P9-DNC-866

ENCYCLOPEDIA OF INTEGRATED CIRCUITS:
A Practical Handbook of Essential Reference Data

ENCYCLOPEDIA OF INTEGRATED CIRCUITS: A Practical Handbook of Essential Reference Data

WALTER H. BUCHSBAUM, Sc.D.

Prentice-Hall, Inc., Englewood Cliffs, New Jersey

Prentice-Hall International, Inc., *London*
Prentice-Hall of Australia, Pty. Ltd., *Sydney*
Prentice-Hall of Canada, Ltd., *Toronto*
Prentice-Hall of India Private Ltd., *New Delhi*
Prentice-Hall of Japan, Inc., *Tokyo*
Prentice-Hall of Southeast Asia Pte. Ltd., *Singapore*
Whitehall Books, Ltd., *Wellington, New Zealand*

Library of Congress Cataloging in Publication Data

Buchsbaum, Walter H.
Encyclopedia of integrated circuits.

Includes index.
1. Integrated circuits. I. Title.
TK7874.B77 621.381'73 80-21596
ISBN 0-13-275875-X

Printed in the United States of America

ABOUT THIS BOOK

Integrated circuits are among the most magical ingredients in today's electronics field. Some contain only a few dozen transistors and diodes, while others, the large-scale integrated circuits (LSI), contain tens of thousands of transistors in a single package the size of an ordinary postage stamp. Integrated circuits (ICs) are available in a huge variety of functions, characteristics, and special features. The Electronic Industry Association (EIA) parts-numbering system covers less than half of the thousands of different type numbers produced by more than 35 major manufacturers. All publish data on their own IC types, each in a different format and often using different definitions for their specifications. The extent of the available manufacturer's IC data was demonstrated to this author when the material used as the background for this book took up nine feet of shelf space.

Encyclopedia of Integrated Circuits covers the entire field of ICs, but from the *user's* rather than the manufacturer's point of view. We are not concerned with part numbers or interchangeability because it is easy to secure that information from the manufacturers' data sheets or their cross-reference lists. This book covers ICs in respect to what they do and how they perform in electronic equipment.

Integrated circuits are divided into four broad categories: Analog, Consumer, Digital and Interface. Each category is further broken down into subcategories, such as Amplifiers, Comparators, Converters, Demodulators, etc. A further division into specific IC functions, such as Class A, Differential or Operational Amplifiers, narrows down the functions to the level of actual application in equipment. Where a further subdivision is appropriate, all variations of a particular function are included. In the case of the Operational Amplifiers, for example, nine different types of op-amps are described.

As in any good encyclopedia, a uniform, concise format is used for all entries. Starting with a brief description and a functional block or logic diagram to illustrate the essential features, the format includes an explanation of key parameters together with actual electrical values. Such parameters as input impedance, open-loop gain, temperature stability, propagation delay time, etc., are specified, where applicable, for different IC families, such as TTL, CMOS or ECL. A summary of typical applications and any comments pertinent to the use or selection of this type of IC complete the entry.

Each entry is numbered according to category and subclassification to help you locate different functions and relate them to each other. Because the complete information for each IC subcategory is contained in that entry, you don't have to waste time going back and forth between different parts of the book. Counters, or dividers, for example, are listed in D.3.0. This entry covers counters or dividers in general. The detailed features of a pre-settable up/down counter are found in D.3.5. All the information needed to select such a counter, or to troubleshoot equipment using it, is contained right at D.3.5.

The appendix contains general information such as the physical dimensions of standard IC packages, a list of the major IC manufacturers with complete addresses and phone numbers, and excerpts from manufacturers' data describing the most widely used IC families.

When you use the *Encyclopedia of Integrated Circuits* you will find the extensive index a real delight. In addition to the terminology used in this book, the index also includes most of the less frequently used terms to help you locate a particular IC. A circuit that amplifies digital signals for transmission over a bus, for example, is listed as "line driver," but if you are used to calling it a "transmitter," a "buffer," an "I/O driver," or just an "output amplifier," you will find it listed under all of these terms in the index.

Some of the latest ICs announced by the various manufacturers combine several basic functions in a single package. Others may be faster, use less power, or have better temperature stability. Your *Encyclopedia of Integrated Circuits* does not get obsolete. A shift register, after all, works the same way whether it has eight or 18 stages, whether its clock speed is 0.5 or 50 MHz. And when this shift register is combined with tri-state output buffers, you can just look them up as if they were on a separate IC. Because this book is written from the user's point of view and concentrates on what each kind of IC does, you will find your *Encyclopedia of Integrated Circuits* as useful five years from now as you did when you bought it.

HOW YOU CAN USE THIS BOOK

Each person has his or her own way of using an encyclopedia, and once you become familiar with this book you will find it extremely helpful for any type of electronics work, whether you are a kit builder, an experimenter, a technician, or an engineer. The following examples will give you some idea of the book's usefulness.

> A person building a kit learns from the instructions that a particular IC is a differential amplifier, but also needs to know what a differential amplifier does and what the common-mode rejection ratio (CMRR) means. This book provides a clear and concise explanation of differential

amplifiers and all of its key parameters and even gives typical values of the CMRR.

Another person experiments with his own communications receiver design. He knows that he needs a detector, and he can find the AM/SSB detector and the multimode detector IC in this book. The brief descriptions and block diagrams of each help him decide which type of IC is most suitable for his purpose.

Working on a color TV set we decide that the trouble must be in the color sync or in the demodulator section. The schematic diagram just shows rectangles for these ICs, without any explanation of how they work. A quick look in the Consumer IC category of this book provides an explanation of TV receiver ICs and a detailed description of how the color sync and the color demodulator ICs work. The reader is alerted to the effect of the external components and can then check quickly whether one of the external components or an IC is defective.

When designing the interconnection of a personal computer to a communications link, a person has to decide which kind of interface IC to use. When looking for communications interface ICs in the index of this book, the reader will find three different functions: the High-Level Data Link Controller and Synchronous Data Link Controller (HDLC/SDLC), the Synchronous Communications Controller (SCC), and the Universal Asynchronous Receiver Transmitter (UART). The concise descriptions and functional block diagrams of each, together with the key parameters, allow him to choose the type best suited for his communications interface.

You can see from these examples that the *Encyclopedia of Integrated Circuits* is used, just as is any other type of encyclopedia, to provide information about a particular topic—in this case a type of integrated circuit. All you need to know is the functional name of the IC and you can find out what it does, how it does it, and how it can be expected to perform in an electronic device. With ICs taking over more and more functions in electronic equipment, this book will become increasingly valuable as time goes on.

Walter H. Buchsbaum

ACKNOWLEDGMENTS

This book would not have been possible without the cooperation and assistance of the major IC manufacturers, and I am glad to thank all of those listed below.

American Micro Systems
Analog Devices Inc.
Exar Integrated Systems Inc.
Fairchild Semiconductor
Ferranti Electric Inc.
General Instrument Corp.
Harris Semiconductor
Hughes Microelectronics Prod.
Intel Corp.
Intersil Inc.
Mostek Corp.
Motorola Semiconduct. Prod. Inc.
National Semiconductor Corp.
Nucleonic Prod. Co.
Plessey Semiconductors
Precision Monolithics
Raytheon Semiconductor Div.
RCA Solid State Div.
Rockwell Intern. Microele. Div.
Signetics Corp.
Silicon General Inc.
Siliconix Inc.
Solid State Scientific, Inc.
Sprague Electric Co.
Standard Microsystems
Teledyne Semiconductors
Texas Instruments, Inc.
TRW Semiconductor
Watkins-Johnson Co.
Western Digital Corp.
Zilog Inc.

The following sections of this book were written by Mr. John J. Petrale:

D.11 Memory
D.12 Microcomputer and Microprocessor
D.13 Microprocessor/computer Support Functions
I.4 Memory Interface
I.5 System Interface

John J. Petrale is a Staff Engineer, currently employed by Loral Electronic Systems in Yonkers, N.Y., where his duties include the design of electronic systems. He is the author of numerous articles in the field and is now writing a book on microprocessors. Mr. Petrale is a consultant in systems design and testing and teaches courses on logic design and microprocessors. He is the head of the Loral Management Development Committee and also lectures on engineering management. His inventions include PROM Controllers and Automatic System Test Sets.

I also want to express my appreciation to Mrs. Inge Seymour, who has done a very tedious typing job cheerfully and accurately. Finally, I want to thank my wife for her encouragement and forbearance as well as her work on the index of this book.

W.H.B.

Table of Contents

CONSUMER ICs

DIGITAL ICs

INTERFACE ICs

APPENDIX

Table of Illustrations

ANALOG ICs

CONSUMER ICs

DIGITAL ICs

INTERFACE ICs

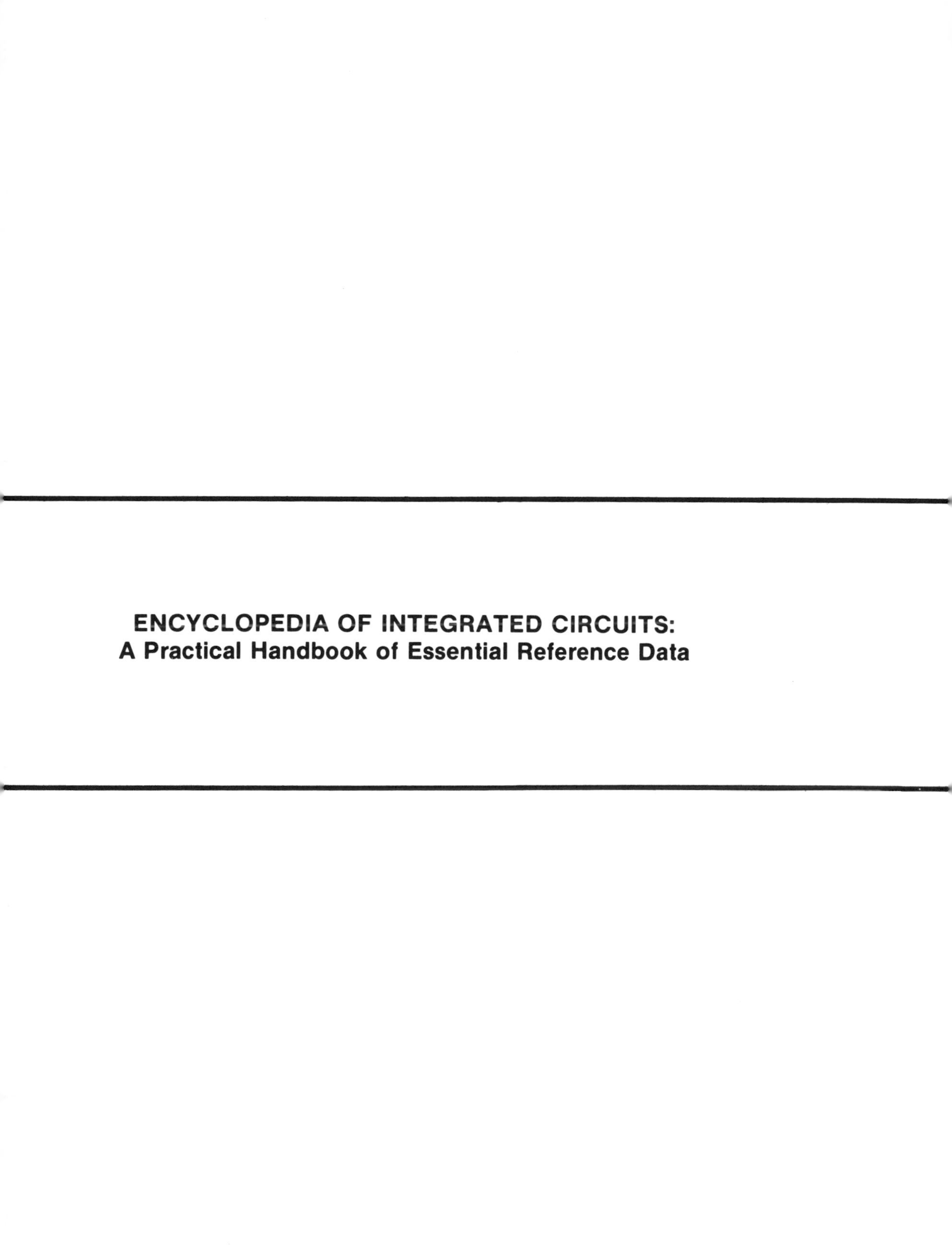

ENCYCLOPEDIA OF INTEGRATED CIRCUITS:
A Practical Handbook of Essential Reference Data

Analog

A. ANALOG ICs

Although some manufacturer's data books call them *linear*, the term *analog* is used in this book to describe those integrated circuits that produce an output that is analogous, or similar, to the input signal. Many of these ICs, such as operational amplifiers, logarithmic amplifiers, function generators, etc., are not really linear circuit functions, and the term *analog* seems a better description

Analog ICs are produced by the use of a variety of semiconductor technologies, including bipolar, field effect, metal oxide, and combinations of these three. In most instances the user has no interest in this aspect of the IC, because he can base his work only on the manufacturer's specifications. In digital IC classification the manufacturing technology is important for the user because digital ICs are used in "family groups," with common electrical characteristics that insure compatibility. Analog ICs are usually selected on an individual basis, and the only compatibility that is important is the power-supply requirement. Even in this respect most analog ICs are available with a range of power-supply options, so compatibility is not a frequent consideration in their use.

The analog ICs listed in Section A are divided into 16 major categories, the largest of which—amplifiers—is subdivided into 11 subcategories. Operational amplifiers (op amps) form such an important, universally used subcategory that we decided to describe the eight major types of op amps as a further "sub-subcategory." The basic op amp is covered in some detail in Section A.1.9.0, and relatively brief descriptions are then presented for the eight major types. Only those special features that distinguish each op amp type from the basic or the general purpose type are discussed in detail. The illustrations would be the same for all op amps and are therefore included for only the basic type and the programmable op amp, which has additional inputs.

The approach used for op amps is also used for such large IC categories as the phase-locked loop (PLL), the voltage regulator, and the analog switch. In each case the basic IC function is discussed in detail, and only the special features of the subcategories are covered fully. For an understanding of the basic function the reader is referred to the first heading describing the particular IC category. This approach is used throughout the book.

A.1.1 Amplifier, Class A (Linear)

DESCRIPTION

In this amplifier the input signal is reproduced, increased in amplitude, in exactly the same waveshape at the output. To achieve this, the quiescent point (Q) is at the center of the collector current (I_C) curve, so that the input signal as well as the amplified output signal operates only over the linear portion of this curve. I_C flows at all times. Class A amplifiers are used wherever the output waveshape must be the same, with a minimum of distortion, as the input signal. Operational amplifiers, and "small signal" amplifiers, such as RF, I.F., preamplifiers, etc., are all basically Class A amplifiers.

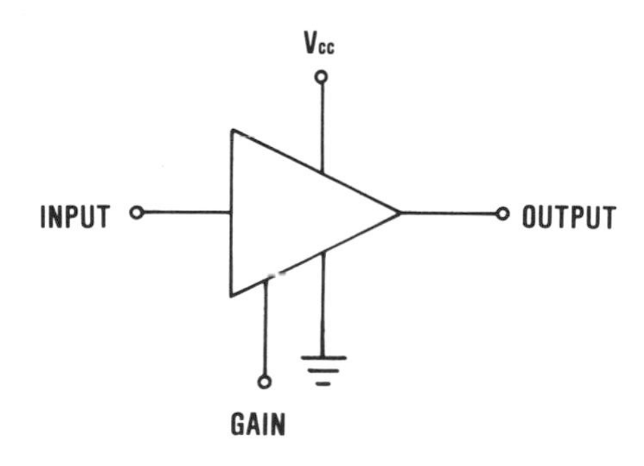

FIGURE A.1.1—1.5

KEY PARAMETERS

a) *Gain*. Depending on the application and type of amplifier, this may be expressed as voltage gain, current gain or power gain, usually in dB. In many types of amplifiers the gain can be electronically controlled by varying DC signals, such as the automatic gain control (AGC), in a receiver. Manual gain control can be applied through a potentiometer connected to the gain terminal on the amplifier.

b) *Frequency (bandwidth)*. Class A amplifiers generally do not have a lower frequency limit, but the upper limit, together with the maximum gain possible at that upper frequency, is an important parameter. This frequency limit is independent of external resonant networks and frequency compensating networks.

c) *Linearity distortion*. This parameter is usually given as the maximum peak-to-peak input signal voltage. In audio circuit applications the percent of output signal distortion, under specific operating conditions, is sometimes stated in the specifications.

d) *Power output.* The maximum power output is stated either as dB with a given input or as the maximum peak-to-peak voltage across a specified load impedance.
e) *Noise.* Stated as signal-to-noise ratio or simply as noise figure, this parameter indicates the amount of noise generated within the amplifier with a specified input impedance.
f) *Power dissipation.* The maximum power dissipation, in milliwatts or watts, not only indicates the power required from the supply but is also a measure of the temperature rise and, if necessary, of the heat sink requirements of the particular amplifier.

APPLICATIONS

Class A amplifiers are used as low-level amplifiers in audio circuits, in the RF and IF stages of receivers of all kinds, and in the video stages of TV receivers and displays. Basic Class A amplifiers can also be connected as operational amplifiers, servo amplifiers, power amplifiers, instrumentation amplifiers, and in any other application where the input signal is amplified in a linear fashion. When used as RF and IF amplifiers, the input and output circuits are generally resonant circuits, tuned to a selected band of frequencies. It is essential in these applications that the frequency or bandwidth characteristics of the IC itself go beyond the frequency band they must amplify. In the case of video amplifiers, low-frequency compensating networks may be necessary to boost the low-frequency components of the video signal that may have been reduced in other parts of the system.

COMMENTS

Some ICs may be labeled by their specific applications, such as audio and video amplifiers, but they can be used for other purposes, such as instrumentation amplifiers, as long as the key parameters listed above are suitable. In some instances Class A amplifiers can be changed to Class AB, B, or even C operation by changing the bias level.

A.1.2 Amplifier, Class AB

DESCRIPTION

In this type of amplifier the quiescent (Q) point is set below the center of the linear portion of the I_C curve. As a result, one half of the output will be the linear reproduction of one half of the input but the second half of the output signal will be partly suppressed. There are two versions—the Class AB1 and the Class AB2. In the Class AB2, the Q point is fairly close to the cutoff point; in the Class AB1, the Q point is about 20% or 30% above the cutoff point. Both versions are used in push-pull circuits, and, by arranging them back to back, the crossover distortion can be minimized. The AB2 version is somewhat more efficient than the AB1 but not as good as the Class B (see paragraph 1.3). Class AB1 and AB2 amplifiers are used widely as push-pull amplifiers to drive speakers to servo motors, applications where linear sinewave amplification with moderate amounts of power is required.

KEY PARAMETERS

See A.1.1, Amplifier, Class A, and A.1.10, Amplifier, Power.

APPLICATIONS

See A.1.1, Amplifier, Class A, and A.1.10, Amplifier, Power.

A.1.3 Amplifier, Class B

DESCRIPTION

In this type of amplifier, the quiescent (Q) point is set exactly at the cutoff of the IC curve, resulting in amplification of only one half cycle of the input sinewave. Class B amplifiers are invariably connected back to back in push-pull circuits. In this arrangement one amplifier is operating during the positive cycle of the sinewave and is cut off while the other amplifier is operating on the negative part of the sinewave. Widely used as audio amplifiers, servo amplifiers, and in similar applications where a linear sinewave output is essential, Class B amplifiers feature excellent efficiency and good second and third harmonic distortion characteristics. Some distortion occurs near the zero crossover point because of the slight nonlinearity of the I_C curve at the point. Where very good linearity is essential, Class AB amplifiers are used in push-pull circuits, as described in paragraph 1.2.

KEY PARAMETERS

See A.1.1, Amplifier, Class A, and A.1.10, Amplifier, Power.

APPLICATIONS

See A.1.1, Amplifier, Class A, and A.1.10, Amplifier, Power.

A.1.4. Amplifier, Class C

DESCRIPTION

In the Class C amplifier the quiescent (Q) point is located at twice the cutoff on the I_C curve. Only one half of a sinewave half cycle is actually amplified at the output. Class C amplifiers are usually used in RF oscillators and, in some instances, in RF transmitters. In these applications the flywheel effect of the resonant circuit supplies the other half of the cycle. High efficiency is the key feature for the Class C amplifier in properly designed and aligned RF circuits.

KEY PARAMETERS

a) *Gain.* In most applications a voltage gain of 20 is adequate.
b) *Frequency.* For application as RF oscillator or transmitter output amplifier, the frequency limit of the device should be at least 10% above the expected resonant frequency.
c) *Power output.* Depending on the actual frequency, power output may vary but is a key design criterion.
d) *Power dissipation.* Class C amplifiers usually operate near the limits of their power dissipation specifications, and the thermal characteristics of the mechanical mounting arrangement may become critical.

APPLICATIONS

Class C amplifiers are usually found in oscillators at frequencies above 100 kHz. Linear sinewave output and freedom from harmonics are usually not important in such applications as walkie-talkies or remote control transmitters. Special high-frequency ICs are used in Class C oscillators, reaching into the microwave band.

COMMENTS

Listed here in order to present a complete picture of amplifiers, Class C amplifiers, as such, are not widely used in IC applications.

A.1.5 Amplifier, Current (Linear Follower)

DESCRIPTION

Current amplifiers are basically Class A amplifiers which usually have a voltage gain of 1 and effectively act as impedance transformers. Their main feature is their ability to develop a substantial output current. Sometimes they are called "linear followers," similar to emitter follower transistor circuits. Current amplifiers are often used in series with an operational amplifier, inside the feedback loop, to provide additional output current.

KEY PARAMETERS

a) *Output current.* Determines the application and useful range.
b) *Slew rate.* Particularly important for servo systems and other motor-driving applications. Slew rates range up to hundreds of volts per microsecond.
c) *Bandwidth.* Conventional frequency range.
d) *Full power bandwidth.* The frequency band over which the maximum output current is available.
e) *Input resistance.* Usually in the megohm range.
f) *Output resistance.* Usually below 50 ohms.

APPLICATIONS

Current amplifiers are used to drive coaxial cables, servo motor systems, precision data recorders, and step-up high-voltage transformers, and they are also useful as audio output amplifiers and in the regulator circuits of power supplies.

COMMENTS

Some manufacturers' literature lists current amplifiers as linear followers, and a wide range of slew rate, output current, and bandwidth is readily available.

A.1.6 Amplifier, Differential

DESCRIPTION

Differential amplifiers have two input terminals which are both isolated from ground by the same impedance. Basically a Class A voltage amplifier, the differential amplifier amplifies only the difference in voltage between its two input terminals. Signals that appear at both terminals are not amplified, allowing the differential amplifier to pick up weak signals in the presence of strong magnetic and electrical interference. This ability to reject signals that are common to both input terminals is expressed in the common-mode rejection ratio.

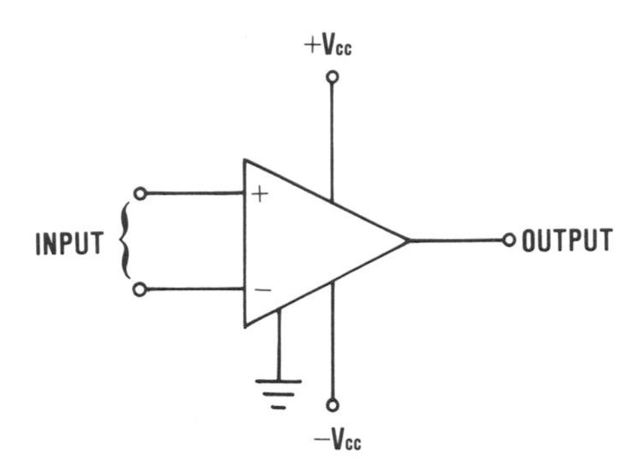

FIGURE A.1.6

KEY PARAMETERS

a) *Common-mode rejection ratio (CMRR).* A measure of the differential amplifier's input characteristic. CMRRs of 100 dB and up are readily available.
b) *Input common-mode range.* The maximum positive and negative voltage which will be rejected by the CMRR at the input.
c) *Differential voltage gain.* Indicates the amplification of the differential input voltage.
d) *Bandwidth.* Indicates the frequency range in a conventional manner.
e) *Input resistance.* The impedance between the two differential input terminals.
f) *Input offset voltage.* Usually measured in millivolts, it indicates the amount of unbalance between the input terminals.

APPLICATIONS

Widely used in all types of magnetic recording pickup head amplifiers, in a variety of instruments in industry, scientific laboratories, and medical applications where small signals must be amplified in the presence of external interference.

COMMENTS

Differential amplifiers are available with differential output or with single output. In many models the gain is selectable by strapping external terminals or connecting outboard resistors. Very high impedance inputs are provided by FET technology, combined with conventional bipolar transistors.

A.1.7 Amplifier, Isolation

DESCRIPTION

Consisting of several stages of amplification, the input amplifier is either electrically or optically isolated from the output. The input amplifier is usually a differential amplifier, the output of which is RF modulated and then transferred through an RF transformer to the second stage, where it is demodulated and filtered. The DC power supply for the input amplifier section must also be isolated so that there is no DC or low-frequency connection between the input amplifier section and the output. Optical isolation amplifiers operate in a similar manner, substituting an optical coupler for the RF transformer. Isolation amplifiers are generally packaged in a single unit and are used wherever very low leakage levels at DC or line voltage are required.

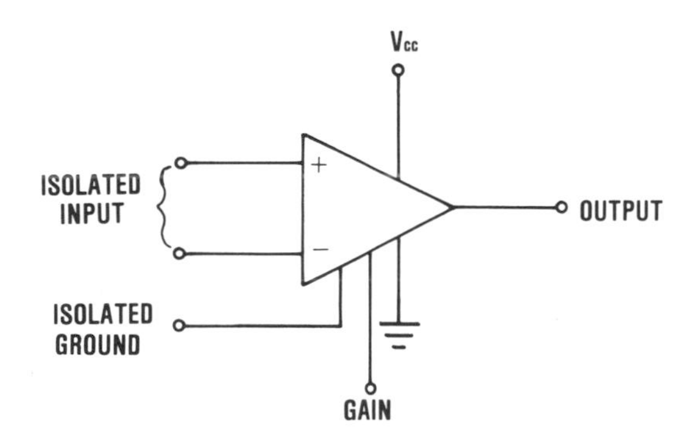

FIGURE A.1.7

KEY PARAMETERS

a) *Maximum safe differential input voltages.* The input voltage limit beyond which the isolation will break down.
b) *Common-mode rejection to guard ground.* This refers to the CMRR of the input amplifier with respect to the isolated ground of the input amplifier.
c) *Common-mode voltage input-to-output.* The amount of external interference (common mode) signal that will be transferred between input and output.
d) *Maximum input leakage current.* Determines the degree of safety for a patient in medical applications.
e) *Overload resistance.* Apparent input impedance when the input amplifier is saturated. This limits any possible fault currents.

APPLICATIONS

Input amplifiers in patient electrocardiograms, electroencephalograms, and other physiological monitors. Isolation amplifiers are also used in nuclear power plant instrumentation and in industrial process controls, wherever there is a problem of electrical safety.

COMMENTS

Isolation amplifiers always require isolated power supplies as well as properly isolated and insulated cables between power supply and the amplifier. In some cases batteries are used to evade the problem of the isolated power supply.

A.1.8 Amplifier, Logarithmic (Antilog)

DESCRIPTION

In this amplifier the output voltage is proportional to the log or to the exponential function (antilog) of the input signal. Most logarithmic amplifier ICs are available with an external strapping option which makes them operable in either the logarithmic or the exponential (antilog) mode, as indicated in Figure 1.8. The antilog element consists of semiconductor stages in which the input voltage produces a current that is an exponential function. To produce a logarithmic output, the antilog element is connected as part of the feedback circuit as shown in Figure 1.8(a), while Figure 1.8(b) shows the connection for producing an exponential output function.

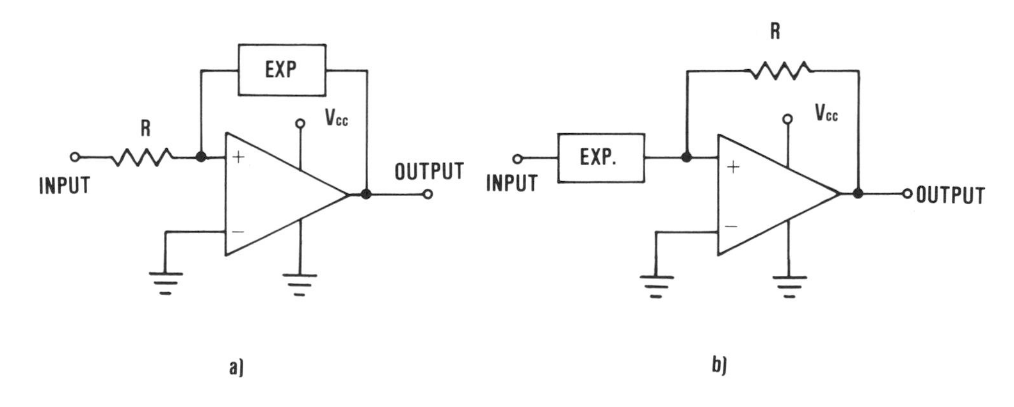

FIGURE A.1.8

KEY PARAMETERS

a) *Log conformity error.* Indicates the deviation of the output from a straight line when plotted on semilog paper, after adjustment for offset voltage, reference voltage and current factors, and other variants.
b) *Offset voltage and current.* Must be adjusted for a minimum in order to obtain a logarithmic function over the entire dynamic range.
c) *Reference voltage and current.* Adjustment of this internally generated voltage and current is essentially the same as in most operational amplifiers.
d) *Gain factor.* External strapping connections are usually available to provide different amounts of amplification and therefore different logarithmic or exponential gain factors.

APPLICATIONS

The compression of any wide-range input signal, analog computing circuits, linearization of exponential sources, such as transducers, certain types of audio circuits, compression and expansion circuits for voice communications.

COMMENTS

A relatively specialized device, a logarithmic amplifier can perform a variety of useful functions, provided the key parameters are selected correctly for the particular application.

A.1.9.0 Amplifier, Operational (Op Amp)

DESCRIPTION

The basic op amp is a DC coupled, high-gain, differential amplifier with external negative feedback. It is characterized by almost infinite open-loop gain, almost infinite input impedance, and almost zero output impedance. Invariably used with an external negative feedback element, the op amp provides a fixed gain, determined entirely by the ratio of the input resistance and the feedback resistance. As the feedback resistance increases, the gain increases as well. When the feedback resistance is zero, the op amp provides unity gain.

Op amps can be used as electronic integrators or differentiators, depending on the R-C network in the input and feedback circuit. Op amps generally have a large bandwidth and low noise, and the key parameters are relatively stable with respect to temperature and time.

A large variety of op amps are available as ICs, ranging from the general purpose op amp to a number of highly specialized types. The basic operation of all types of op amps is the same, and specialized types are distinguished by the emphasis on some key parameters. The major types of op amps are described in the following pages.

KEY PARAMETERS

a) *Open-loop gain.* Usually specified at DC, but a plot of open-loop gain versus frequency is important in many applications. The open-loop gain is an indication of the degree of stability that can be obtained with negative feedback.
b) *Input impedance.* Represented accurately by a parallel R-C circuit across the input terminals, it is most frequently stated as input resistance only.
c) *Input offset voltage.* Measured by using a feedback resistor to set a large fixed gain. The value of the error at the output is then divided by the gain. In most circuits the input offset voltage can be adjusted to zero by a potentiometer.
d) *Input noise.* Frequently given as signal-to-noise ratio, noise figure or input noise current, low-frequency noise and wideband noise are specified separately. Spot noise occurs at certain frequencies and is indicated in detailed manufacturer's specifications.
e) *Common mode rejection ratio (CMRR).* The definition is the same as that given in Section 1.6, Amplifier, Differential, because the input portion of the op amp consists of a differential amplifier.
f) *Maximum common mode voltage.* Same as CMRR.
g) *Drift versus temperature.* Temperature change of offset voltage, bias current, difference current, and other parameters is specified as the temperature coefficient for each parameter, or else is listed in a detailed table or in graph form.
h) *Bias current.* The current from an infinite source impedance applied to either input that will drive the output to zero.
i) *Difference current.* The difference between the bias currents at the two input terminals.
j) *Bandwidth.* The frequency range over which some of the specifications apply.
k) *Full power response.* The sinewave output at the maximum frequency with unity, closed-loop, gain, for a specified distortion into the rated output load.
l) *Rated output voltage.* The minimum peak output voltage at rated current without any nonlinearity.
m) *Rated output current.* The minimum output current at the rated output voltage.
n) *Overload recovery.* The time required for the output voltage to return to the rated value after a 50% overdrive saturation.
o) *Settling time.* The period from the input of a step voltage to the time

the output value reaches the specified error range of its final value. This is usually measured at unity gain, without capacitive loading.

p) *Slew rate.* Usually given in volts per microseconds (V/μs), this indicates the maximum rate of change of the output voltage for an input step-voltage change.

APPLICATIONS

Op amps are used so widely that a list of applications would exceed the space allotted here. Refer to the special purpose op amps in paragraphs 1.9.1 through 1.9.9 for specific applications.

COMMENTS

Op amps were originally developed for use in analog computers but are now used in digital applications as well. While most op amps require a positive as well as a negative power supply source, a whole line of op amps is now available for use with a single power supply source, frequently the same voltage as that used in digital logic ICs. Most DIP IC packages contain several op amps in the same chip. The characteristics of each op amp on a chip are generally identical.

A.1.9.1 Amplifier, Operational (General Purpose)

DESCRIPTION

See Section 1.9.0.

KEY PARAMETERS

The following values for key parameters of general purpose op amps are typical of this device.

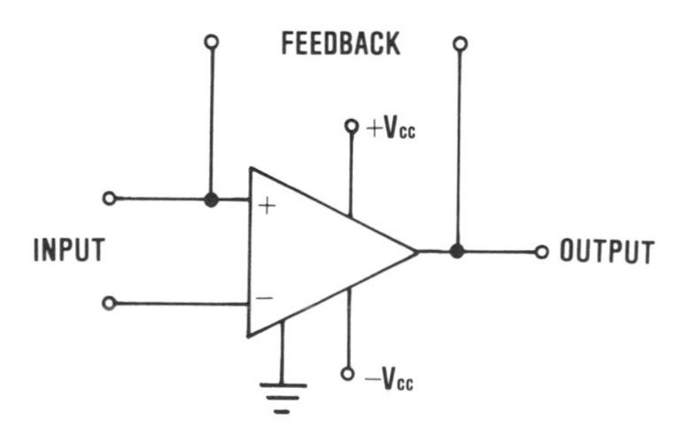

FIGURE A.1.9.1—A.1.9.7 Amplifier, Operational (Op Amp)

a) *Open-loop gain.* 50,000 minimum, 200,000 typical.
b) *Input impedance.* 1 to 10 megohms typical.
c) *Input offset voltage.* 2 to 10 millivolts typical.
d) *Common mode rejection ratio (CMRR).* 90 dB minimum.
e) *Drift versus temperature.* Maximum offset voltage drift is usually about 5 mV/C°.
f) *Bias current.* 50 nA maximum.
g) *Frequency response.* Full power to 10 kHz, unity gain to 1 MHz, typical.
h) *Full power response.* Undistorted output voltage of ±10 volts into a 1,000 ohm load.
i) *Slew rate.* 0.5 V/μs at unity gain.

APPLICATIONS

Op amps find their way into practically all types of equipment, from radio and TV receivers to the most complex computers. General purpose op amps, of course, are the most widely used type, because the vast majority of circuits do not require the special characteristics of some key parameters. The following list of circuit functions is by no means complete: voltage follower, current-to-voltage converter, summing amplifier, integrator, differentiator, active filter, capacitance multiplier, simulated inductor, voltage-to-current converter, precision rectifier, analog-to-digital converter, sample-and-hold circuit, voltage comparator, servo amplifier.

COMMENTS

For more detailed information on op amps and their applications the reader is referred to these two books:

Operational Amplifiers: Theory and Servicing, by E. Bannon, Reston Publ. '75.
Operational Amplifiers and Linear Integrated Circuits, by R. Coughlin and F. Driscoll, Prentice-Hall, Inc. '77.

A.1.9.2 Amplifier, Operational (High Output Current)

DESCRIPTION

This device is basically the same as a general purpose op amp, but its requirement for high output current sometimes results in a special construction for more effective heat dissipation. Most op amps of the high output current type also have short-circuit protection at the output and feature wide bandwidth.

KEY PARAMETERS

Essentially the same as for general purpose op amps except for the following:

a) *Rated output current.* From 100 mA at ±10 volts up to 2 A at 30 volts.
b) *Bandwidth.* Stated as the highest frequency for rated output current, with a typical value of 100 mA at 8 MHz.
c) *Slew Rate.* Ranges from 3 to 1,000 V/μs depending on the type of unit and the total output current capability.

APPLICATIONS

The combination of substantial output current and relatively high-frequency response makes this type of op amp ideal for digital-to-analog current converters, analog-to-digital input amplifiers, video pulse amplifiers, coaxial cable drivers, and CRT deflection amplifiers. When the frequency response is not as high, the op amp can be used successfully for audio output stages and servo motor drivers.

A.1.9.3 Amplifier, Operational (High-Voltage)

DESCRIPTION

The operation of this type of op amp is the same as for a general purpose unit, but a special construction, using a number of semiconductors in series at the output stage, permits operation at higher voltages. Short-circuit protection is usually provided for the output of this type of op amp.

KEY PARAMETERS

Essentially the same as for general purpose op amps except for the following:

a) *Output voltage swing.* ±30 volts up to ±200 volts are typical.
b) *Maximum voltage input.* ±35 volts (typical).
c) *Maximum supply voltage.* ±40 volts up to ±225 volts (typical).
d) *Bandwidth.* Up to 5 MHz (typical).
e) *Slew rate.* From 3 to 200 V/μs (typical) with faster slew rates available.

APPLICATION

High-voltage op amps are used in industrial controls, power supplies, as regulators for high-voltage circuits, signal conditioning circuits, and to provide excitation for resolvers in servo systems.

COMMENTS

The power supplies required for high-voltage op amps must be able to deliver a voltage higher than the maximum output swing, with sufficient current and adequate regulation.

A.1.9.4 Amplifier, Operational (Low-Power)

DESCRIPTION

Designed to operate over a wide range of power-supply voltages, the low-power op amps consume only microwatts when in the standby mode, but are capable of peak output power in the order of milliwatts.

KEY PARAMETERS

Essentially the same as for general purpose op amps except for the following:

a) *Supply voltage*. Ranges from as little as ±0.8 to ±20 volts.
b) *Supply current*. Ranges from 1 mA to 2 mA.
c) *Bias current*. Ranges from 1 to 50 nA.
d) *Bandwidth*. Up to 10 MHz.
e) *Slew rate*. Ranges from 1 to 20 V/μs or higher.

APPLICATIONS

Because of the low power consumption, this type of op amp is generally used in battery powered equipment and is particularly suitable for such applications as active filters, oscillators, amplifiers of all types.

COMMENTS

Many of the low-power op amp models are programmable in their key parameters by external connections. Their gain-bandwidth performance is usually stable over most of the supply voltage and current range.

A.1.9.5 Amplifier, Operational (High-Speed, Wide-Band)

DESCRIPTION

The basic difference between general purpose op amps and op amps of this type is their relatively high-frequency performance. This also usually means that they have a high slew rate and fast settling time. In some of these high-speed, wide-band op amps, the slew rate and settling time are emphasized, while in others the high frequency response is the key feature.

KEY PARAMETERS

Essentially the same as general purpose op amps except as follows:

a) *Slew rate*. Ranges from 100 to 1,000 V/μs.
b) *Settling time*. Ranges from as little as 300 ns to 1.5 μs.
c) *Bandwidth*. Generally stated as 50 to 200 MHz at unity gain, which may mean as much as 100 MHz at full power gain.

APPLICATIONS

Used primarily in high-frequency or high-speed circuits, these op amps are suitable for fast D/A converters, analog data buffers, analog multiplexers, and for amplifiers of video or pulse signals, driving capacitive loads. They are also frequently used in sample-and-hold amplifier circuits.

A.1.9.6 Amplifier, Operational (Low Bias Current)

DESCRIPTION

Most of the op amps in this category use field-effect transistors (FET) in the input stage with a conventional monolithic op amp following. The main result of the low bias current is the availability of very high input impedance, a characteristic particularly important in some instrumentation applications.

KEY PARAMETERS

Essentially the same as general purpose op amps except for the following:

a) *Bias current*. Ranges from 0.1 to 50 pA.
b) *Input impedance*. Ranges from several hundred up to 10,000 megohms.
c) *Offset voltage*. 0.5 to 50 mV.
d) *Slew rate*. Ranges from 0.3 to 10 V/μs.

APPLICATIONS

Used in low current, high input impedance preamplifiers for transducers, photomultipliers, flame detectors, pH electrode cells, radiation detectors, and similar instruments.

COMMENTS

In most instrument applications the drift characteristics play an important part, and warm-up drift, especially, must be considered. Low bias current op amps are also available in low-power and wide-band versions.

A.1.9.7 Amplifier, Operational (Low Drift)

DESCRIPTION

To obtain the low drift characteristic, internal compensation for temperature changes is usually combined with low bias current and high input impedance. Op amps that rely on external compensation generally have greater bandwidth, higher gain, and faster slew rates.

KEY PARAMETERS

Essentially the same as low bias current op amps except for the following:

a) *Drift versus temperature*. Ranges from 0.5 to 5 mV/C°.
b) *Offset voltage*. Ranges from 25 μV to 1 mV.
c) *Bias current*. Ranges from as low as 3 pA to 100 nA.

APPLICATIONS

High accuracy instrumentation is the major application area for low drift op amps. This includes low-level transducer bridge circuits, precision voltage comparators, active filters, and applications in analog computation systems.

COMMENTS

Where low temperature drift is an important parameter, the temperature stability of the immediate environment of the op amp is very important, and sources of temperature variation, such as high-power devices, should not be located near low drift op amps.

A.1.9.8 Amplifier, Operational (Programmable)

DESCRIPTION

Some general purpose op amps are available with an external tab which can be connected to a "set" or programming current. This external terminal

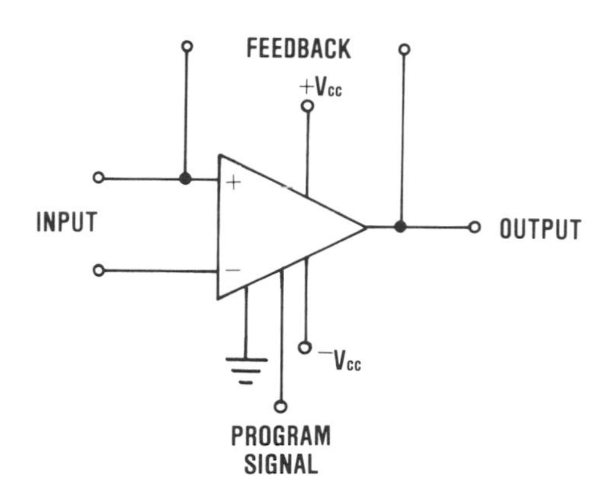

FIGURE A.1.9.8

can control such important parameters as the power dissipation, bandwidth, slew rate, output current, input impedance, and input noise. While the programming current controls some of these parameters, other parameters will remain fixed. Externally supplied programming current can be in the form of a fixed resistor, a potentiometer, or a current source. If the "set" or programming current varies rapidly over a fixed range, the programmable op amp can be used as modulator, current controlled oscillator, or variable frequency active filter. In addition to general purpose op amps, some types of low-power and high-frequency op amps are also programmable.

KEY PARAMETERS

Essentially the same as general purpose or low-power op amps except the following:

a) *Set current range.* 1 to 100 mA.
b) *Bandwidth.* 5 kHz to 10 MHz (typical) variation over the set current range.
c) *Slew rate.* 0.06 to 6 V/μs (typical) variation over the set current range.
d) *Input resistance.* 50 to 5 megohms (typical) variation over the set current range.

APPLICATIONS

Programmable op amps are particularly useful for the experimental design phase, where final specifications are yet to be determined and a range of adjustment of key parameters is important. They are also found in variable frequency active filters, current controlled oscillators, and modulators.

COMMENTS

The manufacturer's detailed specifications should be carefully reviewed for details of the "set" or programming current requirements and the parameter changes obtainable with set current changes.

A.1.10 Amplifier, Power

DESCRIPTION

I.C. power amplifiers generally range up to 10 watts, available in single and dual versions. Most of these amplifiers are designed for Class B operation, requiring that two stages be used in push-pull circuit, such as is commonly found in audio output amplifiers. IC power amplifiers generally have an external heat sink to dissipate the relatively large heat rise.

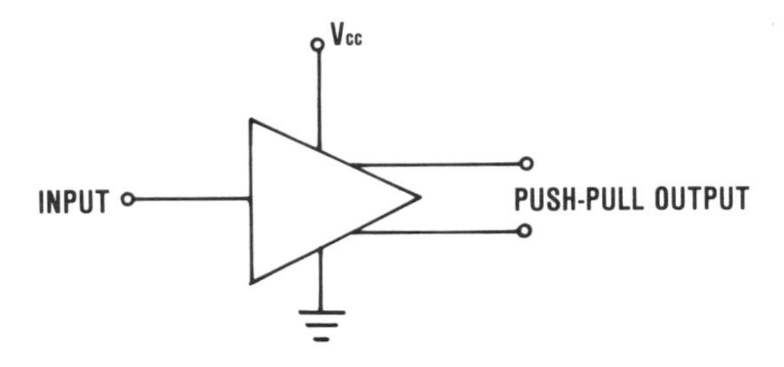

FIGURE A.1.10

KEY PARAMETERS

Essentially the same as for Class A amplifiers, except for the following:

a) *Output power*. 1 to 15 watts.
b) *Maximum power dissipation*. 2 to 25 watts.
c) *Thermal resistance*. The junction-to-case temperature variation as a function of power dissipation changes. A typical value will be 4°C per watt.

APPLICATIONS

Most frequently found in audio output amplifiers as the final stage to drive a loudspeaker, this type of IC is also used in some servo motor-driving applications.

COMMENTS

The heat sink mounting and heat dissipation of an IC power amplifier are its most critical performance characteristics. Many power amplifiers have automatic shut-down circuits to protect the IC when the temperature limit is reached. Short-circuit and overvoltage protection are also often included in the power amplifier IC.

A.1.11 Amplifier, Sample-Hold

DESCRIPTION

This amplifier IC generally consists of two op amps with an electronic switch connected between them and a charging capacitor usually as an external component. As illustrated in Figure 1.11, the amplifier operates in the sampling or tracking mode when the switch is closed. The output signal follows the input signal, usually with unity gain. When the switch is opened, the output signal remains at the previous value until the closing of the switch changes back to the sample mode. The sampling capacitor is usually connected either as shown here or as the series feedback element of the op amp closest to the output.

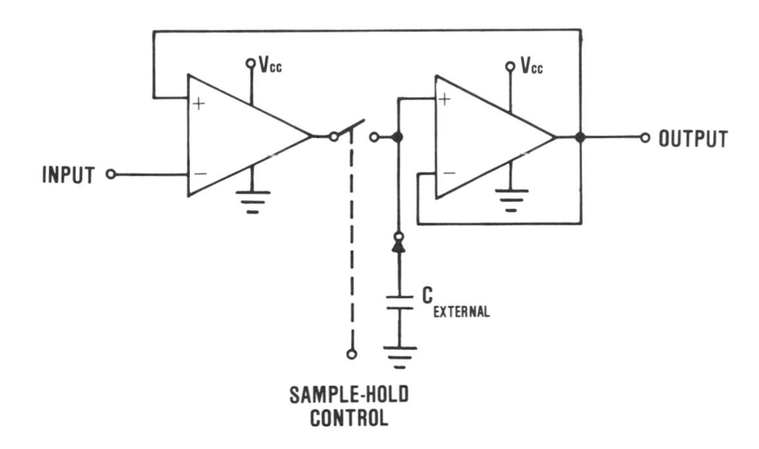

FIGURE A.1.11

KEY PARAMETERS

a) *Acquisition time.* The time required to reach the final output value after the "sample" command. A value from 0.5 to 5.0 μs is typical.
b) *Aperture or delay time.* The time required for the switch to open after the "hold"command. A typical value is from 150 to 500 ns.

c) *Aperture or delay jitter.* The range of aperture time variation. 0.25 to 5 ns is typical.
d) *Charge transfer or offset step.* The effect of stray capacitance during the "hold" mode. A value of 0.025% is typical.
e) *Droop.* The loss in output voltage during the "hold" period, due to leakage current. Typical values are 50 pA to 10 nA leakage current.
f) *Feedthrough.* The amount of AC signal coupled through the switch in the "hold" mode.

APPLICATIONS

Sample-hold amplifiers are used in data acquisition and data distribution systems, in input stages of analog-to-digital converters, and for peak measurements of analog values.

COMMENTS

Sample-hold amplifiers can be constructed of two op amps and an analog switch, but the monolithic construction of the same amplifiers in a single IC provides the advantage of better temperature drift performance, and also requires much less space. Sample-hold ICs are available for high-speed applications.

A.2.1 Compandor (Expandor)

DESCRIPTION

This IC provides compression or expansion of the amplitude dynamic range of audio signals. As indicated in the block diagram of Figure 2.1, a signal current used as gain reference is applied through R1 to a full wave rectifier, and the rectified signal is filtered by the external capacitor C. The DC voltage from the rectifier is applied to the variable gain section, where it controls the gain of an input signal connected to R2. In the variable gain section an internal reference voltage, V_{REF}, is generated at a summing node, and this is also connected to one of the inputs of the op amp. At the trim terminal an external potentiometer can be connected to nullify the internal offsets for minimum output distortion. The gain of the output op amp can be controlled either by an external resistor or by connecting one end of R3 to the output. Depending on the input connections, this IC can serve either as compandor or expandor.

KEY PARAMETERS

a) *Dynamic range.* The amplitude variations in dB which can be compressed or expanded. 110 dB is a typical dynamic range.

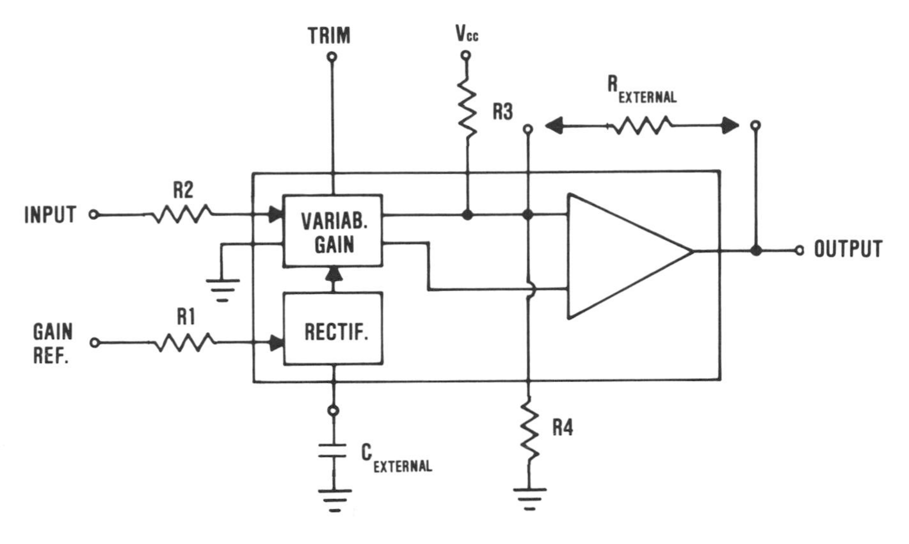

FIGURE A.2.1

b) *Tracking error*. The variation in gain compression at different levels of input current to the rectifier section. Other parameters, such as the supply voltage, power dissipation, output current, etc., are the same parameters as apply to any kind of amplifier.

APPLICATIONS

Telephone trunk and subscriber circuits, voice-controlled microphones, high-level limiters, voltage-controlled amplifiers, dynamic active filters, dynamic noise-reduction systems.

COMMENTS

The value of the external capacitor will determine how well the rectified control signal is filtered, but if the capacitor is too large, the compandor will not be able to follow rapid changes in dynamic range.

A.3.1 Comparator

DESCRIPTION

Essentially a ve. v fast-acting, high-gain differential amplifier, a comparator senses the difference between two voltages and, by rapid amplification, provides maximum output immediately. Designed for low-input currents, most comparators are essentially voltage devices which interface with digital logic circuits. The various types of comparators differ

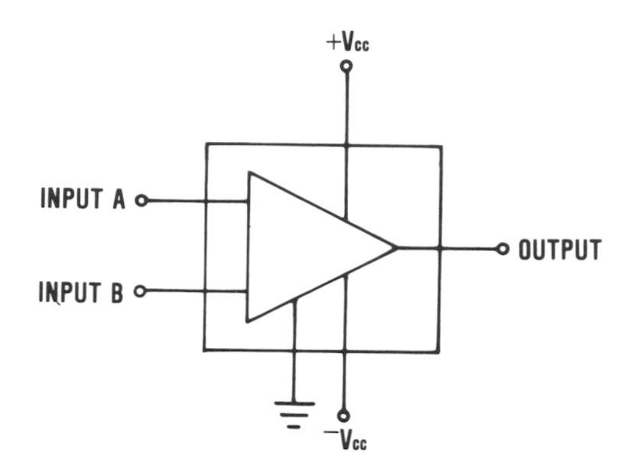

FIGURE A.3.1

primarily in the emphasis of some key parameters, such as low power consumption, fast operating time, high input impedance, and stability over a range of temperatures.

KEY PARAMETERS

a) *Response time.* The length of time required from the appearance of a voltage in excess of the comparator threshold to the time that the output has reached its maximum, the logic level. 50 to 500 ns typical.
b) *Off-set voltage.* The difference between the two input terminals. Typical values range from 0.5 to 10 mV.
c) *Offset current.* The input current difference between the two input terminals. Typical values range from 10 to 100 nA.
d) *Voltage gain.* The difference between input and output, stated either in overall ratio or in volts output versus millivolts input. Typical voltage gains are 200 to 500,000.
e) *Output voltage level.* For compatibility with different types of logic families, specific logic levels to indicate a "zero" and "one" are given.
f) *Output current.* The source or sink current for compatibility with specific digital logic families.
g) *Power supply.* Many comparators require positive and negative power-supply voltages, but there are some on the market that operate from a single voltage and are therefore more easily compatible with the logic families they are supposed to interface with.

APPLICATIONS

Comparators can be used as a threshold detectors, zero-crossing detectors, level changers for different logic levels, detectors for magnetic transducers, window detectors, high stability oscillators, and various analog interface circuits for digital logic systems.

COMMENTS

Most voltage comparators are available in two or four per DIP IC. In some instances medium-power transistor-driven amplifiers are included in the IC to allow the output of the voltage comparator to drive a number of different logic elements.

A.4.1 Converter, Frequency-to-Voltage

DESCRIPTION

As illustrated in Figure 4.1, the input signal is compared in the high gain op amp (1) with a small fixed voltage. Each input pulse or sinewave peak generates a pulse in the one-shot circuit, causing a current pulse, via the current source, to be applied to the output op amp (2). The external R-C network turns this op amp (2) into an integrator, and the output voltage is therefore analogous to the input frequency.

KEY PARAMETERS

a) *Output ripple.* Varies with the input frequency. Typical value is 80 mV RMS at 10 kHz and 35 mV at 100 kHz.

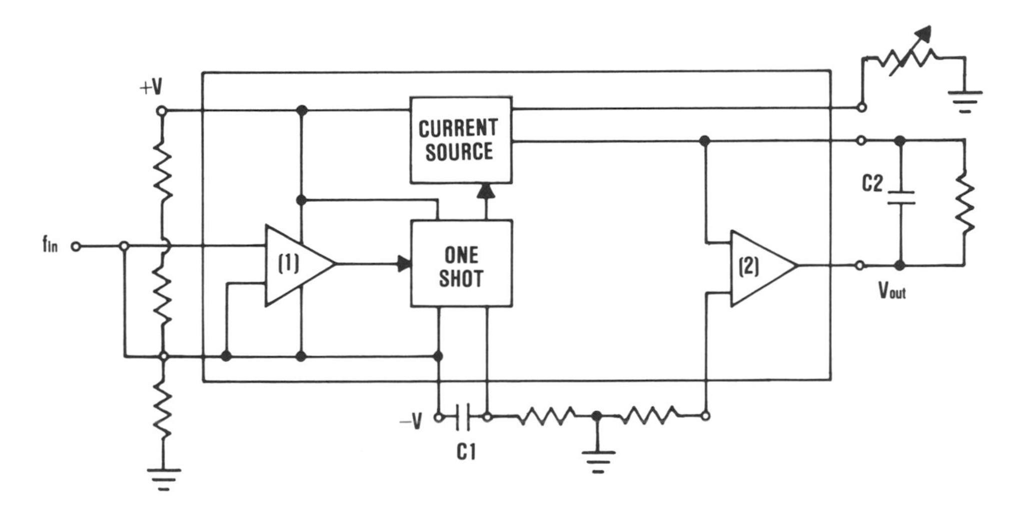

FIGURE A.4.1

b) *Threshold.* This is the minimum voltage level of the input frequency which will cause the one-shot to operate. 1.4 V is a typical value.
c) *Overall temperature coefficient.* Excluding the external components, ±30 ppm/°C is typical.
d) *Linearity.* Exclusive of external components, typical linearity ranges are 0.01% at 10 kHz up to 0.1% at 100 kHz.

APPLICATIONS

Motor control and speed indicators, wow and flutter measurements on tape recorders, stabilizing circuits for voltage-controlled oscillators.

COMMENTS

A voltage-to-frequency converter can be connected as a frequency-to-voltage converter by using a standard PLL (phase-locked loop) arrangement.

A.4.2 Converter, Voltage-to-Frequency

DESCRIPTION

The analog voltage is applied to op amp (1) which converts it into a drive current to the NPN transistor follower. Collector current from this transistor controls an astable multivibrator which has an external capacitor to set the center frequency. The squarewave output signal from the multivibrator (current-to-frequency) passes through buffer amplifier (2) and the open collector output transistor. The external collector to V_{cc} connection

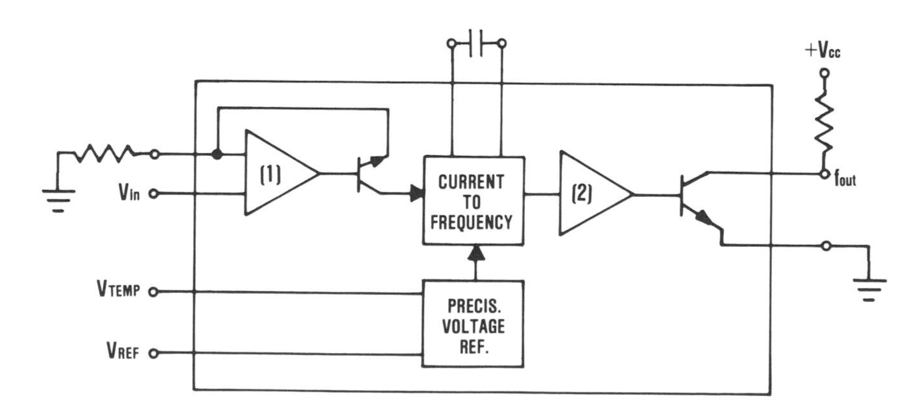

FIGURE A.4.2

can be used to assure compatibility with various logic families. The reference voltage and bias for the multivibrator are supplied by the precision voltage reference generator, which also provides a temperature proportional output, V_{TEMP}, together with the reference voltage output, V_{REF}, allows the IC to be used for temperature-to-frequency conversion, with a variety of temperature scales determined by external resistors.

KEY PARAMETERS

a) *Full scale frequency.* Is determined by F = V/10 RC. Typical full scale frequency is 100 kHz.
b) *Maximum input voltage.* Usually slightly less than V_{cc}. A typical value is V_{cc} -4 volts. V_{cc} may go as high as 36 volts.
c) *Overall temperature coefficient.* Excluding external components, a typical value is ± 30 ppm/°C.
d *Input impedance.* Up to 25 megohms is typical
e) *Linearity.* Decreases at higher frequency. Typical values are .01% at 10 kHz and 0.1% at 100 kHz.

APPLICATIONS

A/D converters, two-wire digital transmissions systems in high noise environments, DVMs, and pulse-width modulators. This IC is also used for signal conversion from various transducers and for DC motor speed control. As indicated in 4.1, a voltage-to-frequency converter can also be connected to act as frequency-to-voltage converter.

COMMENTS

Voltage-to-frequency converters are also available with special features such as pulse output, opto-isolation, and a certain degree of programmability.

A.5.1 Delay Line—Analog

DESCRIPTION

Also called a bucket brigade delay line, this IC consists of a series of "tetrode" transistors with very small capacitors, and is driven by two squarewave clock signals of opposite phase. In effect, the clock signals shift the analog input voltage through the series of capacitive storage units, providing a time delay to the analog signal.

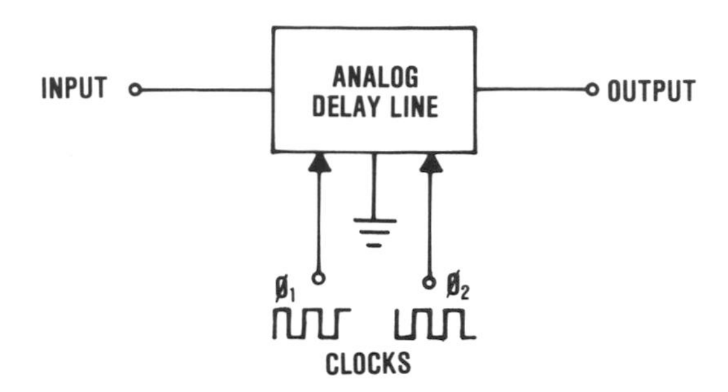

FIGURE A.5.1

KEY PARAMETERS

a) *Signal delay*. Depends on clock frequency and number of stages. Typical delays range from 0.5 to 50 milliseconds.
b) *Number of stages*. Up to 512 stages.
c) *Clock frequency*. Minimum is usually 5 kHz with maximum up to 500 kHz.
d) *Maximum signal frequency*. Highest frequency that can be passed through delay line. 45 kHz is a typical value.
e) *Insertion loss*. The maximum attenuation between input and output. 5 to 8 dB is typical.
f) *Signal-to-noise ratio*. 75 dB is typical.

APPLICATIONS

Particularly useful for delaying audio signals for such purposes as voice-operated control circuits, reverberation, vibrato and chorus effects. Equalizing speech delay in PA installations. Speech scrambling for encryption purposes. Also used in some analog industrial control systems to delay specific signals.

COMMENTS

Analog delay lines are very specialized products and the manufacturer's instructions should be carefully followed.

A.6.1 Decoder, Tone

DESCRIPTION

This IC is designed to decode or demodulate tone signals from their higher frequency carriers by using a phase-locked loop (PLL) and a

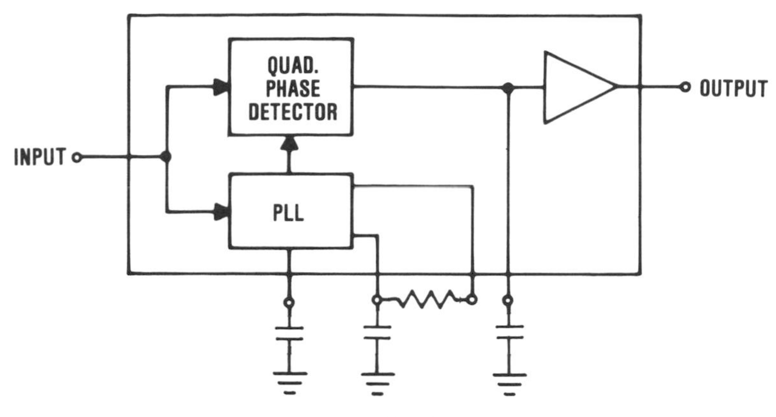

FIGURE A.6.1

quadrature phase detector. The PLL provides the carrier reference to the phase detector, and the resultant tone signal is then amplified, as illustrated in Figure 6.1. This IC usually operates below 10 MHz and requires external capacitors and resistors to set the center frequency of the carrier and, to some extent, the bandwidth. Two separate low-pass filters are provided by these external components: one is the PLL filter and the other, connected between the phase detector and the output amplifier, removes the carrier frequency components from the data.

KEY PARAMETERS

a) *Center frequency range.* Adjustable by external resistor and capacitor from 0.01 Hz to 500 KHz, typical.
b) *Center frequency stability.* Usually stated in percent change per hour or over longer time periods. Frequency stabilities better than 1% are typical, depending on the stability of the external components.
c) *Bandwidth adjustment range.* The frequency band above and below the center frequency over which the PLL and phase detector will still detect or demodulate tone signals from their carrier. Up to 15% of center frequency is typical.
d) *Out-of-band signal rejection.* Detailed characteristics vary between manufacturers. 45 dB is typical.
e) *Noise characteristics.* Indicates the noise contribution due to the IC, stated either as noise figure or signal-to-noise ratio.
f) *Output signal current.* Intended to be compatible with some of the standard logic families because tone decoders frequently work with digital logic. Capability of sinking 100 mA is typical.

APPLICATIONS

Tone decoders are used in a variety of industrial and scientific telemetry systems, remote radio control devices, radio paging systems, ultrasonic remote controls, carrier tone transceivers, tone signaling over the AC powerline, and as "touch-tone" detectors in telephone sets.

COMMENTS

Dual-tone decoders on a single IC are particularly useful for full duplex communications systems. To assure center frequency stability, the IC and its external components should be located away from any strong heat source.

A.6.2 Demodulator, FSK (Frequency Shift Keying)

DESCRIPTION

FSK is a particular type of modulation in which two carrier frequencies are used to indicate the presence of a logic "zero" (**0**) or logic "one" (**1**). All FSK demodulators, like many tone decoders, depend on a phase-locked loop (PLL) to perform the detection. As shown in Figure 6.2 the PLL consists of a voltage-controlled oscillator (VCO), a phase detector, and a loop filter. A second phase detector and lock detector filter provide an output to indicate that the VCO has locked onto the input carrier signal. A data filter precedes the FSK output amplifier, and an internal voltage reference source assures accurate operation of the VCO. The three filters indicated in Figure 6.2 and several other bypass capacitors are external components which must be selected to cover the frequency range of a particular application. The FSK demodulator IC can also be used as a tone decoder, but then there is no need for the lock detector output.

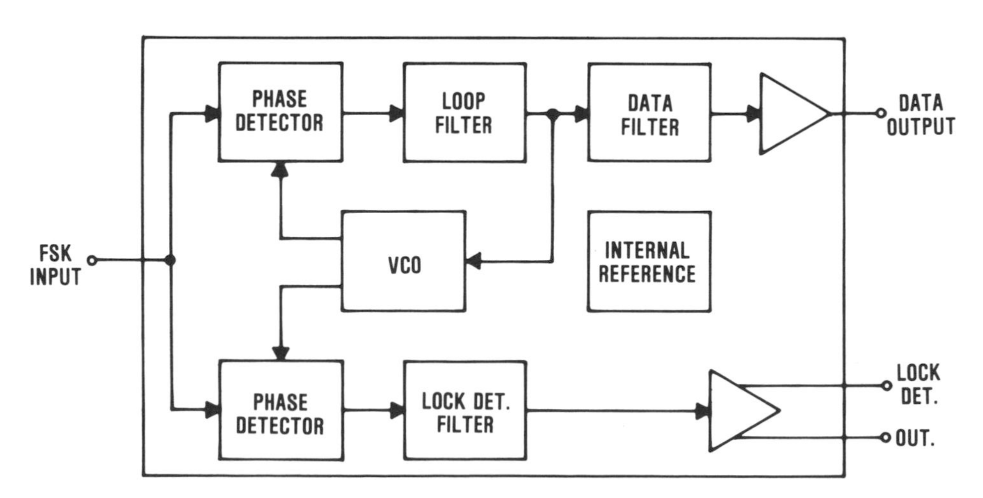

FIGURE A.6.2

KEY PARAMETERS

a) *Center frequency range.* Adjustable by external RC network from 0.01 Hz to 300 KHz, typical.
b) *Center frequency stability.* Depending on external components, and varying manufacturers, excellent performance such as 20 ppm/°C can be obtained.
c) *Adjustable tracking range.* The frequency bandwidth over which the FSK demodulator will work. It can be adjusted from ±1% up to ±80% of the center frequency.
d) *Input dynamic range.* Permissible variations in input signal, 2 mV to 4 V RMS, typical
e) *Supply voltage range.* This determines, to some extent, the output current and the compatibility with different logic families. 4.5 to 20 V is typical.
f) *Output signal current.* Essentially compatible with the logic family for which the supply voltage has been selected. Up to 5 mA driving current is available.

APPLICATIONS

In all types of communications systems that use frequency shift keying as modulation. This IC can also be used for tone decoding (see Section 6.1), FM signal detection, detection of the presence or absence of a carrier, and various types of data synchronization. Telemetry systems, remote control systems, and commercial comunications systems are the main applications.

COMMENTS

The precision and stability of any FSK demodulator will depend on the quality of the external components. For applications requiring very high precision, crystal-controlled PLLs can be applied as FSK demodulators.

A.6.3 Detector, AM/SSB

DESCRIPTION

Intended specifically for single-sideband AM receivers, this IC includes a unity-gain amplifier (1) which drives the AM detector. Unfiltered audio output is available, and two separate terminals are brought out for a potentiometer which will set the automatic gain control (AGC threshold). A variable amplifier (2) amplifies the detector output, controlled by the external threshold potentiometer. Phase correction can be made by another external control through amplifier (3). An internal bias generator circuit provides additional stability over a wide range of temperature and V_{CC} variations.

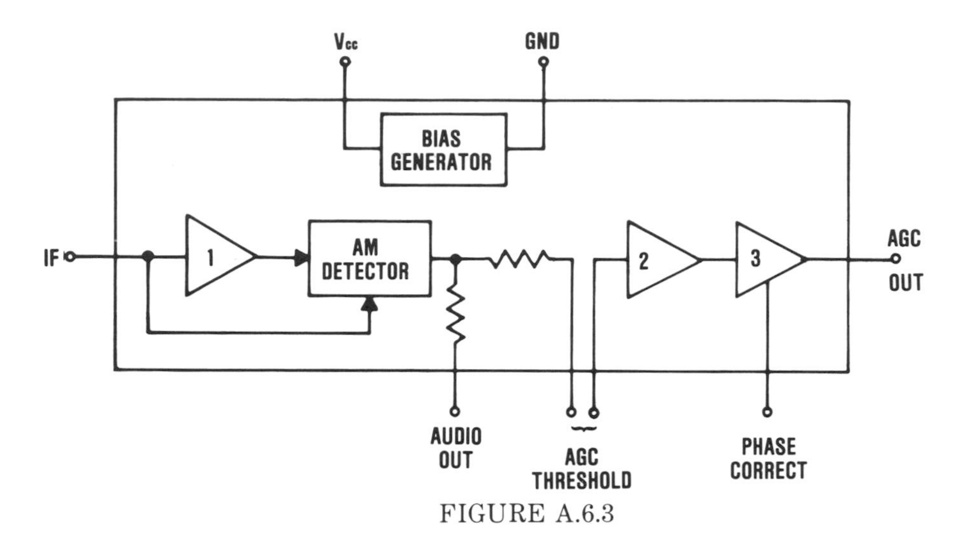

FIGURE A.6.3

KEY PARAMETERS

a) *Maximum frequency.* The highest IF frequency, usually up to 30 MHz. This circuit can work up to 120 MHz, though with some deterioration in performance.
b) *AGC range.* The input level change required to change the AGC output voltage by 2:1. A typical AGC range value is 5 dB.

APPLICATIONS

This IC is specifically designed for use in portable or mobile SSB/AM receivers.

COMMENTS

In some versions of this IC, an audio amplifier following the AM detector is included.

A.6.4 Detector, Multimode

DESCRIPTION

This IC can be used as detector for AM, FM, SSB and CW signals. Depending on the application, it can be connected to operate as synchronous detector, quadrature detector, or product detector with built-in oscillator. For product detection, inputs A and B are used. For FM detection, the input to amplifier (2) is used because that amplifier provides limiting. When limiting

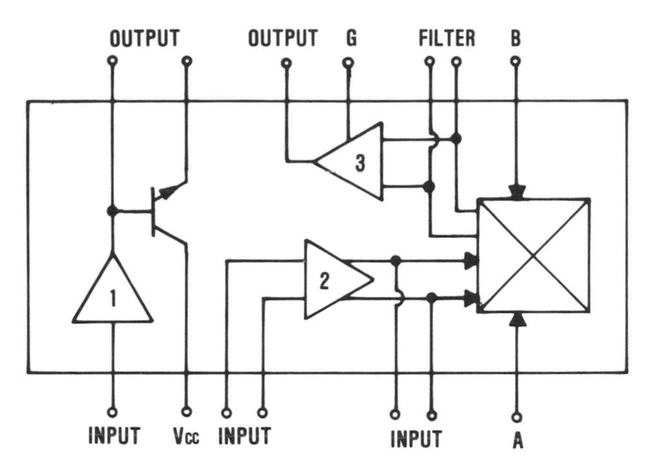

FIGURE A.6.4

is not required, the input to the detector itself can be used. There is a gain-controlled output amplifier (3) with the gain control connected at G. An external filter can be applied at the input to that amplifier. This particular IC also contains a separate audio amplifier (1) for maximum versatility.

KEY PARAMETERS

a) *Maximum frequency.* The highest carrier frequency which the input amplifier (2) will accept. 50 MHz is a typical value. When amplifier (2) is not used, considerably higher frequencies are possible.
b) *Conversion gain.* Depends on the setting of the amplifier gain control terminal (G).

APPLICATIONS

This IC is used in portable and mobile transceivers of all types.

COMMENTS

When used as AM detector, this IC is less noisy and less susceptible to broad-band IF interference than a simple diode detector.

A.7.1 Filter, Active

DESCRIPTION

Active filter ICs usually require external components to determine the particular filter frequency and the type of filter desired. In the example

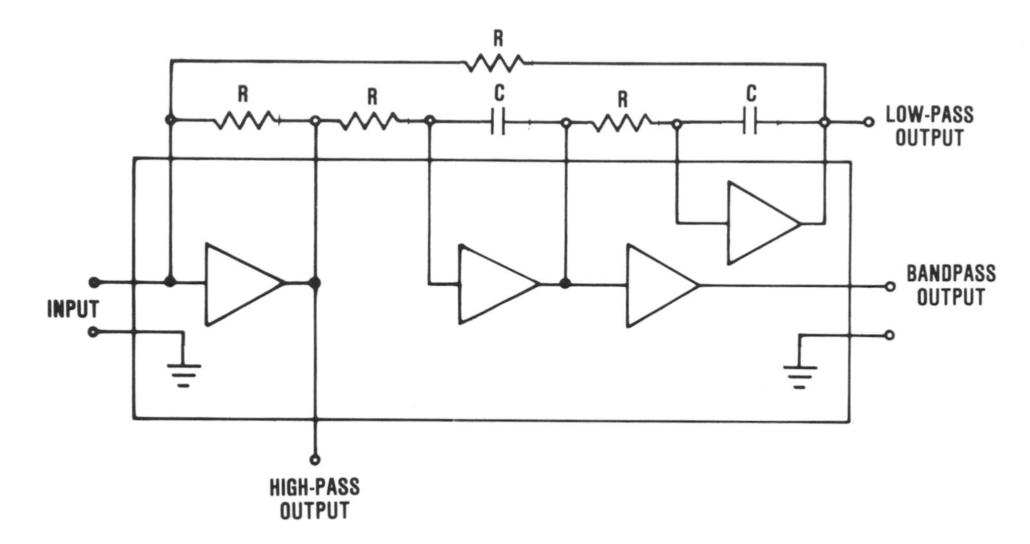

FIGURE A.7.1

illustrated, four separate amplifiers, contained on a single IC, are combined to provide a two-pole active filter. It is made of the combination of three amplifiers, one with purely resistive feedback and the remaining two with capacitive feedback, with overall resistive feedback from the first to the last amplifier. An isolating amplifier is provided for the bandpass filter output signal. Because all amplifiers are made on the same IC, their performance characteristics will be uniform, resulting in a highly stable filter. Figure 7.1 is typical, but many variations of this kind of IC are used in active filters.

KEY PARAMETERS

a) *Frequency range*. Depends on the values of R and C selected. In a typical circuit the frequency equals $1/2\pi RC$ and can range up to 1.0 MHz, typical.
b) *Q range*. Same as the Q factor for inductors, active filters can have Qs up to 1,000.
c) *Frequency accuracy*. Exclusive of the RC accuracies, the IC itself provides up to ±5%.
d) *Frequency temperature coefficient*. For the IC alone, 100 ppm/°C is typical.
e) *Amplifier voltage gain* Up to 110 dB is typical.
f) *Signal-to-noise ratio*. 50 dB is a typical value.

APPLICATIONS

Active filters can be used in the frequency range from audio to 1.0 MHz, and can perform low-pass, high-pass, bandpass and other filter functions.

Analog

COMMENTS

Most ICs intended for active filters will provide a two-pole filter based on the state variable filter principle to implement a second order transfer function. By changing external connections, it is possible to obtain notch, band-stop and all-pass filter functions. Several active IC filters can be combined to provide higher order functions.

A.8.1 Function Generator

DESCRIPTION

As illustrated in Figure 8.1, feedback from the Schmitt trigger circuit to the current source forms an effective voltage-controlled oscillator. Its frequency is determined by the actual resistor and capacitor and can be adjusted over a 10:1 range. The modulating signal input to the current source provides frequency modulation over a 10:1 range. Buffer amplifiers are used for the square-wave and triangular wave outputs. This function generator can be used without the modulating input to generate a fixed frequency signal, or, with modulation, for a swept-frequency, FM, FSK, or other frequency-changing scheme oscillator.

KEY PARAMETERS

a) *Maximum frequency*. Highest possible operating frequency. 1.0 MHz is typical.
b) *Frequency temperature coefficient*. Excluding the external components, 200 ppm/°C is typical.

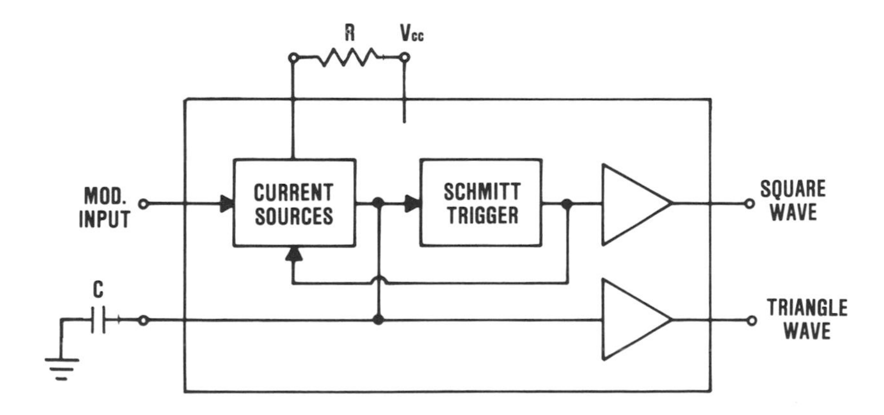

FIGURE A.8.1

c) *Frequency stability with V_{CC}.* Frequency change as supply voltage changes. 1% is typical.
d) *Modulation input impedance.* 1 megohm is a typical value.
e) *Maximum modulation rate.* The highest modulation signal frequency; 1.0 MHz is typical.
f) *Maximum modulation range.* The frequency range over which the center frequency can be varied by the modulation input.

APPLICATIONS

This IC can be used as part of a signal generator, sweep-frequency generator, function generator, or as PSK or FM modulator.

A.8.2 Function Generator, Precision Waveform

DESCRIPTION

This IC is capable of producing sine, square and triangular waveforms in a simultaneous mode. Its frequency can be varied over a wide range by the selection of an external capacitor, and a separate adjustment controls the duty cycle. This changes the square wave to pulse signals and the triangular wave to a sawtooth signal. A modulation signal can be used to vary the output frequency over a large range by controlling the first current source. The oscillator itself is a flip-flop, receiving feedback from the second current source and controlling, through an electronic switch, the modulated current

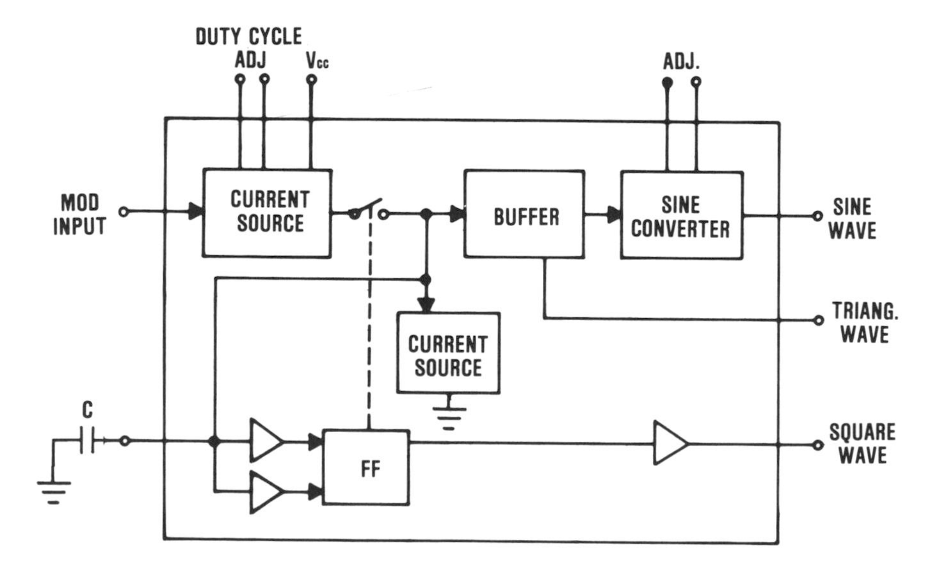

FIGURE A.8.2

source. A separate sine-wave converter with external balancing adjustments provides the desired sine-wave output.

KEY PARAMETERS

a) *Maximum frequency.* The highest possible operating frequency. 1.0 MHz is typical.
b) *Frequency temperature coefficient.* Change in frequency as temperature varies. The effects of the external components are excluded and typical stabilities are about 50 ppm/°C.
c) *Variable duty cycle.* The range of adjustment by the external potentiometer over the duty cycle. Typical range is from 2% to 98%.
d) *Maximum modulation range.* The frequency range over which the center frequency can be varied by the modulation input. 10:1 of the center frequency is typical.

APPLICATIONS

This IC can be used for precision waveform generators, tone generators, sweep-frequency generators, and as clock generators for PLLs. Many applications are in test instrumentation because of the precision available.

COMMENTS

The basic operation of this function generator is the same as the one described in Section 8.1, but the sine-wave converter and the use of two current sources instead of one add greater versatility and precision.

A.9.1 Gyrator, Integrated

DESCRIPTION

This IC uses a number of transistors, capacitors, and resistors to perform the same electrical function as an inductor. A simplified, four-transistor gyrator circuit is illustrated in Figure 9.1. The external components determine the equivalent inductance value. In the actual application, C1 may be part of a coupling network, and the combination of the two resistors and capacitors C2 can simulate inductances of up to 1 millihenry.

KEY PARAMETERS

a) *Efficiency.* Based on the signal power divided by the supply power required. 1.5% efficiency is a typical value.

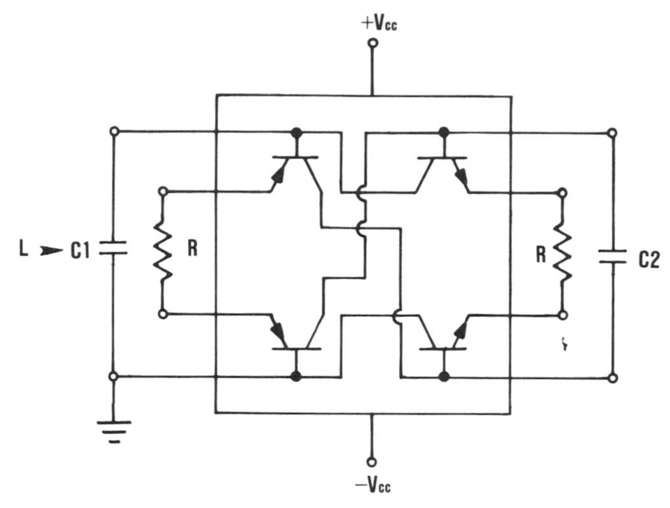

FIGURE A.9.1

b) *Q*. Depending on the external components, an inductor with a Q from 500 to 5,000 can be simulated by this IC.
c) *Input offset voltage*. The unbalance voltage between the two input terminals. 25 mV is typical.
d) *Input offset current*. The current accompanying the input/offset voltage. 10 μA is typical.

APPLICATIONS

This IC is used to replace iron core inductors in telecommunication filter circuits. Also suitable for phase-compensation networks.

COMMENTS

This IC can be cascaded and, combined with capacitors and resistors, forms a variety of highly stable filters and other frequency-compensation networks in the audio range.

A.10.1 Modulator, Balanced

DESCRIPTION

This double-balanced modulator replaces diode ring modulators in AM double-sideband, suppressed carrier operations. Able to operate with great

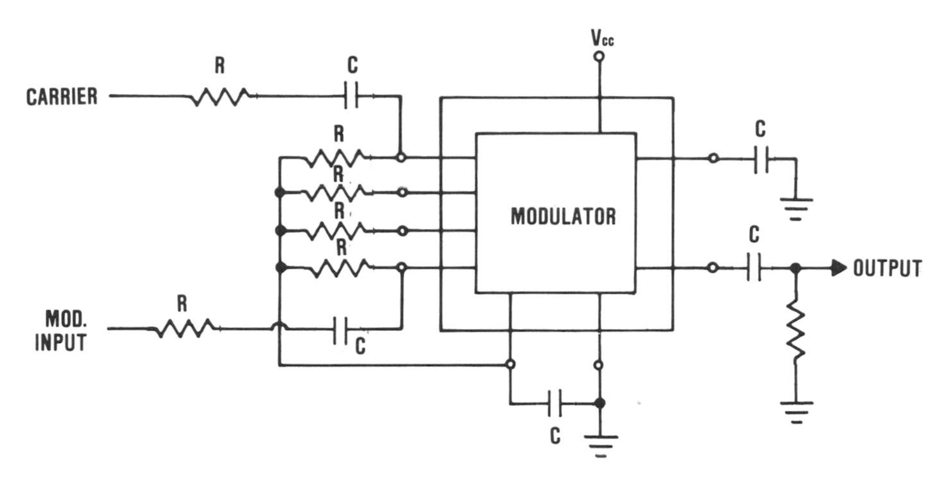

FIGURE A.10.1

stability up to 1.0 MHz, this IC includes internal amplitude limiting of the modulating signals. All resistors illustrated in Figure 10.1 are 600 ohms, the conventional telecommunications line impedance. Capacitors are chosen according to the input frequency.

KEY PARAMETERS

a) *Carrier suppression*. The ratio of carrier-to-signal amplitude at the output. 50 dB is typical.
b) *Conversion gain*. The ratio of output and input signals. 5 dB is typical.
c) *Second harmonic suppression*. The ratio of the fundamental to the second harmonic of the carrier frequency. 75 dB is typical.
d) *Intermodulation products*. The effects of heterodyne action. -60 dB is a typical value.

APPLICATIONS

This type of IC is used as modulator in telephone transmission systems, as synchronized detector, and as analog multiplier.

COMMENTS

A variety of different modulators are available, and it is essential that detailed manufacturer's data be consulted. See also 10.2, Modulator/ Demodulator.

A.10.2 Modulator/Demodulator

DESCRIPTION

Basically a double-balanced modulator, this IC uses "tree" configuration multipliers, an internal bias generator, and a two-stage, common collector output to provide particularly low output impedance. As illustrated in Figure 10.2(a), it can be used with transformer-coupled input signals for the signal and/or carrier. It is also possible to connect the carrier in an unbalanced mode, as illustrated in Figure 10.2(b), and apply the signal through an R-C network to provide the appearance of a balanced input.

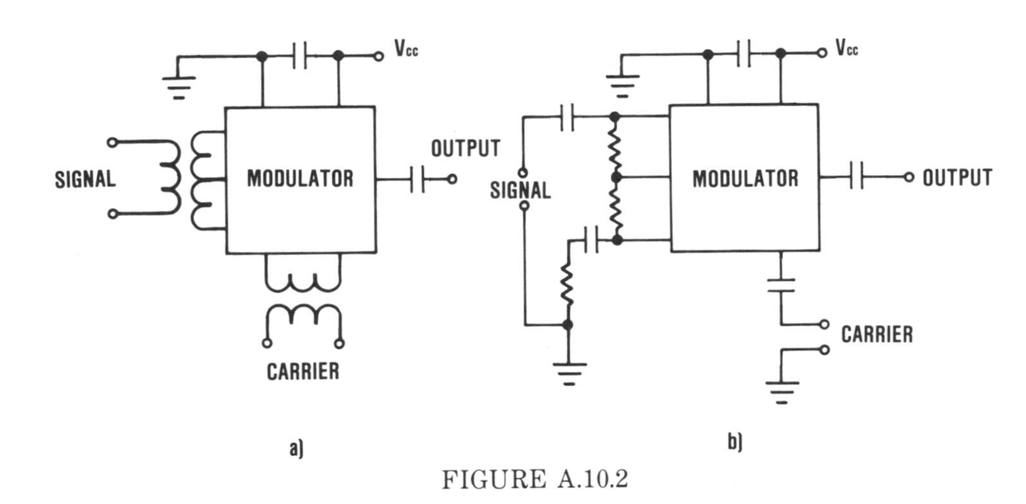

FIGURE A.10.2

KEY PARAMETERS

a) *Carrier suppression.* Ratio of carrier-to-signal amplitude at the output. 50 dB is typical.
b) *Conversion gain.* Ratio of output to input signal. Unity gain is typical.
c) *Second harmonic suppression.* Ratio of the fundamental to the second harmonic of the carrier frequency. 40 dB is typical.
d) *Output impedance.* 10 to 30 ohms typical.
e) *Noise level.* With transformer input -110 dB is typical.
f) *Intermodulation products.* The effects of heterodyne action. -58 dB is a typical value.

APPLICATIONS

This IC is designed for telecommunications transmission systems. It is also used as synchronous detector, FM detector, and phase detector.

COMMENTS

While similar in application to the balanced modulator of 10.1, this IC can operate in a variety of different modes and some of its key parameters are different. A careful check of manufacturer's literature is recommended.

A.11.1 Multiplier, Operational

DESCRIPTION

This IC is based on a four-quadrant multiplier which provides separate, external, gain adjustment for the X and the Y input signal. This means that both the X and the Y input signals can have positive and negative values.The basic multiplier is a variable gain amplifier and one input signal controls the gain of the other signal, so that the output is the product of the two inputs. External X and Y gain adjustments can be used to add scale factors to X and Y. In addition to the direct output signal, there is also a separate high-frequency amplifier output which extends the frequency range that would otherwise be available.

KEY PARAMETERS

a) *Voltage gain.* The maximum gain for each input signal. 25 is a typical value.
b) *Frequency response.* Up to 500 MHz at -3 dB is available in some units.
c) *Phase shift bandwidth.* The highest frequency for a specific phase shift. A typical value may be 3° phase shift at 10 MHz.
d) *Linearity.* The minimum amplitude distortion for the output. 0.3% is typical.
e) *Gain—temperature stability.* The change in gain expected over a given temperature range. A typical value is 3 dB from 0 to +70°C.

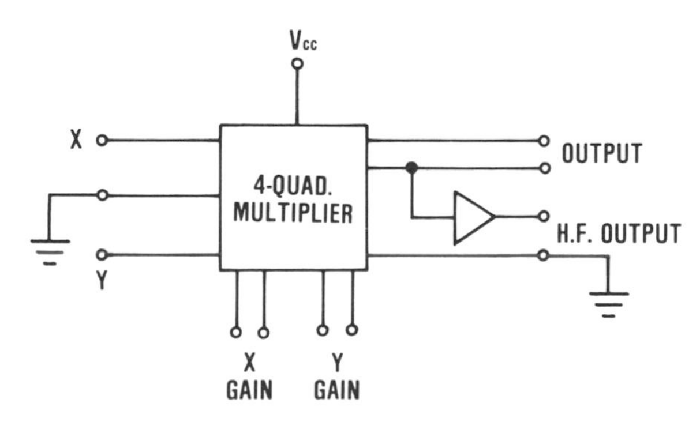

FIGURE A.11.1

APPLICATIONS

Analog computation circuits, certain signal processing circuits, triangle-to-sine-wave converters.

COMMENTS

Variations of this IC often include separate op amps for additional gain at the output. A different version of multiplier usually provides a hybrid IC, uses two logarithic amplifiers and a summing amplifier to provide multiplication in terms of log A + log B = log A B.

Operational multipliers can also be used as dividers by inverting one of the input signals.

A.12.1 Oscillator, Precision

DESCRIPTION

This oscillator is essentially a high-gain amplifier with positive feedback, and operates like an R-C phaseshift oscillator. External components are used to set the frequency. A buffer amplifier is included to keep the oscillator independent of the output loading. Bias voltage variations can be used to vary the center frequency.

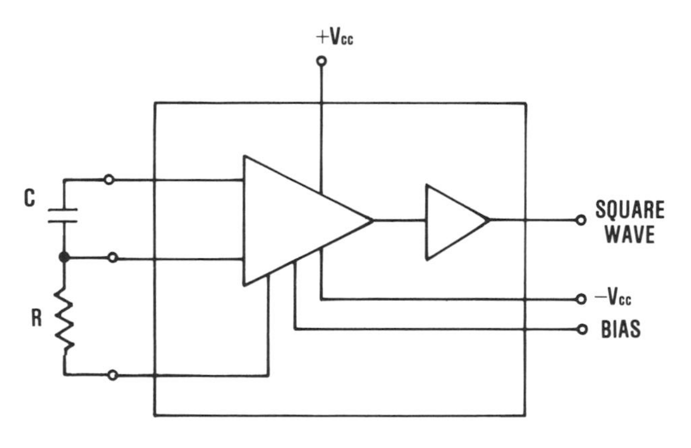

FIGURE A.12.1

KEY PARAMETERS

a) *Frequency range.* Depending on external components, a typical value is from 0.01 Hz to 1.0 MHz.

b) *Frequency temperature coefficient*. Change of frequency with changes in temperature. Excluding the external components, a typical value is 20 ppm/°C.
c) *Supply voltage sensitivity*. Frequency change as a result of supply voltage variations. 0.15% per volt is a typical value.

APPLICATIONS

Because of its stability, this type of IC is often used as clock frequency oscillator or as a very stable PLL oscillator.

COMMENTS

A variety of IC amplifiers can be connected to produce a square-wave signal but only specific oscillator ICs will provide the key parameters indicated above. Other versions of precision oscillators eliminate the effects of the external components by using a crystal as the tuning element.

A.12.2 Oscillator, Voltage-Controlled (VCO)

DESCRIPTION

The VCO generates basic periodic signals which are converted into triangular and square-wave outputs by two separate amplifiers. Because square waves are often used in logic circuits, an open collector and open emitter transistor are used at the output stage to permit the user to select the desired output voltage level. An external timing capacitor sets the basic VCO frequency. A key feature of this VCO is the binary frequency control. Two binary inputs permit the selection of four separate, discrete frequencies which are determined by external timing resistors connected to the current switches. Other versions of the voltage-controlled oscillator have an analog voltage input in place of the two binary inputs shown in Figure 12.2.

KEY PARAMETERS

a) *Frequency range*. The maximum frequency range with any combination of capacitors and resistors is usually from 0.01 Hz to 1.0 MHz.
b) *Frequency accuracy*. Exclusive of external components, a typical value is ±3%.
c) *Frequency temperature coefficient*. 30 ppm/°C is typical.
d) *Power supply versus frequency stability*. Changes in frequency as supply voltage varies. 0.15% per volt is typical.

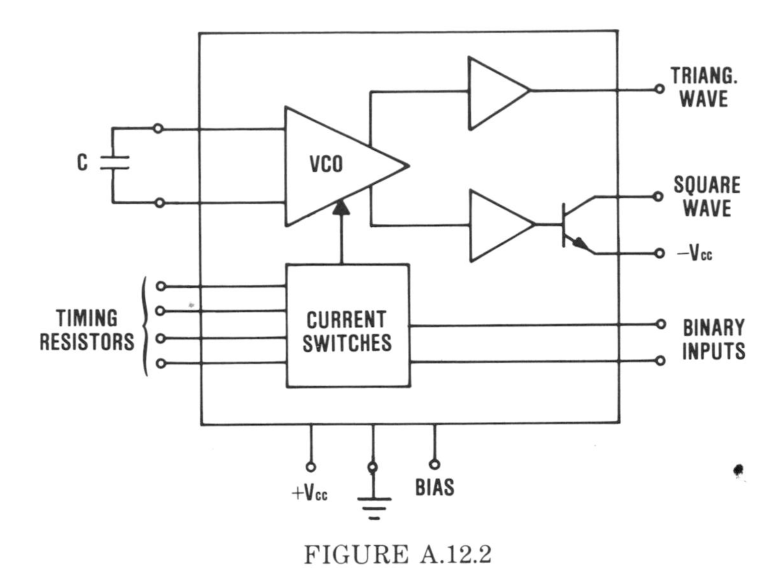

FIGURE A.12.2

e) *Sweep range*. The range over which frequencies can be changed by changing only one of the resistors. 1,000:1 is typical.
f) *Sweep linearity*. The linearity of frequency change versus resistor change. Typical values are 2% linearity over 10:1 frequency range and 5% linearity over 1,000:1 frequency range.

APPLICATIONS

Stable PLL (phase-locked loop) oscillator, FSK modulator, voltage or current-to-frequency conversion. This circuit can also be used as sweep signal generator.

COMMENTS

In many PLL applications the voltage-controlled oscillator (VCO) is included in the PLL IC. This has all of the advantages of monolithic construction, saves parts, and provides better temperature and supply voltage stability.

A.13.1 Phase-Locked Loop (PLL)

DESCRIPTION

The basic feedback circuit of a PLL is illustrated in Figure 13.1. Without input signal, the voltage-controlled oscillator (VCO) operates at its free-running frequency f_o, and the error voltage V_e, as well as its filtered version

V_d, is zero. The phase comparator compares the frequency and the phase of the input signal and of the VCO, and generates either a positive or a negative error voltage which is filtered and applied, as V_d to the VCO. If the input signal is within the capture range of the PLL, the VCO will lock onto the input frequency and generate a signal of the same frequency as f_s with a small, fixed phase difference. Once the VCO is locked onto the input signal, its output frequency f_o will vary over the locking range as f_s varies. This locking range is usually greater than the capture range. In most applications of the PLL, external components are necessary to set the free-running frequency of the VCO and the provide the capacitive component for the low-pass filter. These external components effectively determine the frequency and the range of the PLL.

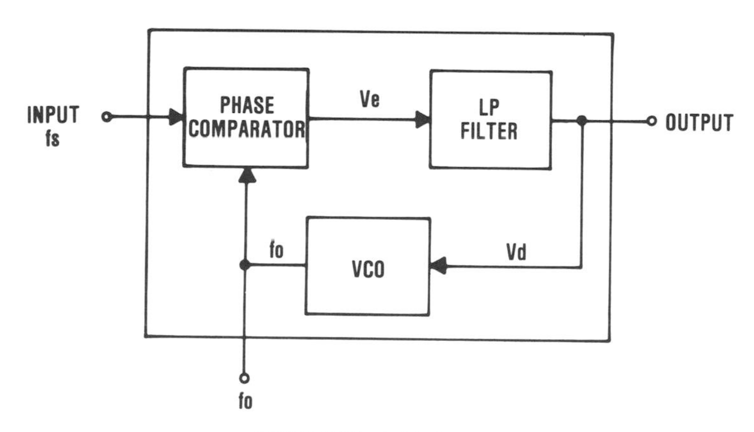

FIGURE A.13.1

KEY PARAMETERS

a) *Lowest operating frequency.* The lowest frequency at which the entire PLL IC can operate in a stable manner.
b) *Highest operating frequency.* The highest frequency at which the entire PLL IC can operate in a stable manner.
c) *Frequency accuracy.* The accuracy with which frequency can be maintained, exclusive of the characteristics of external components.
d) *Frequency temperature coefficient.* The variation in frequency as the temperature changes.
e) *Frequency power-supply stability.* The variation in frequency as power-supply voltage varies.
f) *Input impedance.* The impedance for the external frequency f_s input.
g) *Dynamic range.* The variation in input signal amplitude that the PLL can accommodate.
h) *Output voltage swing.* The maximum output signal, either in peak-to-peak or in R.M.S. voltage.

i) *Tracking (locking range).* The percentage of the free-running frequency f_o over which the PLL will retain frequency lock.
j) *Signal-to-noise ratio.* Noise characteristics of the VCO output signal.

APPLICATIONS

Specific PLL characteristics make some ICs suitable for particular applications. In general, PLLs can be used in FM demodulation, FSK demodulation, frequency synthesizers, data synchronization, synchronous detection of AM signals, tone detection, FM stereo decoding, signal conditioning, tracking filters, and motor speed controls. Specific applications, based on key parameters, are described in more detail in the following pages.

COMMENTS

In some IC PLLs the interconnections of each functional blocking are brought out to allow the user maximum freedom in arranging these blocks for particular applications. The majority of monolithic PLLs, however, are designed for specific applications with a minimum of external leads.

A.13.2 Phase-Locked Loop (PLL), General Purpose

DESCRIPTION

The general purpose PLL usually contains a phase comparator, a low-pass filter with some gain, and a voltage-controlled oscillator (VCO) which also has some adjustable gain. To provide for maximum versatility, a differential amplifier, as shown in Figure 13.2, which can also be used as op amp, is usually included on the IC. External adjustments are provided for the bias of the phase comparator, the capacitive elements of the low-pass filter, one side of the output amplifier, and both range selection and gain control of the VCO. In many general purpose PLLs a VCO sweep input, as well as a VCO output, is provided.

KEY PARAMETERS

a) *Lowest operating frequency.* Lowest frequency at which the entire PLL can operate. 0.5 Hz is a typical value.
b) *Highest operating frequency.* The highest frequency at which the entire PLL can still operate. A typical value is 30 MHz.
c) *Frequency accuracy.* ±5%.
d) *Frequency temperature coefficient.* The variation in frequency as the temperature changes. 250 ppm/°C is typical.

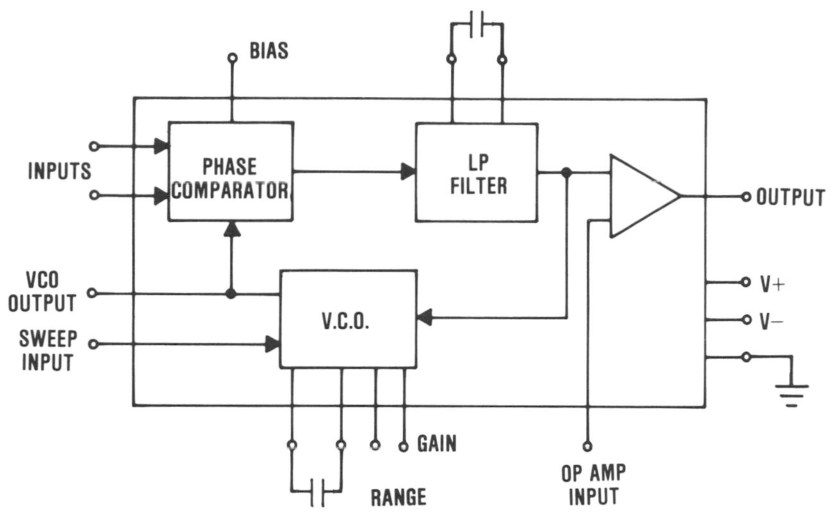

FIGURE A.13.2

e) *Frequency power-supply stability*. The variation in frequency as power-supply voltage changes. A typical value is 0.1%/V.
f) *Input impedance*. The impedance at the input to the phase detector. 1.0 megohm is typical.
g) *Dynamic range*. Input signal amplitude variations that the PLL can accommodate. 300 mV to 3 V typical.
h) *Output voltage swing*. The maximum variation of output voltage. A typical value is 10 V peak to peak.
i) *Tracking (locking) range*. A percentage of the free-running VCO frequency over which the PLL will remain locked. In a typical general purpose PLL this range is adjustable from ±1% to ±50%.
j) *Signal-to-noise ratio*. 65 dB is a typical value.

APPLICATIONS

FM demodulation, FSK demodulation and modulation, tracking filter, data synchronization, voltage-to-frequency conversion, FM or FSK or sweep generation, frequency synthesizer.

A.13.3 Phase-Locked Loop (PLL), High-Frequency

DESCRIPTION

The high-frequency response of a general purpose PLL can be improved by adding an input preamplifier (1) and a special output circuit. As illustrated in Figure 13.3, the special output circuit consists of an amplifier and a two-input Schmitt-trigger circuit which provides the output signal. By

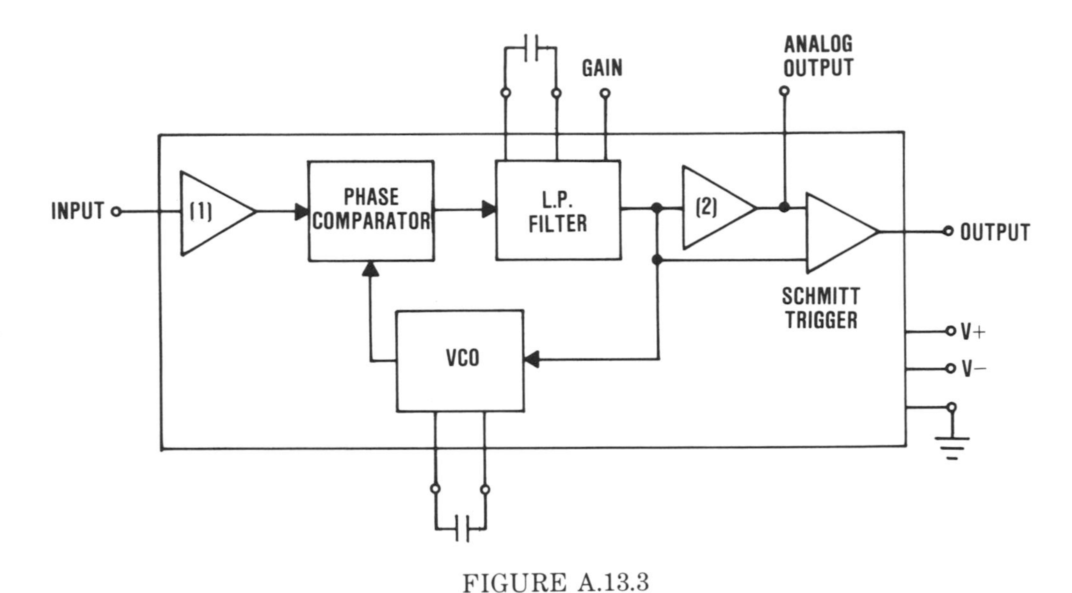

FIGURE A.13.3

using an integrating amplifier (2), an analog output signal is obtained as well.

KEY PARAMETERS

Essentially the same as for the general purpose PLL except for the following:

a) *Highest operating frequency*. The highest frequency at which the entire PLL can operate in a stable manner. 50 MHz is a typical value.
b) *Frequency temperature coefficient*. Variation of frequency as temperature changes. 400 ppm/°C is typical.
c) *Frequency power-supply stability*. Variation in frequency as power-supply voltage changes. A typical value is 3%/V.
d) *Output voltage swing*. The maximum output voltage available. 15 mV R.M.S. is typical.
e) *Signal-to-noise ratio*. 40 dB typical.

APPLICATIONS

High-speed modems, FSK receivers and transmitters, frequency synthesizers, and similar applications at frequencies above 30 MHz.

A.13.4 Phase-Locked Loop (PLL), Precision

DESCRIPTION

As shown in Figure 13.4, an ultra-stable voltage-controlled oscillator (VCO) is combined with isolating amplifiers at the input and output, and a special internal reference generator adds to high precision. The overall operation is the same as for a general purpose PLL.

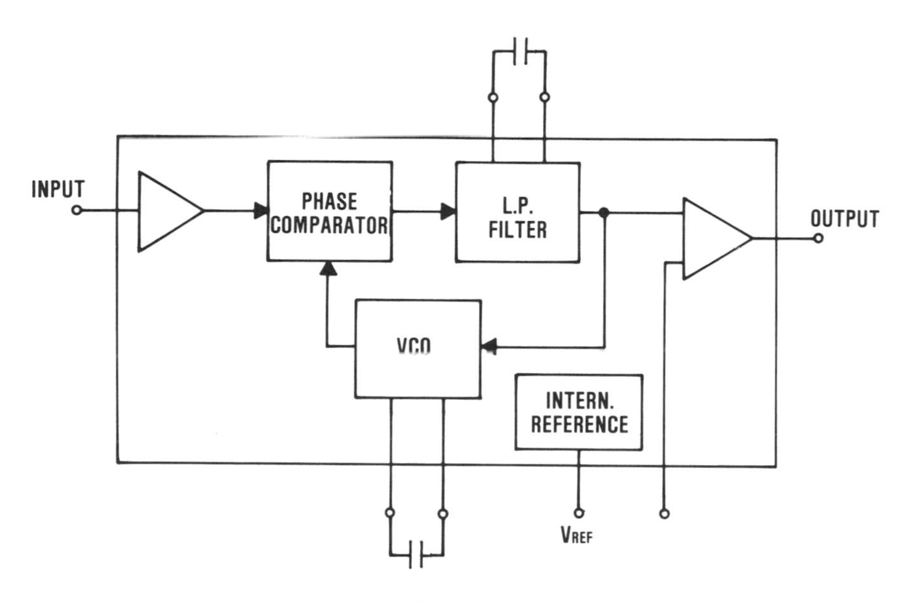

FIGURE A.13.4

KEY PARAMETERS

Essentially the same as for the general purpose PLL except for the following:

a) *Lowest operating frequency*. Lowest frequency at which the entire PLL operates in a stable manner. 0.01 Hz typical.
b) *Highest operating frequency*. Highest frequency at which the entire PLL operates in a stable manner. 300 KHz typical.
c) *Frequency accuracy*. Excluding external components, a typical value is ±1%.
d) *Frequency temperature coefficient*. Variation in frequency as temperature changes. 200 ppm/°C is typical.
e) *Frequency power-supply stability*. Variation in frequency as power-supply voltage changes. 0.05%/V.

APPLICATIONS

High-precision frequency synthesizers, data synchronization systems, and tracking filters.

COMMENTS

The external components must have equal or better electrical stability than the precision desired of the overall PLL.

A.13.5 Phase-Locked Loop (PLL), Programmable

DESCRIPTION

This PLL is essentially the same as a general purpose PLL or any of the other specialized types, with the additional feature that the VCO and phase comparator can be switched between two input signals, effectively time-multiplexing the operation of the PLL. As illustrated in Figure 13.5, the program input terminal on the VCO can be connected to a digital signal. When this signal is high, logic **1**, the PLL operates on input 1, and when that signal is low, logic **0**, the entire PLL operates on input signal 2.

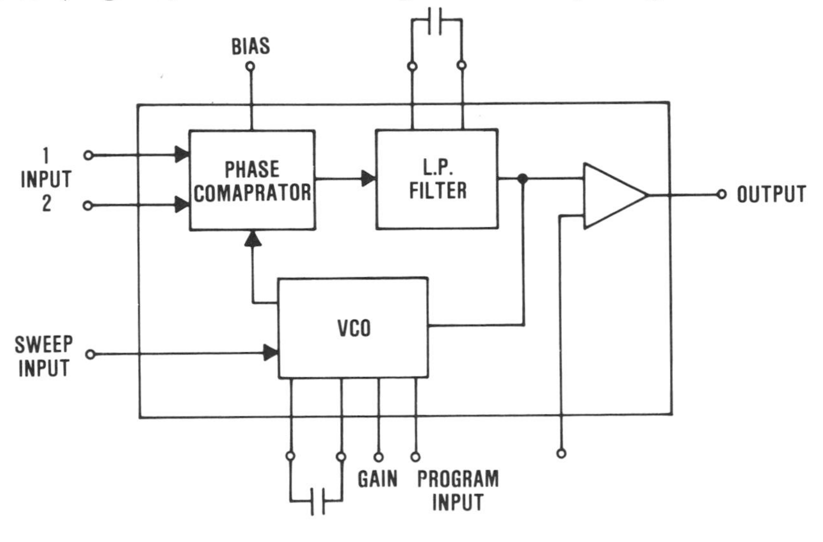

FIGURE A.13.5

KEY PARAMETERS

The same as for the general purpose PLL or other specialized PLLs listed in the previous sections except for the programming capability.

APPLICATIONS

The ability to time multiplex the PLL between two inputs makes it especially useful for telemetry and dual-channel communications systems applications. The actual applications are the same as those shown for the general purpose PLL.

A.14.1 Regulator, Voltage

DESCRIPTION

IC voltage regulators operate basically in the same manner as discrete circuit regulators. As illustrated in Figure 14.1(a), the input voltage, usually the rectified and filtered power-supply output, is applied to the reference circuit and to the op amp. For optimum regulation, the op amp must have high open-loop gain and a fairly large common-mode rejection ratio (CMRR). The reference source determines the accuracy of the regulated voltage output. Reference voltage can be provided by such simple devices as a zener diode or by more complex multitransistor circuits, such as the "band gap" voltage reference. Extremely stable reference circuits depend on automatic cancellation of drift characteristics due to temperature and input voltage changes.

Figure 14.1(b) illustrates a method called "regulator within a regulator," which is used in many IC voltage regulators to provide adjustment of the output voltage and, at the same time, reduce the output impedance of the regulator itself. In this circuit the output voltage of op amp (1) is determined by the ratio of R2/R1. The final output voltage V_o will be the reference voltage, modified by R2/R1 because the feedback factor of amplifier (2) is unity.

IC voltage regulators have a limited power-handling ability, and this can be expanded by using one or more external power transistors. The basic connection for a single external power transistor is shown in Figure 14.1(c), but a variety of other connections can be used. Both NPN and PNP power transistors as well as Darlington pairs and parallel transistors can be controlled by the IC voltage regulator. Special features, such as automatic overload protection, overvoltage protection, automatic thermal shut-down, short-circuit indication, remote control shut-down, etc., are available on different types of IC voltage regulators.

KEY PARAMETERS

a) *Output voltage*. The regulated output voltage maintained over the specified operating range.

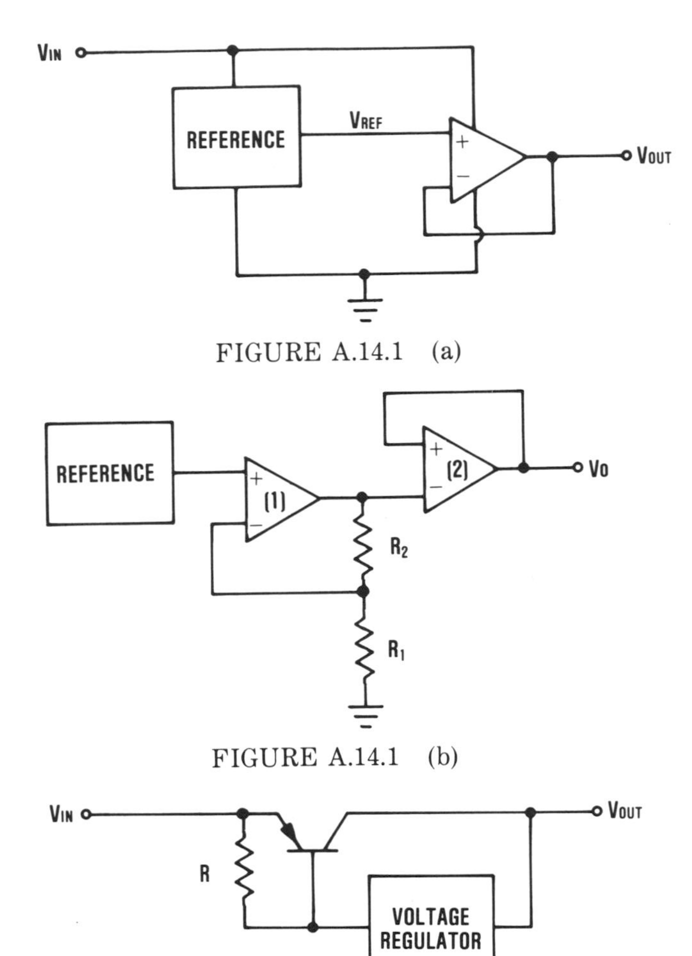

FIGURE A.14.1 (c)

b) *Input voltage.* The range of input voltages for which the output voltage is maintained.
c) *Line regulation.* The change of output voltage for a corresponding change in input voltage.
d) *Load regulation.* The change of output voltage for a corresponding change in output current.
e) *Output voltage temperature coefficient.* The change in output voltage per degree C change in junction temperature.
f) *Output current range.* Minimum output current required to maintain regulation and maximum output current.

g) *Maximum operating junction temperature.* Maximum temperature at which the IC voltage regulator can operate safely.
h) *Maximum power dissipation.* Maximum power that IC can dissipate at 25°C (room temperature) without a heat sink or forced air cooling.
i) *Safe operating area (SOA).* A combination of maximum output current and the difference between input and output voltage. Usually the area under a curve which describes the maximum safe operating characteristics for different values of input and output voltage and maximum output current.

APPLICATION

Wherever the power-supply voltage must be kept at a fixed value, regardless of variations in load current or input voltage. Specific applications will be discussed in the following pages.

COMMENTS

A wide variety of IC voltage regulators, with special features and many different characteristics, is available. Manufacturer's specifications and application data should be studied carefully for special features, mounting instructions, external transistor connections, and thermal requirements. Switching regulators use entirely different operating principles and are described briefly in Section 14.7.

A.14.2 Regulator, Voltage, Fixed Output

DESCRIPTION

This basic three-terminal device is available as either positive or negative voltage regulator for a large range of fixed output voltages. In Figure 14.2(a) small capacitors are shown at the input and output of the voltage regulator, because in many applications such capacitors are required to assure stable operation. Ordinarily we can obtain voltage regulators of the required polarity but it is also possible, as illustrated in Figure 14.2(b), to use a negative voltage regulator to provide a regulated positive output voltage. Similar connections can be used for positive voltage regulators to provide negative output voltages. Optimum regulation performance is obtained when the correct polarity voltage regulators are used. Fixed output regulators are frequently used with external transistors to handle larger regulated output currents.

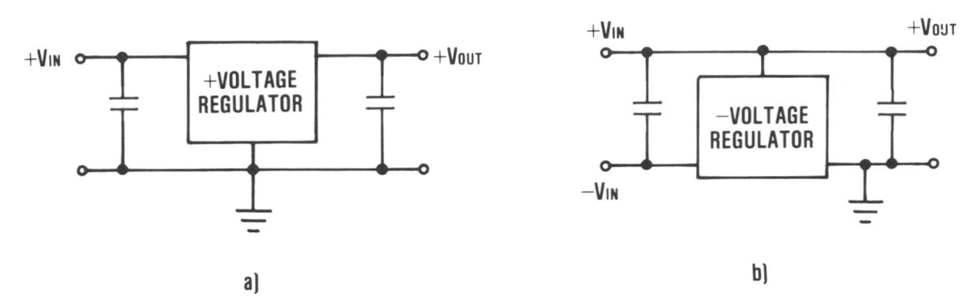

FIGURE A.14.2

KEY PARAMETERS

a) *Output voltage.* The regulated output voltage maintained over the specified operating range. Popular values are 5, 12, 15, 18 and 24 volts.
b) *Input voltage.* Range of input voltages for which the regulated output voltage is maintained. Depending on output voltage, 7 to 30 volts are typical values.
c) *Line regulation.* The change of output voltage for a corresponding change in input voltage. Often stated in millivolts for 1-volt changes, a typical line regulation is 0.1% per volt.
d) *Load regulation.* The change in output voltage for a corresponding change in output current. A typical value is 0.1% over the specified range of load current.
e) *Output voltage temperature coefficient.* The change in output voltage per degree change in junction temperature. Sometimes stated as millivolts per degree C change. Typical values are ±0.002% of output voltage per degree C.
f) *Output current range.* The minimum output current to maintain regulation and maximum output current. Actual values depend on the specific IC, but a range of 100:1 is typical.
g) *Maximum operating junction temperature.* The maximum temperature at which the IC voltage regulator can operate safely. Most commercial ICs are rated for 150°C.
h) *Maximum power dissipation.* Maximum power the IC can dissipate at 25°C (room temperature) without a heat sink or forced air cooling. Commercial ICs can handle from 100 milliwatts to 50 watts. More than 2 watts usually requires an external heat sink.
i) *Safe operating area (SOA).* The area under a curve which describes the maximum safe operating characteristics. For fixed output voltage regulators the SOA is usually not as important as for variable output devices.

APPLICATIONS

As voltage regulators where the output voltage is fixed and usually required in standard values. Digital systems using standard logic families often require simple, fixed output voltage regulators at various points in the system.

COMMENTS

It is possible to get some adjustment of the regulated output voltage even with a fixed output voltage regulator. Some external circuitry, including an op amp and some resistors, can be connected to provide some adjustment, but it is generally more cost effective to use an adjustable type voltage regulator.

A.14.3 Regulator, Voltage, Adjustable Output

DESCRIPTION

The basic circuit is essentially the same as for fixed output voltage regulators, but there is provision for an external resistive divider or potentiometer to adjust the output voltage. Positive or negative polarity output voltage can be obtained by the same type of connections described in Section 14.2. Most of the adjustable type voltage regulators also contain one or more protective features such as overvoltage and overtemperature shutdown. Manufacturers' data contain specific resistance values and/or potentiometer connections to provide the desired output voltage for a specified input voltage.

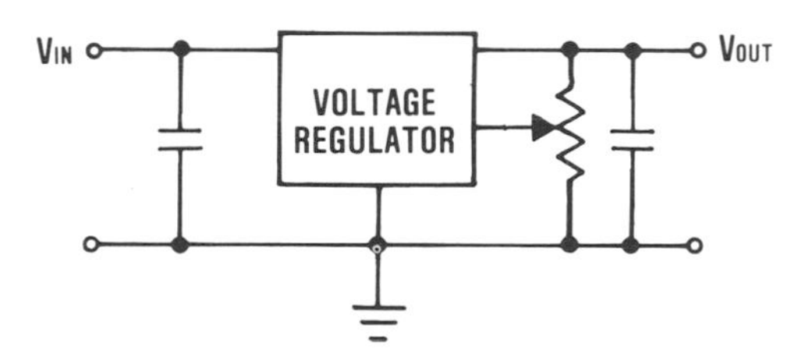

FIGURE A.14.3

KEY PARAMETERS

Essentially the same as for the fixed output voltage regulator, except for the following:

a) *Safe operating area (SOA).* Combination of maximum output current and the voltage difference between input and output voltage. Manufacturers' data contain coordinates and a curve which must be used by the designer to locate the input and output voltage range, and output current, within the safe operating area.

APPLICATIONS

In many types of analog equipment the regulated power-supply output voltage must be carefully adjusted for optimum circuit performance. Photo tubes and transistors are typical of devices for which adjustable, regulated, power supplies are required.

A.14.4 Regulator, Voltage, Tracking

DESCRIPTION

In effect, a tracking voltage regulator performs the function of two separate voltage regulators, positive and negative, with but a single adjustment to track both voltages accurately. As shown in Figure 14.4, the voltage from either output to ground will remain the same regardless of load unbalance. Adjustment of the external potentiometer or resistive network will vary both positive and negative voltages in exactly the same way. Most tracking voltage regulators provide equal positive and negative voltages, but it is also possible to obtain tracking voltage regulator ICs for which the positive and negative voltages are different.

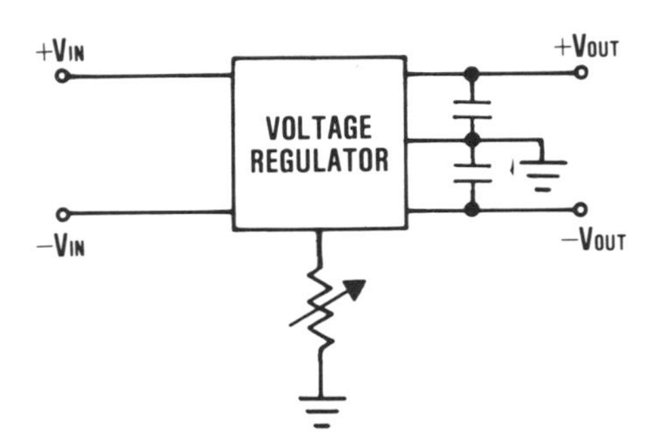

FIGURE A.14.4

KEY PARAMETERS

Essentially the same as for the fixed output or adjustable voltage regulator, except for the following:

a) *Output voltage.* The regulated output voltage maintained over the specified operating range. Typical values are ±10, ±12, ±15, ±18V.
b) *Maximum power dissipation.* Maximum power that the IC can dissipate at 25°C (room temperature) without a heat sink or forced air cooling. Includes both the positive and negative loads. Typical values range up to 3 watts.

c) *Safe operating area (SOA).* Combination of maximum output current and the difference between the input and output voltage. Both positive and negative loads must be added to locate the operating point within the safe operating area curve.

APPLICATIONS

As voltage regulator for power supplies required for op amps, bridge circuits, and other applications where positive and negative voltages must track each other.

A.14.5 Regulator, Voltage, Floating

DESCRIPTION

Used for regulated voltages above 40 volts, the floating voltage regulator requires at least one external transistor, at least one zener diode, and several external resistors. As illustrated in the simplified diagram of Figure 14.5, R1 limits the input current to the regulator IC. The reference voltage is applied to two resistive dividers. The tap from the grounded resistive divider is returned to the non-inverting input of an internal op amp, while the tap going from the reference voltage to the output voltage is connected to the inverting terminal of that op amp. One of these resistive dividers determines the output voltage, while the other sets the maximum

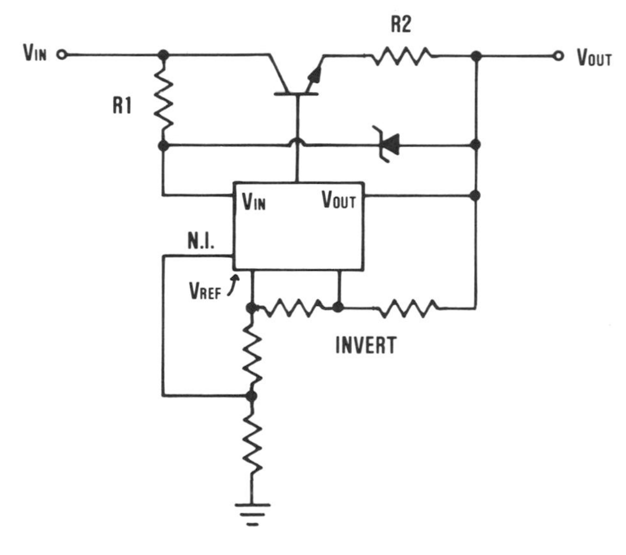

FIGURE A.14.5

output current. A positive voltage output floating regulator is shown, but the same principle applies to negative floating voltage regulator. Additional external components are frequently used to protect the IC regulator against overvoltage and excessive output current.

KEY PARAMETERS

Essentially the same as for the adjustable voltage regulator of 14.3 except for the following:

a) *Output voltage.* The regulated output voltage maintained over the specified operating range. Because of the floating application, several hundred volts are often possible.
b) *Input voltage—output voltage difference.* Instead of absolute input voltage, this parameter limits the application of the floating voltage regulator. Typical difference voltages are in the order of 5 to 10 volts.

APPLICATIONS

To regulate the output of high-voltage power supplies, such as for photo multipliers and other vacuum tubes.

COMMENTS

All of the tolerances and adjustment ranges are generally more limited for a floating voltage regulator, and the manufacturer's application notes should be carefully reviewed for the specific parameters and limitations.

A.14.6 Regulator, Voltage, Precision

DESCRIPTION

High-precision voltage regulators are available for fixed or adjustable output as well as for most positive and negative polarities. The simplified block diagram of Figure 14.6 illustrates the use of a reference amplifier (1) with a special temperature-compensated zener diode conected between its input and output. This, in addition to a highly stable reference source, provides the precision regulation. In all other respects precision voltage regulators operate in the same basic manner as the other, fixed or variable output, regulators.

KEY PARAMETERS

Essentially the same as for fixed and variable output voltage regulators, except for the following:

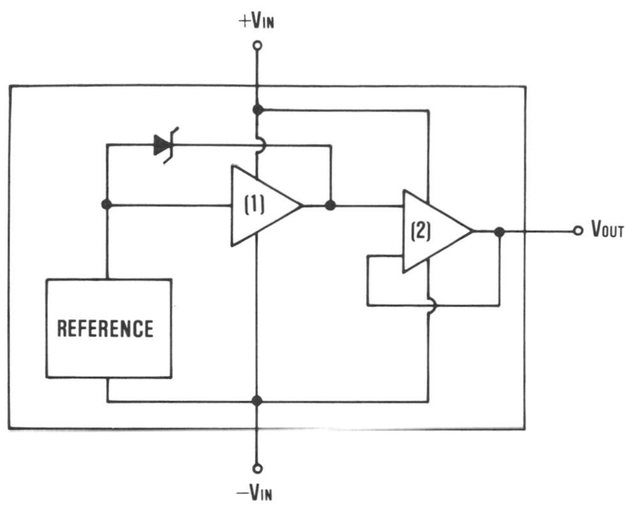

FIGURE A.14.6

a) *Line regulation.* The change of output voltage for a corresponding change in input voltage. Typical value ±0.01%/V.
b) *Load regulation.* The change of output voltage for a corresponding change in output current. ±0.01%/V is a typical value.

APPLICATIONS

In instrumentation and analog computation systems, the precision of the power-supply voltages often requires the use of precision voltage regulators.

COMMENTS

The full precision capability of the IC may be lost when external transistors and resistive networks, with less limited tolerances, are used.

A.14.7 Regulator, Voltage, Switching

DESCRIPTION

Switching regulators are based on chopping the unregulated input voltage with a saturated transistor and then filtering the output with an L-C network, as illustrated in Figure 14.7. Voltage regulation is accomplished by varying the duty cycle of the switching transistor Q1. Compared to the linear voltage regulators, the switching regulator offers much greater efficiency but requires more external components, particularly an iron core inductor or transformer and a filter capacitor. The transient response is limited by the chopping frequency, and the characteristics of the external components, particularly the switching transistor and the diode, can be quite critical. Switching regulators are generally more cost effective than linear voltage regulators above 100 watts or in applications where the 60 Hz components used with linear regulators exceed the space and weight limitations.

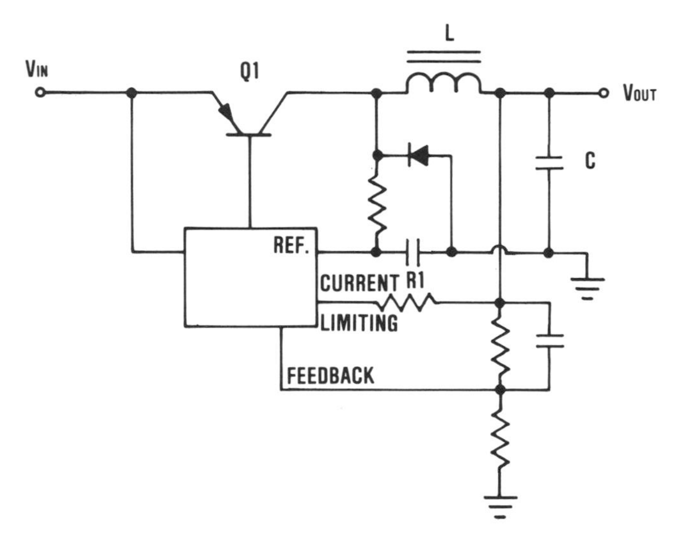

FIGURE A.14.7

KEY PARAMETERS

Essentially the same as fixed output voltage regulators, except for the following:

a) *Output voltage temperature coefficient.* The change in output voltage as the junction temperature changes. Depending on the external transistor and diode characteristic, a 0.3% variation over the allowable temperature range is typical.
b) *Maximum operating frequency.* The highest switching frequency of which the regulator is capable. 300 kHz is a typical value.
c) *Noise characteristic.* High-frequency noise is always generated at harmonics of the switching frequency. Electromagnetic shielding and efficient filtering are required in most applications.

APPLICATIONS

Switching power supplies and regulators are generally used in vehicular, aircraft and space electronics where size and weight limitations exist, and where equipment often has to run from rechargeable battery supplies.

COMMENTS

Switching power supply and regulator design is a highly specialized field and requires special expertise. The problem of electromagnetic interference, caused by switching power supplies, has been the subject of extensive investigation.

A.15.1 Switch, Analog

DESCRIPTION

The basic analog switch, using either MOS, CMOS, FET or JFET semiconductors, is illustrated in Figure 15.1. When the logic control signal is "high," logic **1**, a positive voltage is applied through amplifier (1) to the gate of the N-type transistor. Inverting amplifier (2) applies a negative voltage to a gate of the P-type device. As a result, both transistors are effectively cut off, presenting a very high impedance between the input and output terminals. When the control signal is "low" or a logic **0**, a negative voltage is applied to the gate of the N-type transistor and a positive voltage to the gate of the P-type transistor. This turns both transistors on and they present a very low impedance between input and output terminals. Different

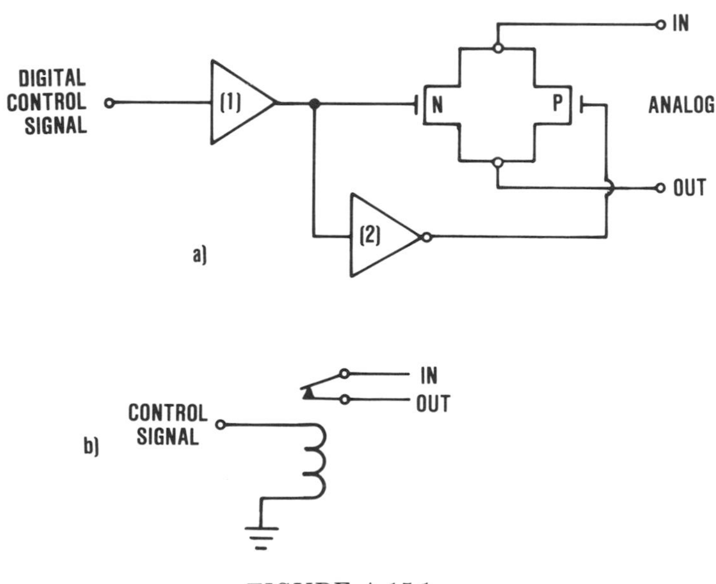

FIGURE A.15.1

versions of this basic analog switch are used to supply the wide range and great variety of analog switches available. General purpose and special feature switches are described in the following pages.

KEY PARAMETERS

a) *Analog voltage range.* The maximum undistorted voltage that can be passed between the input and output terminals when the switch is closed.
b) *Analog current range.* The maximum undistorted current that can pass between the input and output terminals.
c) *"On" resistance.* The resistance between the input and output terminals when the switch is closed.
d) *"Off" isolation.* The attenuation of signals between the input and output terminals when switch is open. This attenuation decreases as frequency increases.
e) *"Off" input leakage current.* The leakage current at the input terminal when the switch is "off."
f) *"Off" output leakage current.* The leakage current at the output terminal when the switch is "on."
g) *Control signal low threshold.* The maximum control voltage at which the switch remains "on."
h) *Control signal high threshold.* The minimum control voltage that opens the switch.
i) *Break-before-make delay.* Essentially the transit time of the switching transistor itself.
j) *Switch "on" time.* The time delay from the leading edge of the control signal to the opening of the switch.
l) *"Off" input capacitance.* Capacitance at the analog input terminal when the switch is "off."
m) *"Off" output capacitance.* The capacitance at the output terminal when the switch is "off."
m) *"On" output capacitance.* The capacitance at the output terminal when the switch is "on."
o) *"Off" input-to-output capacitance.* The capacitance between the input and output terminals when the switch is "off."
p) *Maximum power dissipation.* The total power dissipation, including the digital control signal, V_{CC} and the maximum analog signal.

APPLICATIONS

IC analog switches can be used in many of the same applications as relays and manual switches. The small size, low cost, and high operating speed of IC analog switches have made a host of additional applications possible. These are described in the following pages.

COMMENTS

Many analog IC switches are available in multiple configurations, as, for example, four single-pole, single-throw units on one IC. In some instances common connections of input or output are provided for analog multiplexing and demultiplexing. Logic functions are often included as part of the digital control input. Some of the key parameters listed above vary greatly between different manufacturers and IC technology. Be sure to consult the manufacturer's data for critical switching applications.

A.15.2 Switch, Analog, General Purpose

DESCRIPTION

General purpose analog switches are available in a variety of contact arrangements, as illustrated in Figure 15.2. The most basic contact arrangement is the single-pole, single-throw (SPST) configuration shown in Figure 15.2(a). Note that in this, as well as in all other configurations, a positive and negative supply voltage is required. Although the terminals are labeled "input" and "output," current can flow in either direction, and that makes double-throw switches like the SPDT and the DPDT of Figure 15.2(a) and (b) respectively, quite versatile. The DPST of Figure 15.2(c) uses two SPST switches which are controlled by a single driving source. To

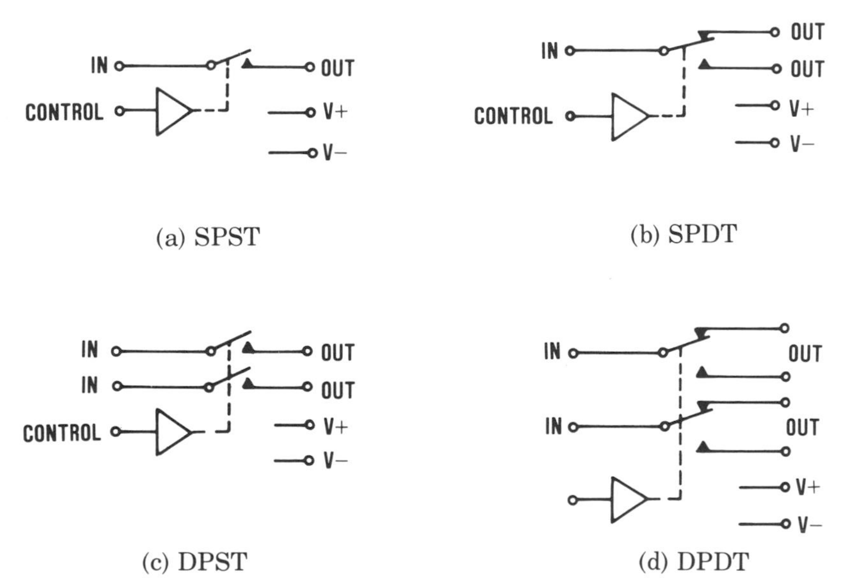

(a) SPST (b) SPDT

(c) DPST (d) DPDT

FIGURE A.15.2

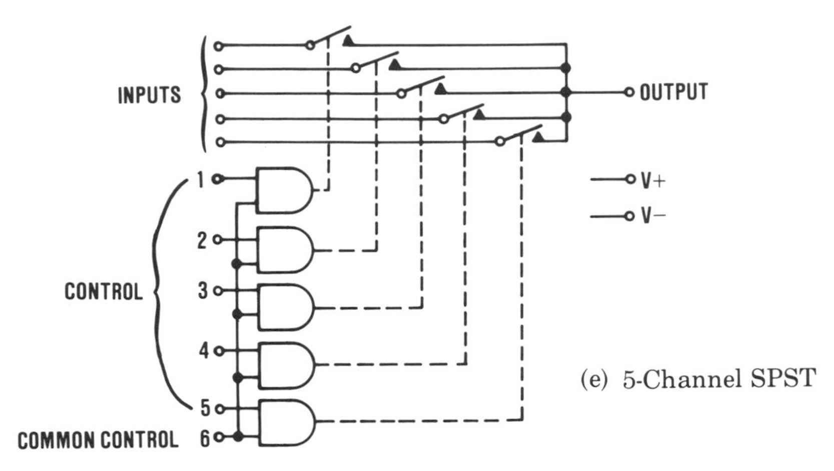

FIGURE A.15.2 (continued)

implement a rotary or push-button switch, the five-channel SPST configuration illustrated in Figure 15.2(e) can be used. Note that in this illustration the five driving sources are AND circuits, with five separate and one common control input. Many different contact arrangements, all based on the fundamentals illustrated here, are available as general purpose analog switches.

KEY PARAMETERS

a) *Analog voltage range.* The maximum undistorted voltage passed through the switch. Typical values are either ±10 volts or ±15 volts.
b) *Analog current range.* The maximum undistorted current passed through the switch. Typical values are 30 to 50 mA
c) *"On" resistance.* The resistance of the switch when it is closed. Typical values range between 10 and 50 ohms.
d) *"Off' isolation.* The attenuation of the signal between input and output terminals when the switch is open. 70 to 80 dB is typical.
e) *"Off" input leakage current.* Leakage current at the input when the switch is open. 5nA is typical.
f) *"Off" output leakage current.* The leakage current at the output terminal when the switch is open. 5 nA is typical.
g) *Control signal low threshold.* The maximum control voltage that closes the switch. 0.8 volts typical.
h) *Control signal high threshold.* The minimum control voltage that opens the switch. 3.0 volts typical.
i) *Switch "on" time.* The delay from the leading edge of the control signal until the switch is closed. 500 ns is a typical value.
j) *Switch "off" time.* The time delay between the leading edge of the control signal until the switch is opened. 250 ns is a typical value.

APPLICATIONS

D/A converters, audio switching, programmable power supplies, and any other circuit where digital control of analog signals is required.

COMMENTS

Consult the manufacturer's literature for the particular logic configuration and contact arrangement best suited to your application. If both the analog switch and the control logic can be obtained on a single IC, temperature stability and overall reliability will be better than if separate ICs are used.

A.15.3 Switch, Analog, High-Frequency

DESCRIPTION

Essentially the same as general purpose switches, the switches capable of handling video or RF must have special, high-frequency characteristics. JFET or CMOS transistors are used because of their small internal capacitance. The upper frequency limit is determined by the "off" isolation, which decreases as frequency increases. Some of the high-frequency analog switches that feature high "off" isolation also have a relatively higher "on" resistance. When the "on" resistance is lower, the attenuation between input and output signals is also lower at the higher frequencies.

KEY PARAMETERS

Essentially the same as for general purpose switches, except for the following:

a) *Analog voltage range.* The maximum undistorted voltage passed through the switch. ±12 volts is typical.
b) *"On" resistance.* The resistance of the switch when it is closed, 30 ohm to 75 ohm, depending on the "off" isolation.
c) *"Off" isolation.* The attenuation of the analog signal when the switch is open. 60 to 50 dB at frequencies from 10 MHz to 100 MHz.
d) *Maximum analog frequency.* The highest frequency with no more than 3 dB attenuation that can pass through the closed switch.
e) *Switch "on" time.* The delay from the leading edge of the control signal until the switch is closed. 150 to 300 ns is typical.
f) *Switch "off" time.* The time from the leading edge of the control signal until the switch is open. 100 to 300 ns is typical.

APPLICATIONS

As switching elements for high-frequency D/A converters, high frequency sample-hold switches, digital filters, and op amp gain switching.

COMMENTS

Remember that the input and output capacitance are the basic limitations of the high-frequency response. External circuitry should minimize capacitive loading.

A.15.4 Switch, Analog, High-Speed

DESCRIPTION

Essentially the same in operation as the general purpose, analog switch described in Section 15.2, high-speed switches use a JFET switching element as well as high-speed control logic. The same characteristics that enable the JFET transistor to switch fast also make it suitable for high-frequency analog signals. Most of the basic contact arrangements described in Figure 15.2 are also available in the high-speed analog switch version.

KEY PARAMETERS

Essentially the same as described for the high-frequency switch of 15.3, except for the following:

a) *"On" resistance*. The resistance of the switch when it is closed. 30 ohms typical.
b) *Switch "on" time*. The time from the leading edge of the control pulse until the switch is closed. 100 to 200 ns typical.
c) *Switch "off" time*. The time from the leading edge of the control pulse until the switch is open. 100 ns typical.
d) *Toggle rate*. The highest frequency at which the switch can be actuated. 4 MHz is a typical value.

APPLICATIONS

High-speed analog switches are useful for sample-hold circuits, high-speed commutators, switching video, and other high-frequency signals that must be turned on and off at great speed.

COMMENTS

Refer to Section 15.3, High-Frequency Analog Switch, for some additional details.

A.15.5 Switch, Analog, Low-Power

DESCRIPTION

This type of analog switch is essentially the same as the general purpose switches described in Section 15.2. Very low standby and operating powers make this family of devices particularly suitable for battery operated equipment or other applications where low-power dissipation is an important characteristic. The basic contact arrangement illustrated in Figure 15.2 can be also obtained as low-power switches.

KEY PARAMETERS

Essentially the same as for general purpose analog switches, except for the following:

a) *Analog current range.* The maximum undistorted analog current passed through the closed switch. 10 mA is typical.
b) *"On resistance.* The resistance between the input and output when the switch is closed. 50 ohms is a typical value.
c) *"Off" isolation.* The attenuation of the analog signal when the switch is open. A typical value is 60 dB at 500 kHz
d) *Standby power.* Power consumption per contact configuration, with the switch open. 0.06 mW is a typical value.
e) *Operating power.* Power required per contact configuration when all contacts are closed. 7.5 mW is a typical value.

APPLICATIONS

Sample-hold, audio, IF, and other communications signals switching in portable, battery operated equipment.

A.15.6 Switch, Analog, Multiplexer (MUX)

DESCRIPTION

Multiplexers are essentially an array of SPST switches, connected to function like a single-pole, multiposition switch. While the eight-channel configuration shown in Figure 15.6 is the most widely used, other arrangements up to 16 channels are available. Most analog multiplex ICs include a digital decoder so that only a binary input is required to select the desired switch position. When used as commutator, the binary input must change at the commutation rate. Differential multiplexers use an array of DPST switches going to two outputs, permitting direct connection to the differential amplifier inputs.

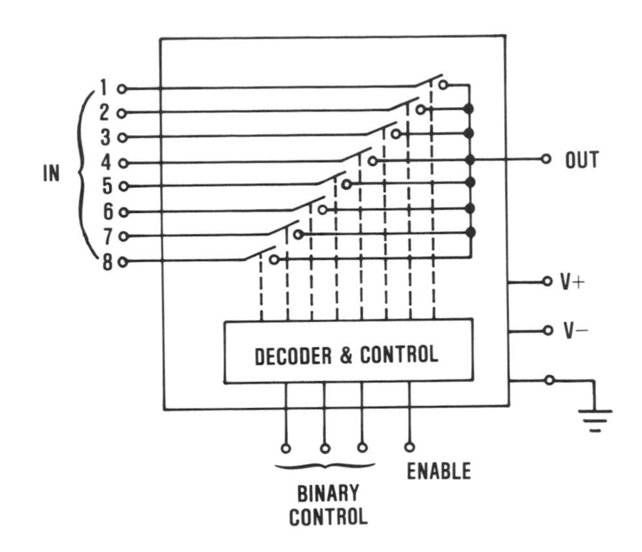

FIGURE A.15.6

KEY PARAMETERS

Essentially the same as for corresponding analog switches described in 15.2 through 15.4, except for the following:

a) *Break-before-make-delay.* Essential to avoid overlap in the output. 100 ns typical.
b) *Channel separation.* The attenuation of adjacent input signals. A typical value is 62 dB at 1 kHz.
c) *Channel switching time.* The minimum time per channel. 30 ns is typical.

APPLICATIONS

Data acquisition systems, telemetry, industrial controls, programmable power supplies, and automatic test equipment. Multiplex front end for A/D converters.

COMMENTS

Because multiplexers usually operate with relatively fast binary pulse signals, manufacturer's data must be consulted to assure that the digital control signals meet the specifications concerning rise time, fall time, duration, and logic levels.

A.16.1 Timer, General Purpose

DESCRIPTION

Most general purpose timer ICs can operate either as a monostable (one-shot) or an astable (multivibrator) circuit. As illustrated in Figure 16.1, the block diagram of one of the most popular timers is connected in the monostable mode. When the circuit is "off," the output of the control FF is **O** which makes Q1 an effective short circuit across the external capacitor C. A negative pulse applied to the trigger input causes the output of comparator 2 to go to **1** which sets the control FF. This in turn cuts off Q1 and allows the external capacitor C to charge, through external resistor R, up to +V. Note that the threshold input to comparator 1 is connected to the junction of the external R-C circuit. The other input to comparator 1 is connected to the voltage divider from +V to ground. When the threshold voltage exceeds the other input of comparator 1, the control FF will reset, turning Q1 on again and shorting out the external capacitor C. The time period during which control FF is "on" therefore depends on R-C and the internal voltage divider. The output of this circuit will be a positive pulse, due to the inverter after the control FF, and the duration of the pulse will depend on R-C.

To operate this circuit in the astable, or free-running, mode, it is only necessary to connect another resistor between the discharge and the trigger terminal, and then connect the trigger terminal to the threshold. In this mode, the external capacitor will charge through the two resistors and will discharge through the resistor between the discharge and trigger terminals. The duty cycle can be set precisely by the ratio of these two resistors, making it possible to obtain pulses rather than square waves and sawtooth rather than triangular waves.

KEY PARAMETERS

a) *Maximum free-running frequency.* The highest frequency at which astable operation is possible. A typical value is 100 KHz.

b) *Turn-off time.* The time delay from the leading edge of the reset signal until the output of the FF goes low. Typical values are 2.0 to 4.0 ms.

c) *Initial accuracy (monostable).* The accuracy of the time interval, exclusive of external component effects. 1% is typical.

d) *Initial accuracy (astable).* Accuracy of the free-running mode, exclusive of external R-C effects. 2% is a typical value.

e) *Frequency temperature coefficient.* The change in frequency as the temperature changes. For monostable operation 50 ppm/°C is typical while 150 ppm/°C is typical for astable operation.

f) *The frequency power-supply stability.* Change in frequency as a result of changes in the power-supply voltage. Typical values are 0.1%/V for monostable and 0.3%/V for astable operation.

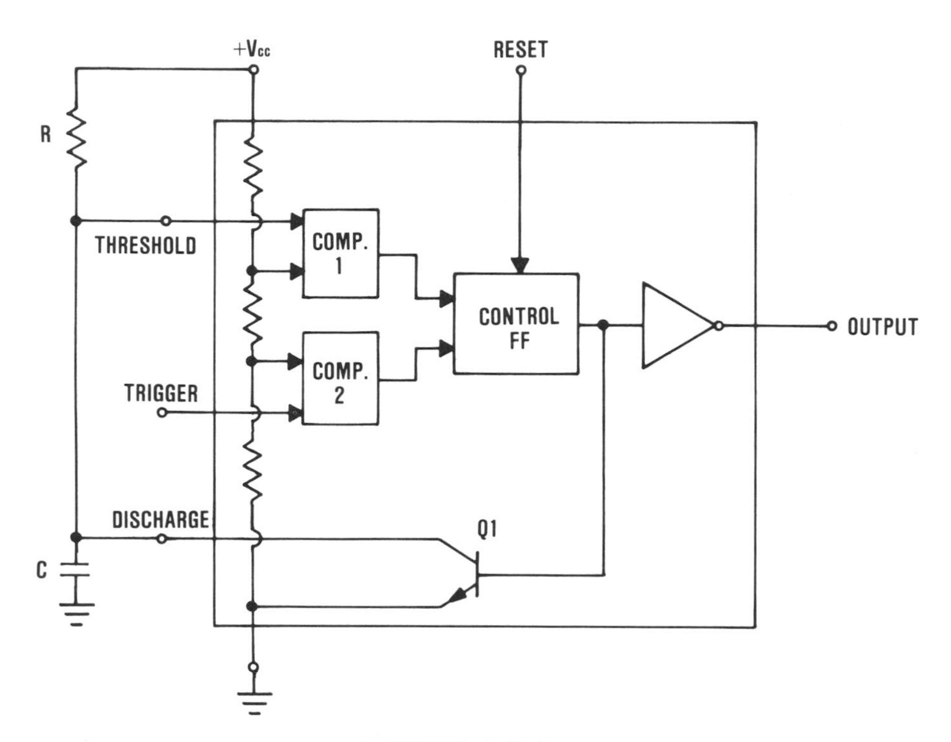

FIGURE A.16.1

g) *Trigger voltage*. Polarity and minimum trigger level for a given V_{CC}. Typical value is 0.3 V_{CC}.
h) *Trigger current*. The minimum current required by the trigger input terminal. 1μA is a typical value.
i) *Reset voltage*. The minimum voltage level to reset the FF when the trigger voltage is high. A typical value is 0.7 V for a V_{CC} of 15 V.
j) *Reset current*. The current required by the reset input. 0.5 mA is typical for V_{CC} of 15 V.
k) *Output voltage*. The nominal output voltage into the specified load. Typical values are 0.1 to 2.5 V for logic **0** and 12.5 to 13.5 V for logic **1**. This assumes V_{CC} = 15 V.
l) *Power dissipation*. The maximum DC power dissipated for an IC is either monostable or astable mode. A typical value is 600 mW.

APPLICATIONS

Accurate timing, sequential timing, time-delay generation, pulse generation, pulse-width modulation, pulse-position modulation, missing pulse detection, and clock applications of all types.

COMMENTS

The most widely used general purpose timer is the 555. A host of application notes, magazine articles, and even kits are on the market for the many different applications of this IC. This and other timers are frequently available in dual and quadruple configurations.

A.16.2 Timer, Precision

DESCRIPTION

Basically the same as the general purpose timer, the precision timer, as illustrated in Figure 16.2, contains an internal reference source which controls V_{CC} very accurately. Exernal access to V_{Ref} and V_{Adj}, the internally set voltage divider, permits external adjustment of this voltage and thereby a variation in the timing ratio of R-C. Just like the general purpose timer, the precision timer can operate in either the monostable or astable mode, but the accuracy is much greater.

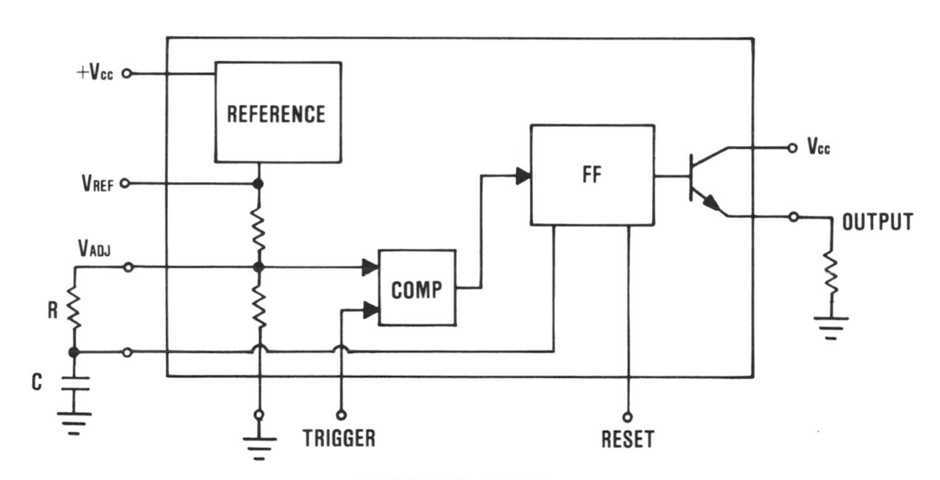

FIGURE A.16.2

KEY PARAMETERS

Essentially the same as the general purpose timer except for the following:

a) *Initial accuracy*. The accuracy of the timing signal exclusive of the external component effects. 0.1% is typical for both monostable and astable operation.
b) *Frequency temperature coefficient*. The change in frequency as temperature changes. 30 ppm/°C are typical for either monostable or astable operation.

c) *Frequency power-supply stability*. The change in frequency as power-supply voltage changes. 0.05%/V are typical values for either monostable or astable operation.

APPLICATIONS

High-precision timing, time-delay generators, sequential timing, or any other timing application requiring more precision than the general purpose model can provide.

A.16.3 Timer, Programmable

DESCRIPTION

In this IC the output of a basic timer, operating in the astable mode, drives a digital counter. The frequency of the oscillator, actually a multivibrator, is determined either by the R-C network or by an external modulating signal to which it is synchronized. Such external signal may be a crystal-controlled clock or, for example, a pulse derived from the 60 Hz powerline. Any of the counter outputs can be programmed to stop the oscillator by connecting that output to the reset terminal of the control FF. The count is started by applying a control signal to the trigger input. The use of the digital counter makes it possible to obtain relatively long delays from the basic timer circuit. To improve the accuracy and stability, an internal voltage regulator is included in most programmable timers.

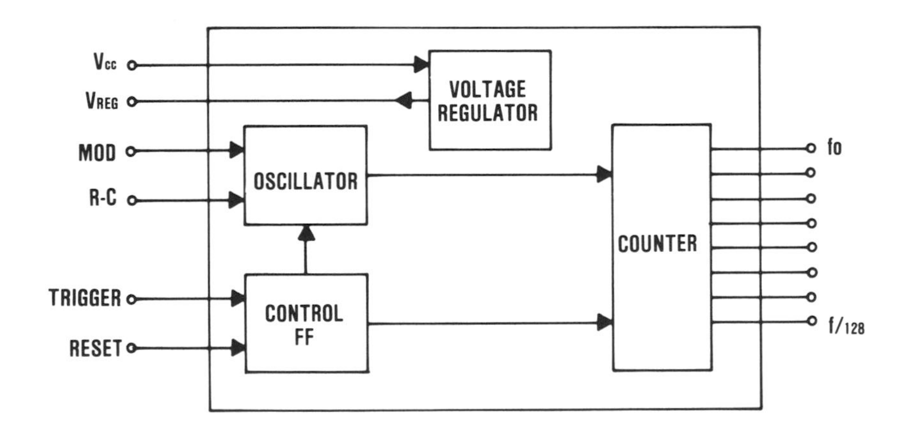

FIGURE A.16.3

KEY PARAMETERS

Essentially the same as for general purpose timers, except as follows:

a) *Size of counter.* The number of counter stages with externally available terminals determines the flexibility of application and the maximum number of periods over which the timer can be programmed.
b) *Counter output circuit.* The type of output circuit provided at each counter output terminal determines the capability to drive different types of logic. Many programmable timers have open collector outputs, requiring external collector resistors which can be matched to be compatible with most logic families.

APPLICATIONS

Programmable timing in process, appliance, darkroom, and other timing controls. Precisely determined time delays can be obtained by connecting various outputs together through combinational logic and returning the output of that logic to the trigger or reset terminals of the control flip-flop.

COMMENTS

Programmable timers are available in which the counter output is set for 59 so that a direct connection to the 60 Hz powerline will generate one-second pulses. Three timers can be connected in series to provide a real time clock.

C. CONSUMER ICs

Consumer

The integrated circuits listed in this category include all those offered by the manufacturers for use in equipment that is generally classified as "consumer electronics." Clearly, the ICs used in digital wristwatches, smoke detectors, TV sets and calculators fit into this category. The ICs used in appliance timers may be the same as those used in industrial timers, and the microprocessor used to control a microwave oven or operate an electronic game could also be listed as consumer IC. This classification problem is compounded by the fact that in any given function, such as a wristwatch IC, a calculator or electronic game IC, there are many different models, some sold only to the manufacturer of the consumer device, some available from electronic distributors. Some of these ICs are so unique that no specifications are ever published, and some are custom designed for a particular calculator, watch or game. ICs used in automated cameras, for example, seem to be almost entirely in that category.

Only a few manufacturers publish data on their consumer ICs, and only for a few selected types. The vast majority of ICs used in mass market consumer equipment apparently are custom made and, if replacements are needed, only the original equipment manufacturers have them in stock.

Consumer ICs are almost always large-scale (LSI) types and often contain both analog and digital circuits. In this section the ICs are listed according to the type of consumer equipment in which they are used. Each IC listed is a representative example of both standard or custom-designed units which perform a particular function. In each instance that function is sometimes combined with others, a variation in features may be available or some other minor differences may exist, but the essential function described here applies to that type of IC.

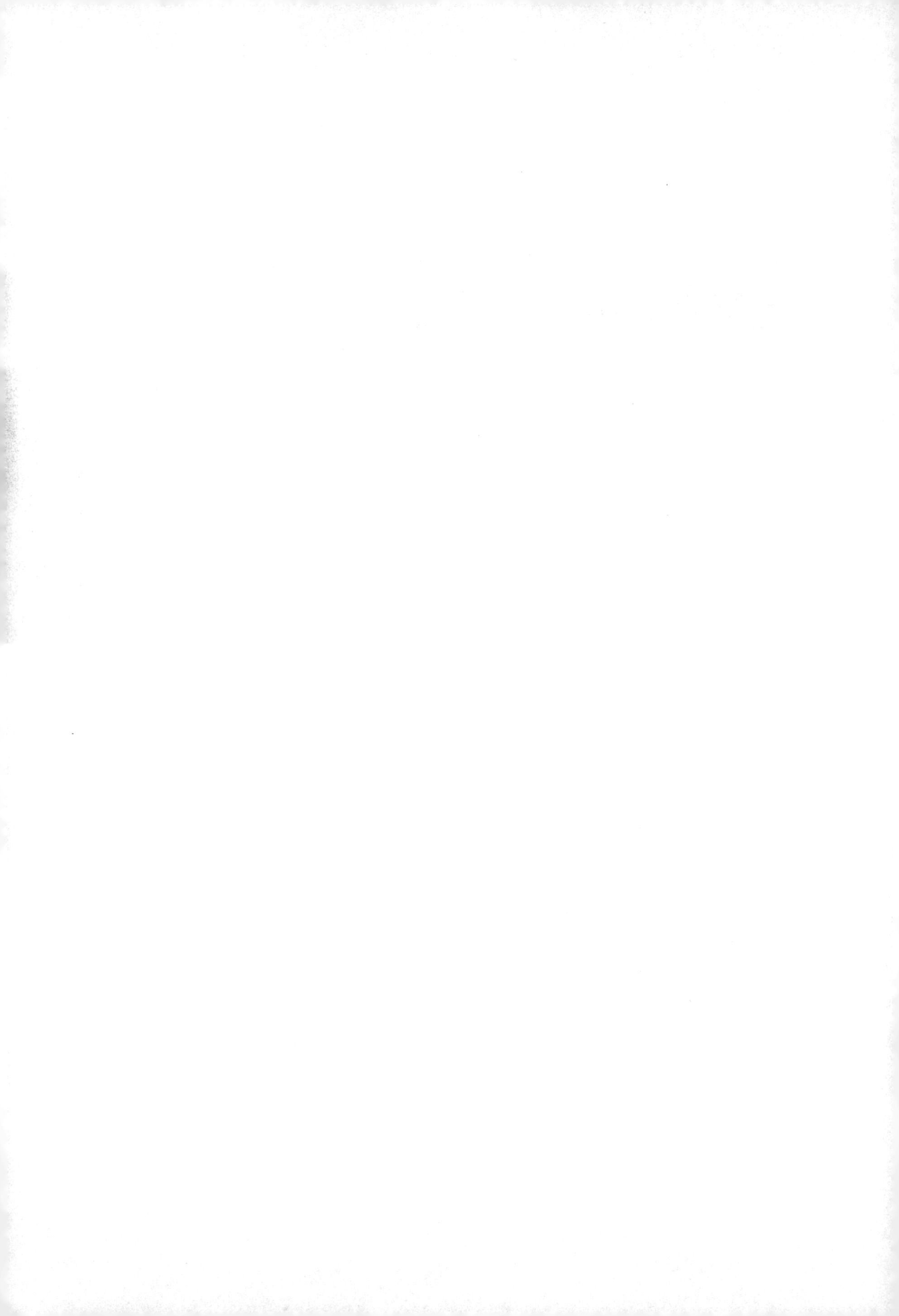

C.1.0 Alarm Circuit

DESCRIPTION

This IC provides all of the functions required for burglar alarms, temperature, moisture, and other types of security systems. Positive and negative input signals are accommodated together with an input noise-suppression signal. One of the features of this IC is its ability to sense the low battery voltage. The output current can be adjusted to drive either a horn, a speaker, or some other audible or visual indication. Separate switch inputs are available to set the alarm and to override the alarm. These switches are usually key operated. Because this type of IC is usually battery operated, the minimum possible power requirement is desirable.

KEY PARAMETERS

a) *Standby power-supply current.* The maximum current required when there is no alarm condition. 5 to 7 μA is typical.
b) *Operating power-supply current.* The maximum current required when the IC is in the alarm condition. Typical values are 5 to 15 mA.
c) *Input voltage threshold.* The level of either positive or negative input voltage which will trigger the alarm. Typical values are 3.0 to 3.4 V.
d) *Low battery threshold detector level.* The voltage at which the low battery alarm will start to indicate the alarm condition. Typical values are 1.7 to 2.0 V.
e) *Maximum output current.* Output current is adjustable in this type of IC to assure correct interface with external logic or indicator devices. Typical maximum current is 15 mA.

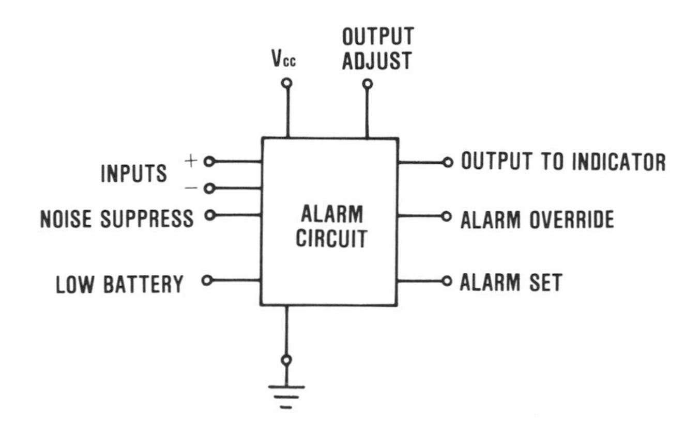

FIGURE C.1.0

Consumer

APPLICATIONS

Alarm circuits can be used with a variety of security systems and other systems where physical parameters, such as temperature, air flow, pressure, illumination, etc., are continuously monitored. A substantial change in the electrical analog of the parameter being monitored will then actuate the alarm device. Because of the low battery voltage-sensing feature, this IC is particularly useful in battery type applications.

COMMENTS

The basic principles of this alarm circuit are also found in the more elaborate smoke detector circuits described in Section C.10.0.

C.2.0 AM Receiver System

DESCRIPTION

As indicated in Figure C.2.0, all of the active portions of a typical AM receiver are contained on a single IC. Only the resonant networks have to be provided externally. This IC includes the RF converter, IF amplifier, detector and AGC circuit, a built-in regulator zener diode, and the audio preamplifier stage. In some AM receiver system ICs the RF amplifier is also included and the tuning meter or audio preamplifier is omitted.

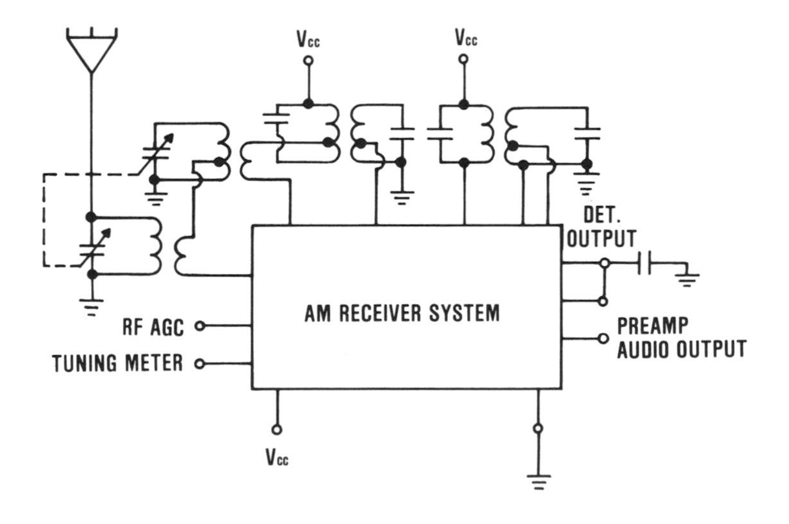

FIGURE C.2.0

KEY PARAMETERS

a) *Sensitivity.* Overall receiver sensitivity based on a particular RF and IF coil arrangement, usually at 1 MHz, with 30% AM modulation, at an audio frequency of 400 Hz, for a specific output voltage level. A typical sensitivity for 10 mV output would be 3.0 μV.
b) *Signal-to-noise ratio.* Measured under the same conditions as in (a) above; a typical value is 45 dB.
c) *Maximum power dissipation.* Generally measured at room temperature. Typical AM receiver system ICs will dissipate 600 mW.

APPLICATIONS

This IC is used in miniature and subminiature AM receivers of the broadcast, weather, and other types.

C.3.0 Appliance Timer

DESCRIPTION

While appliance timers vary in their flexibility of application, a typical timer such as the IC shown in Figure C.3.0 can be used with 50 or 60 Hz power lines and can operate either on a 12- or 24-hour basis. An external clock input is required if a 50 Hz power line is used. External input controls set the minutes and hours and turn the timer on or off. In addition there is a "reset" control, which will set the timer back to the original time, the "repeat" control, which permits the timer to repeat the operation as often as this control is activated, and the "cancel" control, which cancels the alarm. The control output has separate "on" and "off" lines which can go either

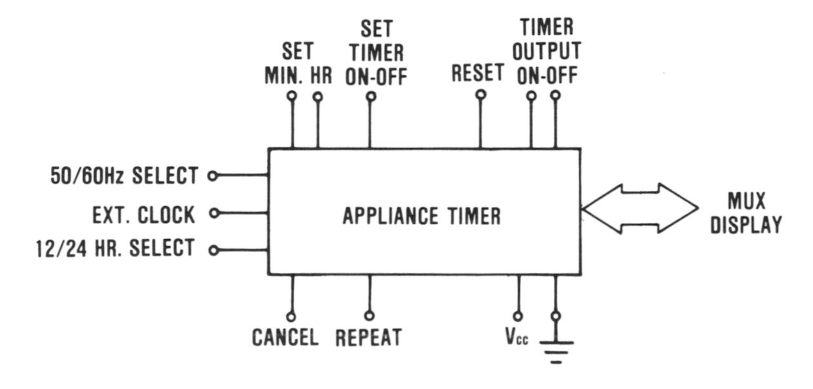

FIGURE C.3.0

directly to the appliance or to a relay to control the AC power to the appliance. Display information is multiplexed by segment and digit in the conventional manner.

KEY PARAMETERS

a) *Control logic levels.* The levels for logic **0** and logic **1** required for any of the input and output controls. Typical values are +0.3 V for logic **0** and -6V for logic **1**. This is based on a supply voltage of -12V.
b) *Display level outputs.* The voltage levels required to turn the display segments on or off. Depending on the type of display used, typical values range from 0 to +5 V for LEDs and -2 to 0 V for fluorescent displays.
c) *Maximum power dissipation.* Depending on the logic family, typical values are 100 mW.

APPLICATIONS

This type of timer can be found in such appliances as microwave ovens, TV tape recorders, electric cookers, stoves, etc.

COMMENTS

Not all of the inputs and outputs of a typical appliance timer are used in every application. It is important that the unused terminals be connected either to logic **0** or to logic **1** levels. Consult manufacturer's instructions.

C.4.0. Dolby Noise-Reduction Processor

DESCRIPTION

This IC is specifically designed to provide the Dolby B-type noise-reduction features for single audio channels. In addition to an internal power supply regulator, it consists of a series of amplifiers and some external R-C networks. One of these networks, containing five capacitors and three resistors, is connected to four external leads, while the second one, forming a feedback path, is comprised of three resistors and three capacitors working with an internal rectifier circuit. These R-C networks are specified in full detail by the manufacturer to assure that the desired Dolby B-type noise-reduction effect is obtained.

KEY PARAMETERS

a) *Distortion.* The maximum distortion contributed by the IC is specified as 0.05% for a 1 kHz, 0 dB level input and as 0.1% for a 10 kHz, 10 dB level input.

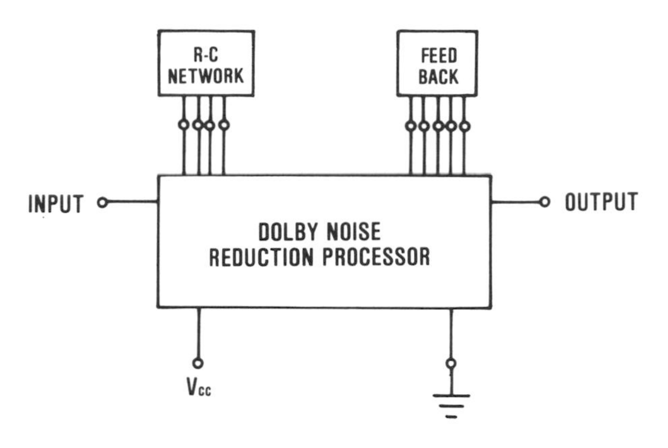

FIGURE C.4.0

b) *Signal handling.* Describes the dynamic range of the signal for 0.3% distortion at 1 kHz. 14 dB is a typical value.
c) *Signal-to-noise ratio.* For the encode mode of operation a typical value is 70 dB, and for the decode mode of operation it is 80 dB.
d) *Input resistance.* Typical values are 65 k ohms.
e) *Output resistance.* Typical values are 80 to 100 ohms.

Consumer

APPLICATIONS

In any kind of hi-fi audio system, recording devices, FM receivers, etc., where the Dolby type of noise reduction is desired.

COMMENTS

External resistor and capacitor values and tolerances must be used to assure obtaining the desired effect.

C.5.0 Calculator, 5-Function

DESCRIPTION

This single IC performs the basic four arithmetic functions and add-on or discount percentages. It operates with a simple keyboard consisting of the C-CE key, ten numerical keys, and six function keys plus a decimal point key. It is typical of the inexpensive, hand-held calculators and contains all of the logic and storage functions in a single, 28-pin IC. A variety of more complex ICs, providing more than eight digits of display, more than the basic five functions, and some limited memory, are used in more advanced calculators, but their basic features are the same.

As illustrated in Figure C.5.0, the nine digit connections are shared

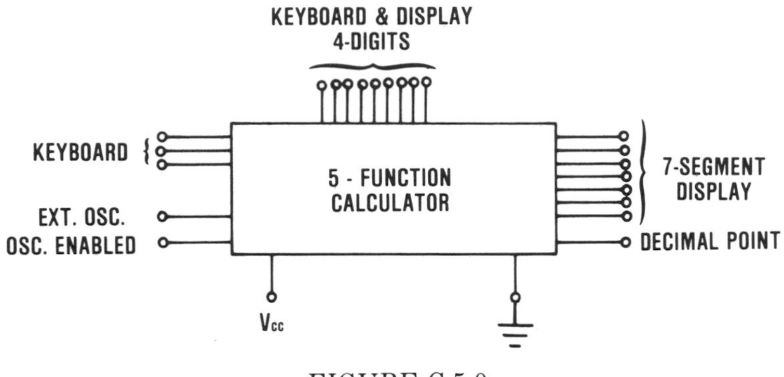

FIGURE C.5.0

between the keyboard and the display. Three lines from the keyboard indicate to the calculator IC which row of keys has been depressed, and this is combined with the digit information. When one of the keys on the keyboard is depressed, the same set of nine digit lines enables one of the eight display digits and the selected segments of the 7-segment numeral are then illuminated. The only other inputs are an external oscillator and the oscillator enable signal.

KEY PARAMETERS

a) *Supply voltage*. This depends on the type of display for which the IC is designed. For fluorescent type displays, -15 V is typical, and for LED type displays, -7.5 V is typical.
b) *Input signal levels*. For 15 V ICs, logic **1** ranges from -15 to -6 v and logic **0** from -1.5 to 0 V. For -7.5 V ICs, logic **1** ranges from -7.5 to -4.0 V while logic **0** ranges from -0.5 to 0 V.
c) *Keyboard input resistance*. Typical value is 1,000 ohms for all types of calculator ICs.
d) *Standby power*. The power used by the IC when all displays are off. For the 15 V IC, 75 μW is typical, and 15 μW is typical for the 7.5 V IC.
e) *Maximum power dissipation*. At room temperature, +25°C, the nominal maximum that most calculator ICs can dissipate is 500 mW.

APPLICATIONS

Calculator ICs are, in effect, custom designed for particular types of calculators.

COMMENTS

When replacing a calculator IC it is important to determine the original IC manufacturer and obtain an exact pin-for-pin replacement.

C.6.0 Clock Circuits

DESCRIPTION

This IC provides all of the functions required in an electronic clock, operated either from the AC powerline, or from an automobile, boat or plane battery. Depending on the application, it can receive inputs from a 3.58 MHz color TV crystal or a 60 Hz powerline. These signals are then counted down to generate minutes, ten minutes, and hour displays. A 3.75 Hz signal is available from the IC for flashing specific numerals or messages. In this IC, separate leads are brought out from the segment drivers to each of the LEDs or fluorescent numeral indicators. Only three control inputs are required. In order to change any particular digit, the "increment" input permits the user either to select hours, tens of minutes or minutes, or to let the clock run. Once the desired state is selected, the digit can be incremented by providing one impulse or pushbutton closing for each advance in the selected digit. The "reset" input automatically resets the clock to indicate 1:00.

KEY PARAMETERS

a) *Power supply voltage*. Nominal voltage of +5 V is typical.
b) *Logic control levels*. For logic **1**, 2.0 to 5.0 V is a typical range. For logic **0**, a typical range would be 0 to 0.3 V.

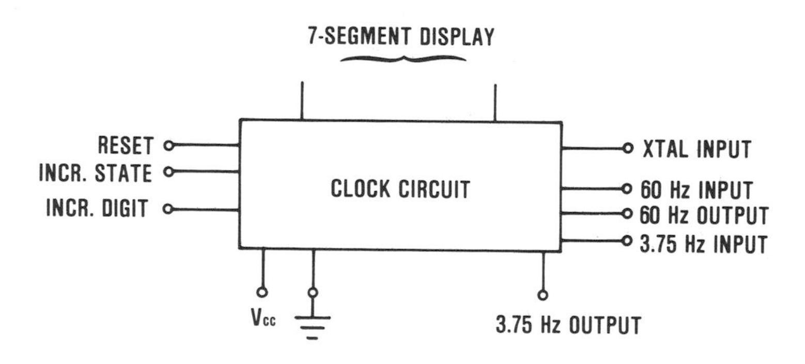

FIGURE C.6.0

Consumer

c) *Maximum power dissipation.* With all segments illuminated, approximately 500 mW will be required.

APPLICATIONS

This IC is used in electronic clocks of all kinds.

C.7.0 FM-IF System

DESCRIPTION

This IC includes a three-stage FM-IF amplifier/limiter arrangement with level detectors for each stage and a balanced quadrature FM detector. As illustrated in Figure C.7.0, the quadrature-tuned circuit requires two external coils, a capacitor and resistor and a single adjustment. Most ICs of this type also include an audio amplifier as well as a driving circuit for the tuning meter output, automatic frequency control (AFC) for the FM tuner, and AGC for the RF amplifier stage. The muting sensitivity control is optional and provides automatic muting (squelching), if desired. An internal power-supply regulator makes it possible to use any supply voltage between 8 and 18 V. The amount of distortion obtained with this IC depends mostly on the phase linearity of the quadrature-detector coil.

KEY PARAMETERS

a) *Input limiting voltage.* The voltage at the input that will cause limiting of the FM-IF signal. 10 μV is a typical value.

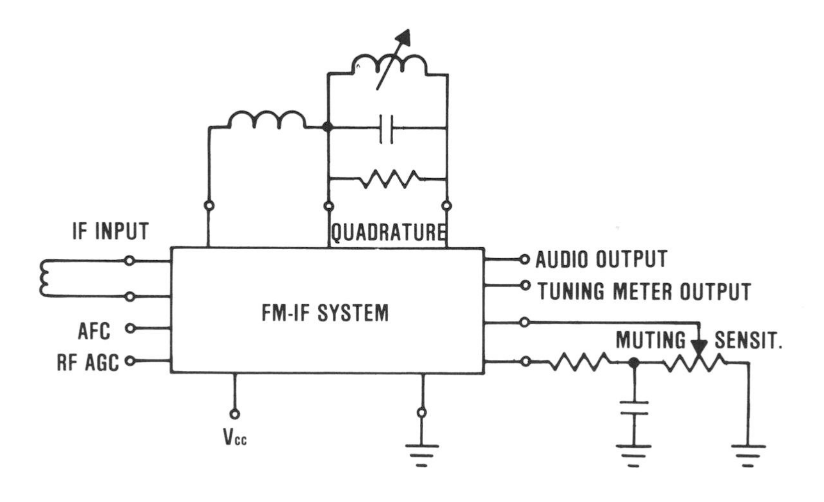

FIGURE C.7.0

b) *AM rejection.* The rejection of amplitude modulation with an input voltage of 100 mV, with 30% AM modulation at 400 Hz. 55 dB is typical.
c) *Recovered audio voltage.* The amplitude of the audio output signal when the input signal reaches the limiting level [see (a) above] with 400 Hz audio, FM modulated at 25 kHz. Typical values range between 300 and 500 mV.
d) *Total harmonic distortion.* Audio distortion at all frequencies with the same input conditions described for (c) above. A typical value is 0.5% with a single-tuned input circuit or 0.1% with a double-tuned input circuit.
e) *Signal-to-noise ratio.* Measured with a deviation of ±75 kHz. 65 dB is typical.

APPLICATIONS

This IC is used in all types of FM receivers and intended for 10.7 MHz IF frequencies.

COMMENTS

To provide sufficient FM-IF selectivity, most manufacturers of FM receivers use a narrow-band filter before the IF input stage. Because of the very high gain of this IC, the layout on the PC board and the location of bypass and decoupling capacitors and grounds can become critical.

C.7.1 Stereo Multiplex Decoder

DESCRIPTION

This IC accepts its input signal from the FM detector and applies it simultaneously to a 19 kHz and a 38 kHz synchronous detector. A 76 kHz local oscillator, part of the IC and controlled by the external circuits illustrated in Figure C.7.1, generates a signal which is counted down to 38 kHz and to two 19 kHz signals in phase quadrature. These signals are then used in the two synchronous detectors. The transmitted pilot tone is compared in one of these detectors. When it exceeds a threshold, the pilot tone sets a Schmitt-trigger circuit which lights the stereo indicator. This enables the 38 kHz synchronous detector and automatically switches from monaural to stereo operation. Internal audio preamplifiers provide the audio signals to the left and right output terminals. External control of stereo "defeat" or "enable" is provided by a separate terminal.

KEY PARAMETERS

a) *Input impedance.* 50 kohm is typical.

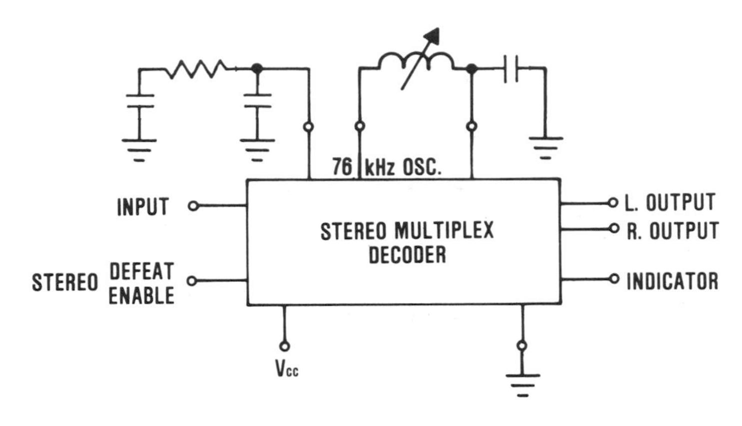

FIGURE C.7.1

b) *Channel separation.* The relative interference between the left and right channels. 0.3 dB is typical.
c) *Channel balance.* In the monaural mode, the difference between the left and right channels. 0.3 dB is typical.
d) *Monaural gain.* The amount of amplification on monaural operation. 6 dB is a typical value.
e) *Stereo/monaural gain ratio.* The difference in gain between monaural and stereo operation. ±0.3 dB is typical.
f) *Capture range.* The deviation from the 76 kHz center frequency over which the IC will operate. ±10% is typical.
g) *Distortion.* The amount of distortion of the second harmonic. 0.2% is a typical value with less than 0.2% at the higher harmonics.
h) *19 kHz rejection.* The reduction in 19 kHz frequency content of the audio output signal. 35 dB is typical.
i) *38 kHz rejection.* The reduction of 38 kHz signal present in the audio signal. 48 dB is typical.

APPLICATIONS

This stereo multiplex decoder is used in many FM-stereo systems and requires one low inductance tuning coil.

COMMENTS

The phase-locked loop (PLL) stereo decoder described in Section C.7.2 is used in more recent FM-stereo receivers and does not require any tuned circuit.

C.7.2 R-C Phase-Locked Loop (PLL) Stereo Decoder

DESCRIPTION

This IC uses a PLL and dividers instead of the 76 kHz resonant circuit of Figure C.7.1 and does not require synchronous detectors. As illustrated in Figure C.7.2, the external components consist of an R-C network for the VCO (voltage-controlled oscillator) which is the heart of the PLL, another R-C circuit for the loop filter, and a single capacitor for the switch filter. It also provides an output of the 19 kHz pilot tone, an indicator and the left and right audio outputs. Its performance characteristics are, in most respects, superior to those of the stereo decoder described in 7.1.0.

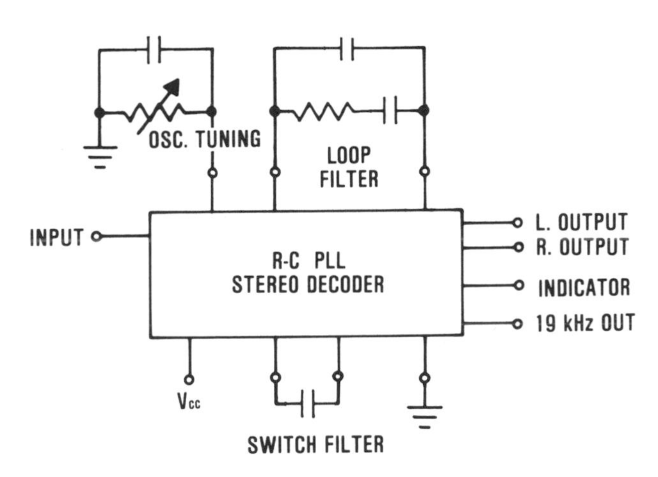

FIGURE C.7.2

KEY PARAMETERS

a) *Input impedance.* 50 k ohms.
b) *Channel separation.* The relative interference reduction between the two audio output channels. 40 dB is typical.
c) *Audio output voltage.* The audio output signal at either stereo channel for full limiting with minimum input signal. 500 mV, RMS.
d) *Channel balance.* The relative amplitude difference between the two stereo channels. A typical value is 0.1 dB with a maximum of 1.5 dB.
e) *Capture range.* The deviation from the center frequency over which the stereo demodulation will take place. ±3.5% is typical.
f) *Total harmonic distortion.* The sum of all distortions at all harmonics. 0.3% is a typical value.

Consumer

g) *19 kHz rejection.* The relative attenuation of this frequency in the audio output channel. 34.5 dB is typical.
h) *38 kHz rejection.* The relative attenuation of this frequency in the audio channel. A typical value is 45 dB.

APPLICATIONS

As stereo multiplex decoder in all types of FM-stereo receivers.

COMMENTS

As indicated in the comments to Section C.7.1, the PLL type of stereo decoder provides less distortion, a simpler adjustment, and generally better performance than the tuned circuit type.

C.8.0 Games, Video

DESCRIPTION

The large variety of dedicated game ICs is quite different from the microprocessor controlled TV games. The IC illustrated in Figure C.8.0 is one of the simplest on the market, yet provides a selection of tennis, soccer, squash, practice, and two rifle shooting games. A second IC can be added to display the same games in color. There are inputs for two players for tennis, soccer and squash, and separate inputs for the rifle game. The game control inputs permit the player to select the size of the racket or the width of the soccer player, the speed of the ball, the rebound angles of the ball, and either automatic or manual ball service. There is an audio output which can produce sounds at the TV set corresponding to the appropriate actions of the game. The sync output provides the horizontal synchronizing signal, and separate video outputs indicate the position of the two players, the ball and an outline of the field and the display of the score. The only electronic input signal required is a suitable clock. When used with a color generator IC, this clock input must be synchronized to 3.58 MHz, the color subcarrier frequency.

This type of IC is frequently used together with a video modulator to generate an RF signal which can be connected to the TV set through the antenna terminals.

KEY PARAMETERS

Manufacturers of game ICs do not usually provide strict electrical specifications. All dedicated game ICs are intended to be battery powered,

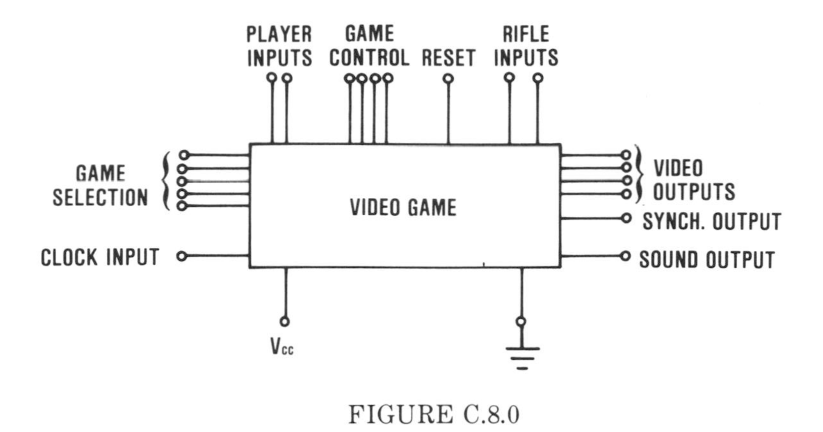

FIGURE C.8.0

operated at room temperature, and to generate audio, video and sync signals that are within the standard American TV signal specifications.

APPLICATIONS

This type of IC is used in dedicated TV games.

COMMENTS

In addition to the simple ball and paddle game IC described above, other ICs are available for such games as roadrace, warfare, wipeout, shooting gallery, battle, motorcycle, etc. Each of these ICs is mass-produced by a particular manufacturer and not usually compatible or interchangeable with any other IC. Consult manufacturer's data for all information.

C.8.1 Game, Video, RF Modulator

DESCRIPTION

This IC consists of two separate circuits, an RF oscillator and the RF modulator. The oscillator requires an external L-C network which is tuned to the desired TV channel, usually Channel 3 or 4, whichever is not in use in the particular area. The output of the RF oscillator serves as the carrier which is modulated by the audio and the composite video signal. Color information is also connected to the video input. The output signal is a double-sideband, video and audio modulated TV signal. A separate, external filter removes one of the sidebands to provide the vestigial sideband transmission standard of U.S. TV channels. TV video modulator circuits are usually

Consumer

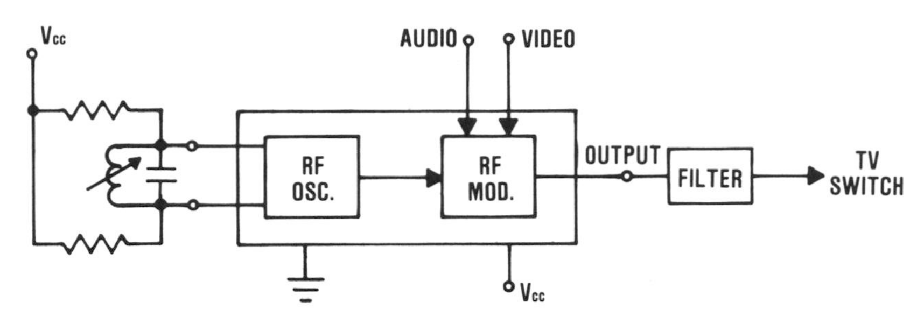

FIGURE C.8.1

contained in a compact 8-pin DIP package or else in the standard metal TO-5 can. Because of the relatively high frequencies involved, the actual circuit layout may be critical.

KEY PARAMETERS

a) *Input dynamic range.* Amplitude range of the input signal. 0 to 1.5 V is typical.
b) *RF output voltage.* Usually stated at approximately 60 to 70 MHz with a video input signal of 1.0 V. 15 mV RMS is typical.
c) *Video linearity.* Degree of distortion of the video input signal at RF output. 2% is typical.
d) *Color linearity.* The amount of distortion of the color information at the modulated RF output. 1% is typical.
e) *Audio input resistance.* 800 ohms typical.
f) *Video input resistance.* 100 k ohms minimum.
g) *Audio or video input capacitance.* Less than 5.0 pF is typical.
h) *RF oscillator stability.* At a frequency of 67.25 MHz and over a temperature range from 0 to 70°C, the RF oscillator is expected to change ±250 kHz typical.
i) *Maximum power dissipation.* At 25°C, typical value is 1.25 watts.

APPLICATIONS

This TV video modulator permits using the TV set as display for video games and personal computers and is used as the output stage on video tape recorders and TV test equipment.

COMMENTS

In order to comply with FCC regulations, the modulated RF video output of any equipment such as video games, personal computers, video tape recorders, etc., must be well shielded and must be submitted to the FCC for prototype testing and approval.

C.9.0 Organ Circuit, Top-Octave Generator

DESCRIPTION

An octave generator produces 12 separate output frequencies, each counted down from the input frequency by an exact integer to form a chromatic scale of 12 notes. The particular octave generated depends on the input frequency. This IC is essentially digital in operation since it uses digital counters to produce pulse trains. Note that the ratios of the output to input frequency correspond to the tones of the chromatic scale. The output pulse trains are usually converted into sine waves in the audio amplifier section of the musical equipment.

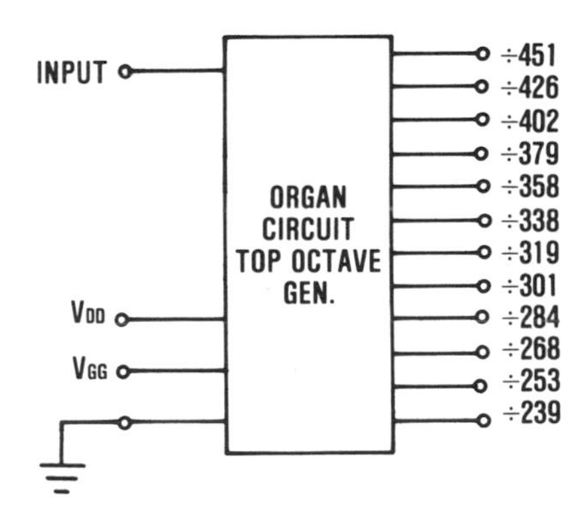

FIGURE C.9.0

KEY PARAMETERS

a) *Input frequency range.* Typical octave generators operate from 250 kHz to 2.5 MHz.
b) *Input capacitance.* The inherent capacitance of the input terminal. 5 pF is typical.
c) *Input pulse width.* The minimum width of the input pulse that the IC can accept. 0.2 μs is typical.
d) *Output pulse rise and fall time.* Characteristic of the output signal. 1 μs is typical with no load attached.

APPLICATIONS

Top-octave generators are used in electronic organs, electronic music synthesizers, and other electronic musical devices.

COMMENTS

Other top octave generator ICs are available which will produce the equal-tempered scale instead of the chromatic scale. Use only exact replacement ICs for this function.

C.9.1 Organ, Electronic Circuit, Rhythm Generator

DESCRIPTION

This IC contains all the logic circuits to generate six sets of rhythm patterns capable of driving up to eight instruments. When connected to a chord generator IC (see Section C.9.2), the rhythm generation can start on the downbeat every time a new chord is started. If multiple patterns are selected, this will result in their combination. Figure C.9.1 shows the external potentiometer which controls the frequency of the internal "tempo" oscillator. It can be adjusted from "largo" to faster than "presto," and its output drives the "tempo light" on the organ. All of the output signals are in the form of square waves.

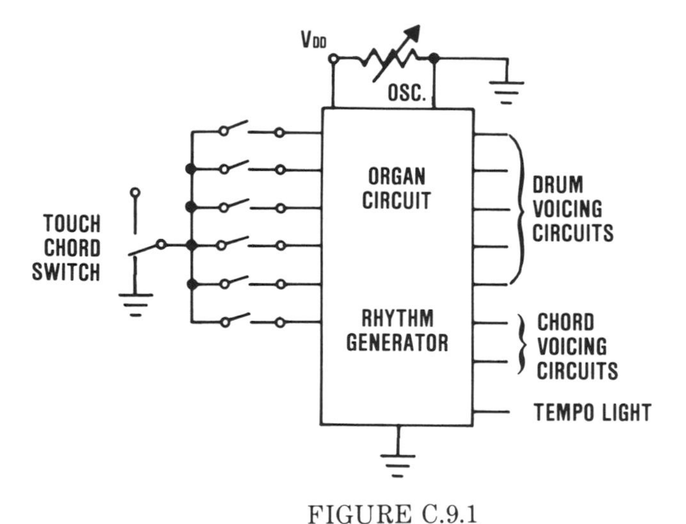

FIGURE C.9.1

KEY PARAMETERS

a) *Selector switch input impedance.* Between 15 and 80 k ohms.
b) *Input logic **0** level.* Between 0 and -2 V typical.

c) *Input logic **1** level.* Ranges from -9 V to the maximum V_{DD} (-12V) typical.
d) *Output logic—**0** level.* -0 to -2 V is typical.
e) *Output logic **1** level.* -9 V to V_{DD} (-12 V) is typical.

APPLICATIONS

To provide the rhythm portion for electronic organs and other electronic musical devices.

COMMENTS

Use only exact replacement ICs for this function.

C.9.2 Organ, Electronic Circuit, Chord Generator

DESCRIPTION

This IC accepts one full octave of 12 basic frequencies and combines them to generate the notes necessary to form the major, minor, and seventh chords. The internal oscillator frequency is determined by an external R-C network and, once selected, remains fixed. In addition, this IC provides the capability of generating such special effects as walking bass, rhythm arpeggio, alternating bass, etc. It also interfaces with a keyboard matrix as illustrated in Figure C.9.2

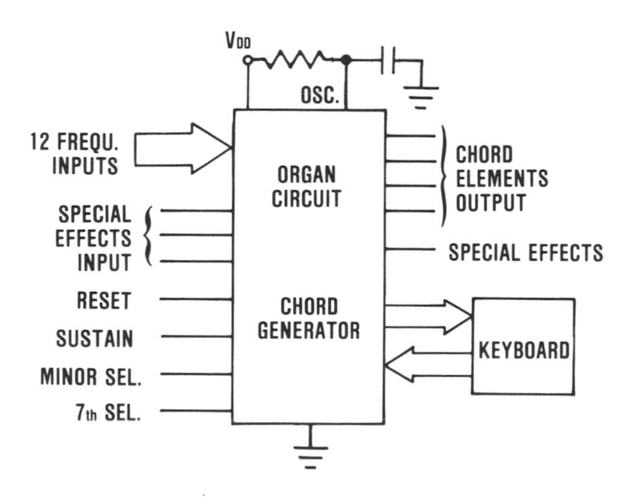

FIGURE C.9.2

KEY PARAMETERS

a) *Clock input frequency range.* From 100 to 1,000 Hz, with a maximum frequency of 100 kHz, is typical.
b) *Logic level for **0**.* V_{DD} to V_{DD} -4 V.
c) *Logic level for **1**.* 0.3 V to -1.3 V.
d) *Internal oscillator frequency range.* Typically from 100 to 1,000 Hz, set by external R-C circuit.

APPLICATIONS

Used to generate chords in electronic organs and other electronic musical devices.

COMMENTS

Manufacturer's data must be consulted concerning interfacing with other music circuitry. Use only exact replacement ICs for this function.

C.10.0 Smoke Detector (Ionization and Photoelectric)

DESCRIPTION

This IC can provide either or both ionization and photoelectric smoke detection. As illustrated in Figure C.10.0, an ionization chamber and a photocell smoke detector can be operated simultaneously, directly from this IC. When either or both indicates the presence of smoke, the IC provides sufficient drive to actuate a horn device directly. Designed to operate from a 9V battery, this IC automatically and continuously checks for proper battery voltage. The external R-C network controls an oscillator which "pulse" loads the battery once every minute or so for a few milliseconds. If the battery voltage is less than 7.5 V, the "low battery" alarm is activated.

Another outstanding feature of this IC is its ability to work in conjunction with other smoke detector ICs. The ground connection and the terminal marked "other units" in Figure C.10.0 are connected to the corresponding terminals on other, remote, smoke detectors. This makes it possible to have a chain of smoke detectors, each of which will sound a horn when one of them detects smoke. The unit actually detecting the smoke will generate a sound that is audibly different from the remote units in the chain, helping in the quick location of the smoke source.

KEY PARAMETERS

a) *Standby current.* The current drawn when no smoke is detected and no alarm is activated. 4μA is a typical value with a 9 V battery.

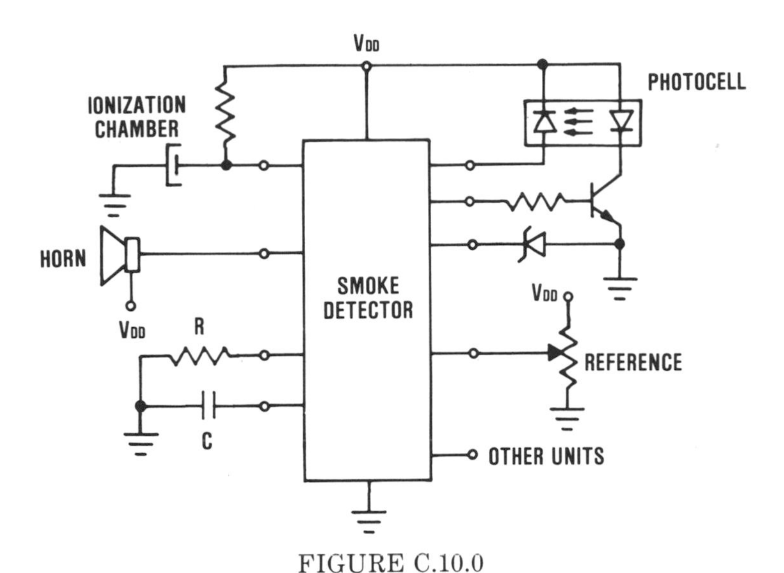

FIGURE C.10.0

b) *"Low battery" current.* The total current drawn when the battery voltage drops to 7.5 V. 5 μA is a typical value, excluding the content of the testing pulse.
c) *"Low battery" oscillator period.* The length of time between pulse tests for low battery voltage. 45 to 60 seconds are typical values.
d) *Low battery test period.* The length of the low battery loading pulse. 10 to 20 ms are typical.
e) *Horn alarm current.* The current drawn from the battery when the alarm horn is activated. 120 mA is typical.
f) *Horn signals.* For local smoke the horn sound is continuous. For "low battery" the horn sounds every 40 seconds for 20 ms, and for remote smoke the local horn sounds every 100 ms for 20 ms.
g) *LED indications.* The LED flashes on for 150 ms once every half second. The same signal is used for local smoke, remote smoke, or low battery.
h) *Priorities.* Since several alarm conditions can occur at the same time, internal IC logic sets the following order of priorities: (1) local smoke, (2) remote smoke, (3) low battery, (4) standby.

APPLICATIONS

In battery operated smoke detectors.

COMMENTS

ICs that perform only photoelectric or only ionization type smoke detection are very widely used. They operate basically in the same manner as described above.

C.11.0 TV Receiver Circuits

DESCRIPTION

The following portions of Section 11 describe ICs that are generally found in modern TV receivers. The TV receiver block diagram of Figure C.11.0 illustrates all of the major functions found in a modern color TV receiver. The power supply portion is omitted because it generally does not employ special ICs. Discrete power transistors and diodes are usually used in the vertical and horizontal output and high-voltage section. All other areas of a TV receiver contain ICs. In some cases a single IC provides several of the functions illustrated here, and in other receivers separate ICs are used to perform a particular function. Automatic color control (ACC), automatic fine tuning (AFT), automatic gain control (AGC), etc., are functions that are not illustrated in Figure C.11.0 and, in most receivers, their implementation is part of an IC performing a larger function.

The TV tuner, in most cases, is an electromechanical device and even where pushbutton, electronic tuning is used, discrete semiconductors are frequently employed.

The RF TV band extends from the VHF into the UHF region, and for this reason VHF or UHF transistors are usually used in the tuner. In the IF section the signals range from 41.25 MHz (the audio IF) to 46 MHz, and this can be easily handled by ICs. The RF TV signal contains the monochrome video information which is amplitude modulated (AM), with vestigial sideband suppression of the RF carrier. The audio signal is frequency modulated (FM) on a subcarrier which is 4.5 MHz higher in frequency than the RF video carrier. In addition, there is the color subcarrier which is 3.58 MHz above the RF video carrier and contains both amplitude (AM) and phase modulation (PM) to provide full color information.

All of the signals are amplified in the IF amplifier and are then split at the detector into the three key frequencies. The FM audio signal is demodulated and amplified to drive a loudspeaker. The monochrome video signal contains the vertical and horizontal synchronizing pulses. During color broadcasts the color synchronizing burst rides on the trailing edge of each horizontal blanking pulse. The sync separator also contains noise-cancelling circuits and, in many instances, automatic gain control (A.G.C.) circuits. It provides a vertical sync signal to the vertical sweep section. In the color section the 3.58 MHz color subcarrier is demodulated with the aid of the color burst, derived from the sync separator. Although a single line is shown in Figure C.11.0, the color section provides the three color signals, red, green and blue, to the color picture tube.

A remote control transmitter, usually operating at ultrasound frequencies, and a remote control receiver are available with most TV sets, and ICs form the key elements in both of these functions.

While there is a large variety of ICs for TV receiver applications, we have, in each case, selected the one IC type that contains the largest number of functions. It is possible, for example, to perform the color demodulation and matrixing function with two or three separate ICs, but in Section 11.4 we describe a single-chip TV chroma processor and demodulator which performs all the functions, including automatic color correction (ACC).

The key parameters listed in the following sections apply to the particular IC, but, because all TV ICs deal with the standard TV signals, these key parameters are generally applicable to similar functional units.

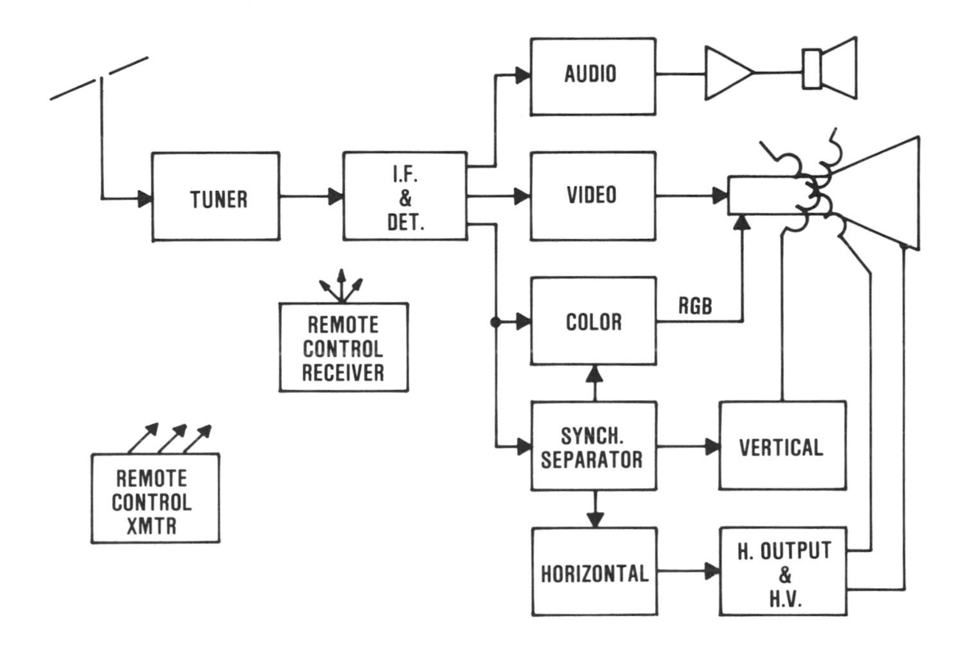

FIGURE C.11.0

Consumer

C.11.1 Remote Control Transmitter

DESCRIPTION

This IC contains a master oscillator, controlled by the 4.4336 MHz external crystal circuit, and counted down to provide ultrasonic outputs in the range of 33.9 to 43.9 kHz. The ultrasound output terminal drives a single transistor stage which then powers the ultrasound transducer. Eleven separate key inputs are available for either touchplate or mechanical keyboard use. Up to 30 separate commands can be encoded by means of these inputs.

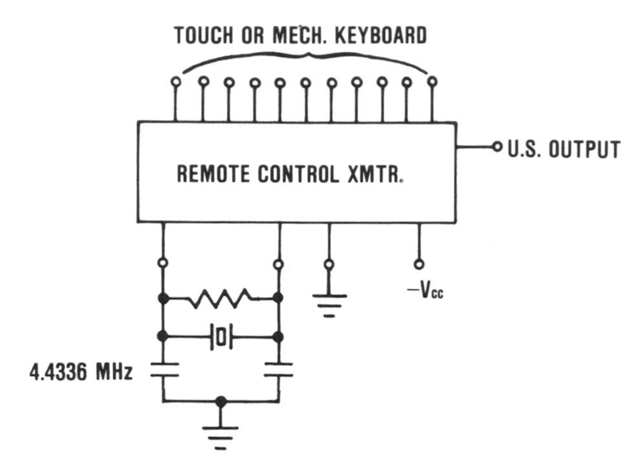

FIGURE C.11.1

KEY PARAMETERS

a) *Clock frequency.* 4.4336 MHz, external crystal controlled.
b) *Output frequencies.* 30 frequencies, separated by 346.4 Hz, are available in the range from 34 to 44 kHz.
c) *Input logic **0**.* -0.5 V typical.
d) *Input logic **1** level.* -4 to -9 V typical.
e) *Standby current drain.* The battery current required when no commands are transmitted. 15 μA is typical.
f) *Operating current drain.* The battery current required when a command is transmitted. 8.0 mA is typical.

APPLICATIONS

For handheld, battery powered ultrasonic remote control of color TV receivers.

COMMENTS

A 30-channel remote control receiver is available to decode the signals transmitted by this IC. The receiver requires a separate capacitor microphone and preamplifiers as well as an external crystal oscillator with the same crystal frequency of 4.4336 MHz. In the receiver module the control channels are used to provide eight codes for the control of three analog outputs, an on-off signal, and 21 independent codes, usually used for TV channel selection. The remote control receiver can also be interfaced directly with a local keyboard.

C.11.2 Automatic Fine Tuning (AFT)

DESCRIPTION

This IC operates from the video IF input and contains an internal shunt regulator and bias generator. In addition to the automatic fine-tuning (AFT) voltage, it also generates an amplified 4.5 MHz intercarrier sound signal. The AFT circuit features a differential peak detector and amplifier, and provides bipolar, positive and negative error signals to the TV tuner. The time constants of the AFT circuit are determined by an external, two-capacitor AFT filter, as illustrated in Figure C.11.2.

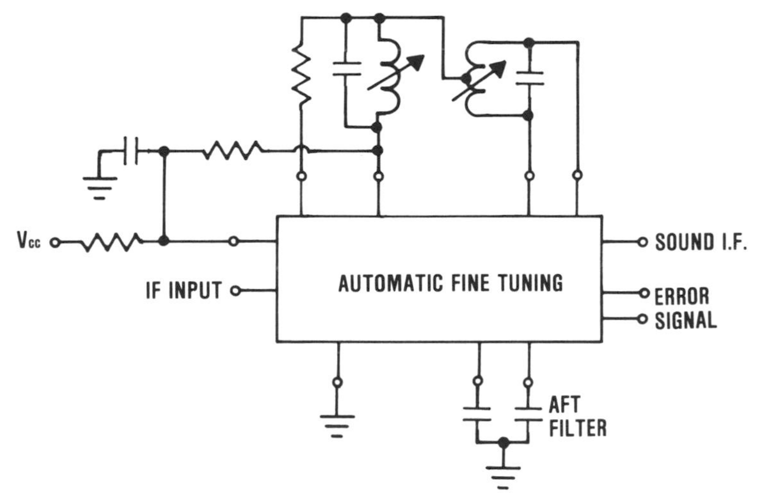

FIGURE C.11.2

KEY PARAMETERS

a) *Signal input.* The amplitude of the IF input required. 15 mV RMS is typically specified.
b) *Error signal amplitude.* Depends on the deviation of the frequency from the desired IF value. When the IF frequency is 44.65 MHz, the error voltage will be, typically, between 2.2 and 4.7 V. When the IF frequency is 46.85 MHz, the error voltage will vary from 9.1 to 12.1 V. At the terminal of the opposite polarity, the error voltage will be opposite, in relation to the IF frequency.
c) *4.5 MHz output voltage.* With an IF input signal consisting of 45.75 MHz and 41.25 MHz (sound carrier), the 4.5 MHz output signal will be approximately 11 mV RMS.

APPLICATIONS

In color TV receivers using automatic fine tuning.

COMMENTS

In typical applications the error signal passes through isolating resistors and bypass capacitors and is applied across a varactor diode, which changes its capacity in response to changes of the voltage across it. This varactor diode is part of the local oscillator resonant circuit in the TV tuner and determines the fine tuning.

C.11.3 IF System, Video

DESCRIPTION

This IC performs the function of a three-stage TV, broadband, IF amplifier, detector and video preamplifier. Automatic gain control (AGC) for the IF section as well as for the tuner is also provided in this IC. A horizontal keying pulse is used with a sample-and-hold circuit so that the AGC bias depends on the horizontal sync pulse amplitude and is independent of variations—and noise—during the video portions. An external adjustment is provided to set the amount of delay for the AGC

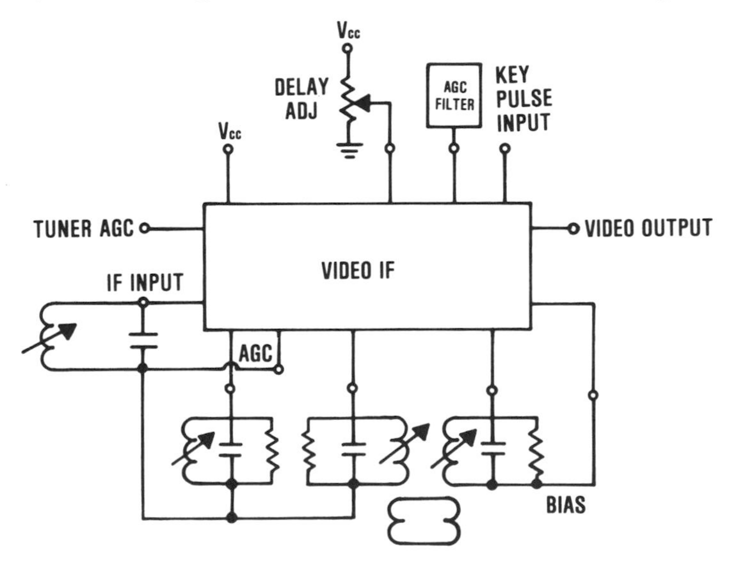

FIGURE C.11.3

signal supplied to the TV tuner. The AGC filter, consisting of resistors, capacitors and a diode, is an external network. One of the most important features of this IC is an internal voltage regulator which assures constant power supply voltage to all stages. The four resonant circuits shown in Figure C.11.3 are typical of video IF bandpass networks but do not include the adjacent channel and the sound IF trap. In most TV receivers these traps are part of a filter connected between the tuner output and the IF input.

KEY PARAMETERS

a) *Nominal IF input signal.* The amplitude of the IF input signal, over the flat portion of the response curve, that will provide a nominal video output of between 1 and 5 V. A typical input value is 400 μV.
b) *Distortion.* The amount of distortion at 50 kHz, with 80% AM and a sync pulse amplitude equivalent to 30 mV, RMS. A typical value is 10%.
c) *Video output level.* Depending on the input signal, a range of 0.9 to 10V output can be expected with 12 V_{CC}.
d) *Horizontal key pulse input.* Through an external 100 k resistor, between 25 and 35 V peak to peak is required.
e) *Maximum power dissipation.* Up to 55°C, 750 mW may be dissipated.

APPLICATIONS

This IC is used in the video IF section of modern monochrome and color TV receivers.

COMMENTS

Because of variations among manufacturers, only an exact replacement IC should be used.

C.11.4 Chroma Processor/Demodulator

DESCRIPTION

One of the most comprehensive color-signal processing ICs, the unit illustrated in Figure C.11.4 accepts the composite video signal from the detector or first video amplifier of an IC, such as described in C.11.3. In addition, the chroma processor/demodulator requires only a horizontal keying pulse to produce the color subcarrier reference signal, amplify the 3.58 MHz color subcarrier, demodulate the subcarrier, and produce the three color difference signals. This IC also contains an automatic color control

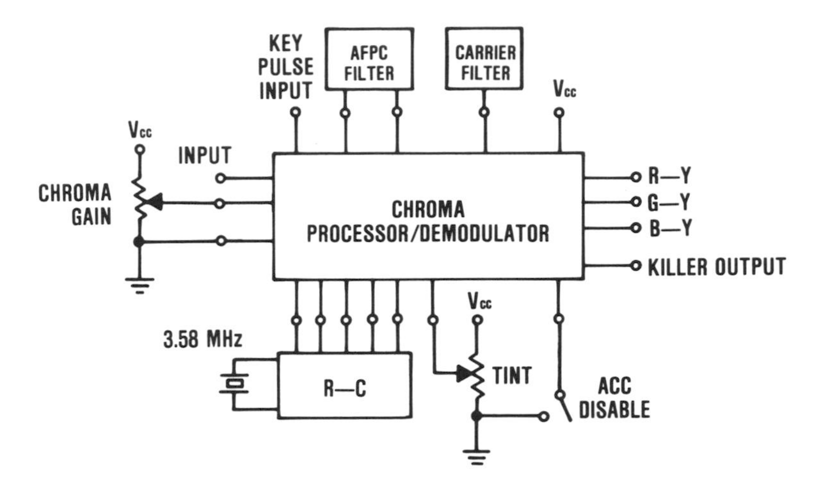

FIGURE C.11.4

(ACC) based on flesh tones, a color overload circuit, and the color-killer detector and amplifier.

As illustrated in Figure C.11.4, the automatic frequency and phase control (AFPC) requires an external filter and a second filter is needed to remove the color subcarrier. The color reference oscillator is based on the 3.58 MHz external crystal and R-C network. The two main color controls, the chroma gain and the tint, are directly connected to this IC. An external switch permits the user to disable the automatic color correction system (ACC) and the overload protection circuit.

KEY PARAMETERS

a) *Maximum power dissipation.* The maximum power that can be dissipated with this IC up to 55°C is 825 mW.
b) *Nominal power dissipation.* The total power dissipated under normal operation. 500 mW is typical.
c) *Minimum oscillator pull-in-range.* The maximum amount of misalignment of the crystal tuning capacitor with which the oscillator will still lock in phase with the reference signal. ±300 Hz is a typical value.

APPLICATIONS

As color-signal processing unit in color TV receivers.

COMMENTS

Only the exact manufacturer's replacement ICs should be used. In some TV receivers two or three ICs perform the functions of the chroma processor/demodulator described above.

C.11.5 Luminance Processor

DESCRIPTION

This IC provides the important functions of equalizing the high- and low-frequency components of the video signal, clamping the signal to the proper black level and mixing the vertical and horizontal blanking signals with the video signal. A power transistor video-amplifier driver is usually required to drive the CRT. Because of the longer time required for processing the color signal, a tapped delay line of approximately 750 ns is used in modern color TV receivers. Three inputs from that delay line are applied to the luminance processor. A peaking amplifier and the main video amplifier on this IC provide for the contrast control and the peaking of the high frequencies. A special black-level clamping circuit requires a horizontal pulse input at the clamp inhibit terminal to prevent the video signal from being clamped to the sync level, which is more negative than the black level. The brightness control sets the level of the video-output amplifier, which is also gated from the vertical and horizontal blanking pulse.

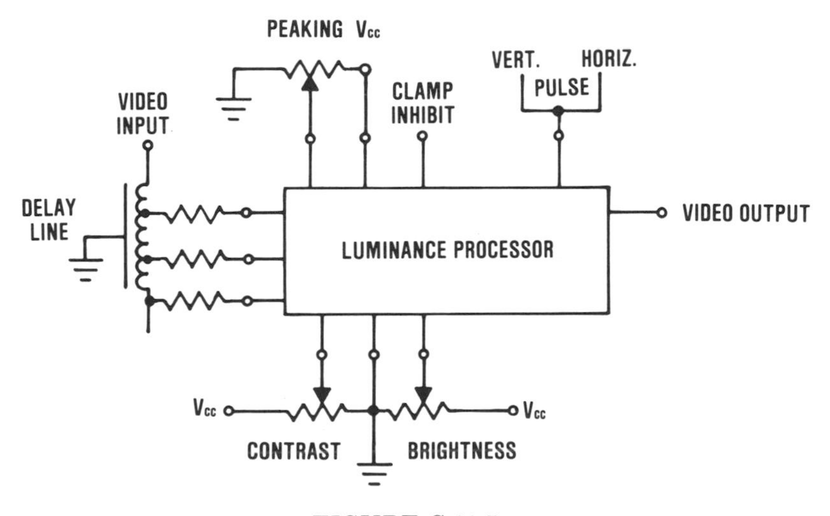

FIGURE C.11.5

Consumer

KEY PARAMETERS

a) *Maximum power dissipation.* 750 mW at 55°C.
b) *Wideband gain.* Amplification over the bandwidth 100 Hz to 3.5 MHz. 8 dB is typical.
c) *Intermodulation distortion.* The amount of distortion due to intermodulation of two or more frequencies. 20% is a typical "worst case" value.

APPLICATIONS

In the video section of color TV receivers this IC performs the video amplification, clamping, and blanking functions.

COMMENTS

Only exact manufacturer's replacement units should be used.

C.11.6 Signal Processor

DESCRIPTION

This IC accepts a composite video signal from the first video amplifier and processes this signal to remove the synchronizing pulses, generate IF and tuner AGC, and reduce the effects of any possible interfering noise. The R-C network illustrated in Figure C.11.6 is a complex filter that operates between the output of the noise inverter and the sync separator itself. A

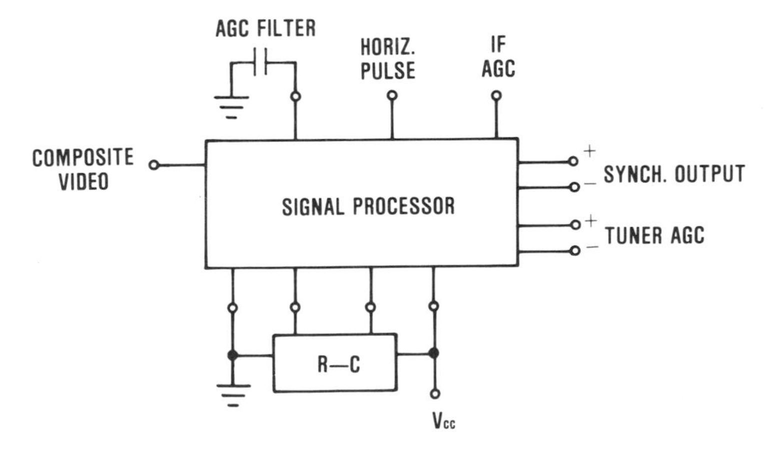

FIGURE C.11.6

horizontal keying pulse is required for this circuit, just as for the video IF section of Figure C.11.3, which also generates its own AGC. If such an IF section is used in the receiver, the AGC capability of the signal processor will not be required. The separated sync pulses are available in both positive and negative polarity. Tuner AGC voltage is also available in a bipolar output.

KEY PARAMETERS

a) *Maximum power dissipation.* This IC can dissipate 750 mW at temperatures up to 55°C.
b) *Video input amplitude.* The nominal amplitude of the composite video signal. 3 V peak to peak is typical.
c) *Sync output levels.* Maximum amplitude is V_{CC} (24 V typical).
d) *Horizontal pulse amplitude.* Ranges from 3 to 6 V.

APPLICATIONS

As sync separator and noise-reduction circuit in monochrome or color TV receivers.

COMMENTS

Use only exact manufacturer's replacements.

C.11.7 Horizontal Processor

DESCRIPTION

The major difference between the signal processor IC illustrated in Figure C.11.6 and the horizontal processor of C.11.7 is that the latter contains the horizontal automatic frequency control (AFC), the horizontal oscillator, and a preamplifier for the horizontal sweep section. Both ICs separate the sync pulses from the composite video signal, provide some noise reduction, and also generate a keyed AGC signal. The section labeled R-C in Figure C.11.7 contains the resistor and capacitor network required for the noise-suppression function. A separate, external AGC filter is required. The horizontal and vertical sync pulse output is connected to the horizontal sync input and to the vertical sweep section which may be on another IC. An external phase-detector network, including resistors, capacitors and two diodes, generates the error signal for the horizontal AFC. The horizontal oscillator itself depends on an external R-C network, and the frequency is adjustable by means of the tuned coil. An internal amplifier provides

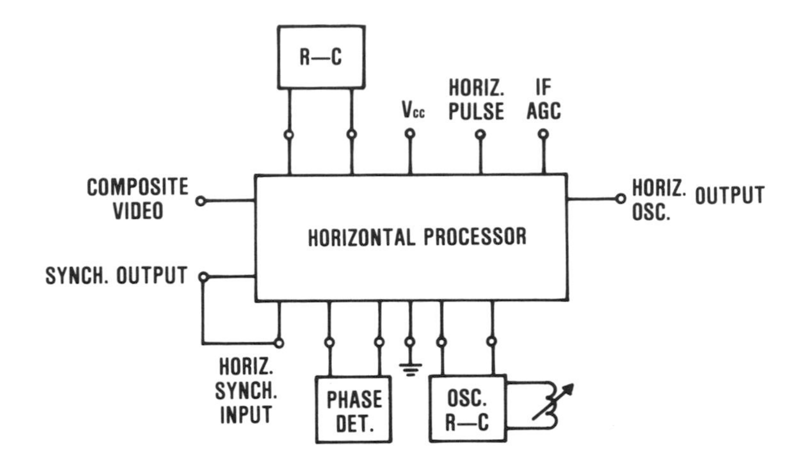

FIGURE C.11.7

sufficient horizontal pulse output to drive the horizontal sweep and high-voltage section.

KEY PARAMETERS

a) *Maximum power dissipation.* Up to 55°C, 750 mW is typical.
b) *Horizontal pulse amplitude.* Typical value of amplitude required is 25 V peak to peak.
c) *Sync signal output.* Amplitude of vertical and horizontal sync signal. 1.5 V peak to peak is typical.

APPLICATIONS

This IC is used in monochrome and color TV receivers to provide sync pulses, AGC and horizontal oscillator output.

COMMENTS

Use only the exact manufacturer's specified replacement.

C.11.8 Horizontal and Vertical Sweep

DESCRIPTION

This IC accepts the horizontal and vertical synchronizing signal from a signal processor, such as described in Section C.11.6, and generates the horizontal and vertical sweep signal.

An unusual technique assures excellent synchronization without vertical or horizontal hold controls. The horizontal flyback pulse input is compared with the horizontal sync input in a standard phase detector which controls the voltage-controlled oscillator (VCO). An external loop filter is required to smooth out the error voltage. The 503.5 kHz ceramic resonator controlled oscillator output is counted down by 32 to generate the 15.75 kHz horizontal sweep signal. Twice the horizontal output frequency is supplied to an internal timing circuit which also receives the vertical sync signal. The vertical R-C network removes the horizontal pulses from the vertical sync pulses. The 60 Hz vertical sweep signal is always locked in with the horizontal sweep, assuring correct interlace on the screen. The horizontal pulse is also used to provide a burst gate which is then applied to a color processing IC to produce the color-sync burst.

An internal regulator assures stability of the power supply voltages.

KEY PARAMETERS

a) *Maximum power dissipation.* Rated at 25°C, 830 mW is typical.
b) *Horizontal output frequency.* With no error voltage applied, typical center frequency is 15,750 Hz.
c) *Horizontal output frequency range.* The normal range of output frequencies, controlled by the resonant 503.5 kHz element, over which the oscillator can be varied by the error voltage. 15, 150 to 16,300 Hz is typical.
d) *Vertical pull-in range.* The frequency range over which vertical synchronizing pulses can be locked in. 58.1 to 67.1 Hz is typical.
e) *Horizontal pull-in range.* Frequency band over which the horizontal oscillator can be locked in. ±600 Hz is typical.
f) *Horizontal static phase error.* With a frequency range of ±600 Hz, the typical phase error of the horizontal output signal should be no more than ±0.5 μs.

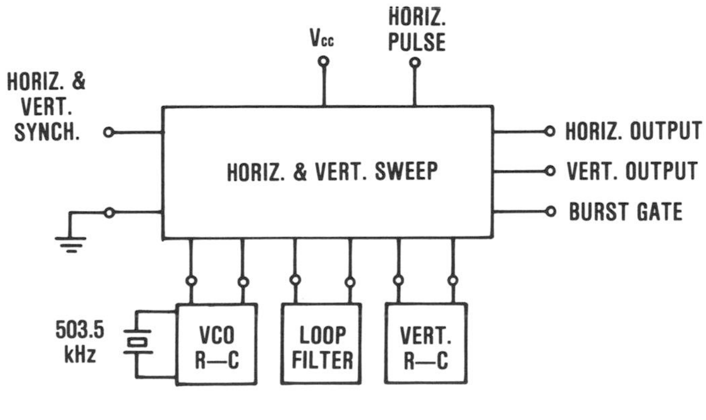

FIGURE C.11.8

APPLICATIONS

To provide horizontal and vertical sweep output signals and the burst gate for color TV receivers.

COMMENTS

Use only manufacturer's specific replacement IC.

C.11.9 Sound IF and Output

DESCRIPTION

This IC accepts the 4.5 MHz sound subcarrier signal through a transformer or other resonant circuit and provides all of the limiting, FM detecting, and audio amplifying functions. As illustrated in Figure C.11.9, the FM detector uses a resonant circuit which must be tuned to the 4.5 MHz center frequency. The volume control is a linear taper DC control which goes to an internal, electronic attenuator circuit. This circuit provides an improved audio taper response and drives the audio power amplifier in connection with the de-emphasis network. Because this IC provides a fair amount of audio output power, a voltage regulator as well as a thermal and current sensing shut-down circuit is contained on it. Another unusual feature of the IC is the fact that it has a copper strap heat sink as part of the IC body.

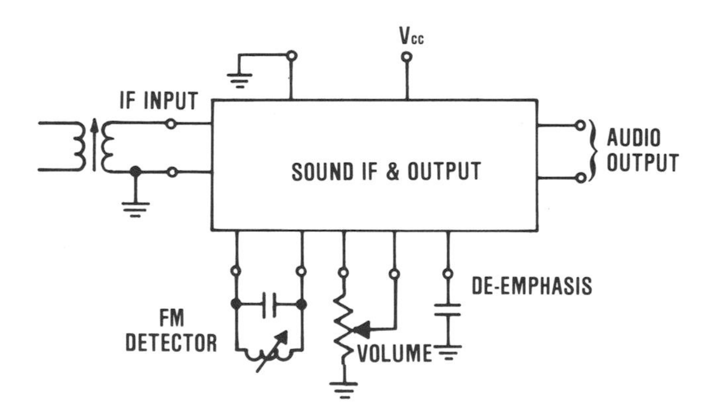

FIGURE C.11.9

KEY PARAMETERS

a) *Maximum power dissipation.* Without external heat sink, up to 25°C, 1.4 W is maximum. With a copper strap heat sink soldered to the IC board, up to 25°C, 3.9 W can be dissipated as maximum.
b) *IF input amplitude for limiting.* Amplitude of the 4.5 MHz IF signal that will cause 3 dB limiting. 200 μV is a typical value.
c) *AM rejection.* The ability of the FM detector to reject amplitude modulation. 50 dB is typical.
d) *Total system harmonic distortion.* The total harmonic distortion from the IF input to the audio output. For a 1 W audio output, 1.5% distortion is typical.
e) *Power output.* The maximum audio power output with a total harmonic system distortion of 10%. Ranges from 1 up to 5 W, with suitable heat sinks.

APPLICATIONS

In the sound section of monochrome or color TV receivers.

COMMENTS

Use only manufacturer's specific replacement ICs. Exceeding temperature limits or maximum power dissipation will cause automatic shut-down of the audio section. When the temperature of the equipment has been reduced, normal operation will be restored. This effect can be mistaken for an intermittent defect.

C.12.0 Watch Circuit, Digital, LCD

DESCRIPTION

This IC is typical of all those used in digital wristwatches. The four-digit liquid crystal display (LCD) indicates the month, date, day of the week, hours, minutes and seconds, and the IC contains all of the counters, memory and control logic necessary for these functions. As illustrated in Figure C.12.0, an external 32 kHz crystal is used to provide the required timing accuracy, including the logic and memory for a four-year calendar. The battery is connected between the ground and V_{CC} and four other leads are used for control functions. The "test" lead is used in factory testing and the "MDR" control permits reversing month-date position. There are only two controls for the user. Pressing the "mode" switch selects, in sequence, the following modes: run (normal watch operation), set month, set date, set day of the week, set hours, set minutes. Once the mode is selected, the "set" switch is used to bring the particular data to the desired reading.

One of the important features of any IC that drives LCDs is the voltage multiplier that increases the battery voltage and provides the required AC drive for the display. The frequency is usually counted down from the crystal oscillator, and the voltage multiplier uses only on-board R-C elements.

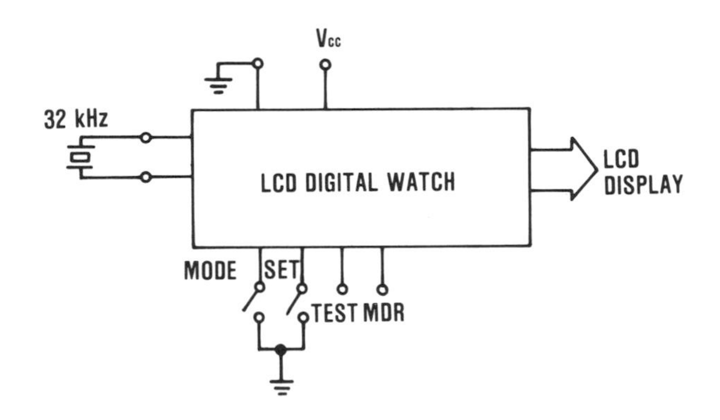

FIGURE C.12.0

KEY PARAMETERS

a) *Maximum power dissipation.* Up to 60°C, 200 mW is typical.
b) *Quiescent IC current.* Total device current, when watch is running normally, with 1.5 V battery. A typical value is 0.1 μA.
c) *Output voltage.* The output from the multiplier that drives the LCDs. Typical values are -4.5 V for logic **O** and 0 V for logic **1**.
d) *Output current, total.* The current delivered to the entire LCD set. 50 μA is typical.
e) *Crystal oscillator frequency.* 32,768 Hz.
f) *LCD drive frequency.* 512 Hz.
g) *Display segment multiplex frequency.* The "update" rate for each LCD segment. 32 Hz is the usual value.

APPLICATION

Used in LCD type wristwatches.

COMMENTS

The same basic IC is available with additional features, such as operation as stop watch or timer, and even with the capability of playing a form of roulette.

D. DIGITAL ICs

By definition, digital circuits deal with signals which can have only one of two possible values. In earlier, discrete, transistor digital circuits the transistor was either cut off, producing a collector output voltage close to the power-supply voltage, or it was in saturation, producing a voltage close to the emitter potential, usually ground. In positive logic systems the level close to ground is considered **0**, and the voltage close to the power supply, if it is positive, is considered logic **1**. The reverse is true in negative logic systems. In this book all descriptions will refer to positive logic and the term logic **1** (or **1**) will refer to the **high** voltage level, while the term logic **0** (or **0**) will refer to the **low** voltage level.

All of the digital ICs, with only a single exception, operate in the manner described above. The exception is the digital IC family using emitter-coupled logic (ECL). This family is used primarily where high operating speed, generally above 2.0 MHz, is required. A typical manufacturer's specifications for ECL digital ICs appear as item F in the Appendix.

The essential digital functions of all ICs are the same, regardless of the particular IC family. An OR gate, a flip-flop, or a whole shift register operates in exactly the same way, whether the IC is implemented with ECL or with MOS technology. Certain key parameters may be different, but the function is identical. An ECL IC may require more power, may operate at higher clock speed, and may have different input and output loading characteristics than its MOS counterpart, but the functions of a shift register or a counter are always the same. In the following pages we have noted the difference in key parameter values in many instances, indicating, for example, how much power a TTL, MOS or ECL IC, performing the identical function, requires. For detailed specifications, however, the reader has to refer to the manufacturer's data. Some typical manufacturer's data sheets for COS/MOS, Low-Power Schottky TTL and ECL IC families are included in the Appendix.

The section on digital ICs is the largest section in this book and represents a very condensed version of the detailed data available from the manufacturer's data books to extract the functional descriptions included here. Some digital IC families implement several microprocessor (MP) or microcomputer (MC) series, each of which might be the subject ot a 500-page book itself. Clearly, this volume of information had to be distilled considerably to bring to the reader concise and accessible data. For this reason, some types of digital ICs may seem to have received short shrift. We have included only 11 categories of IC memory, concentrating on the most widely used, typical examples of their basic operation and key parameters. In the microprocessor (MP) and microcomputer (MC) area we have included only seven basic IC types, because an entire book of this size could not do justice to all of the ICs currently on the market or in the process of coming out. The reader interested in designing equipment with an MP or MC will have to consult the manufacturer's data anyway, and the information in section D will at least provide the basis for setting up preliminary technical requirements. For a full understanding of MPs and MCs, the reader is referred to the many excellent texts on this subject now on the market.

The reader looking for a particular digital logic function may find that it is included in this book under a different name. Many digital functions have several common names, and while only one name is listed in the table of contents, the index includes all names. A *counter*, for example, is often called a *divider*; a *decoder* may be referred to as a *demultiplexer* or a *demuxer*; and nomenclature such as "latch" may apply to a series of flip-flops or a parallel shift register—or vice versa.

While the basic digital IC families usually include TTL, MOS, ECL and IIL, there are some combinations and special trade names that might confuse the reader. RCA, for example, calls its complementary metal oxide semiconductor ICs "COS/MOS," while Motorola calls their version "CMOS" and their ECL family "MECL," the *M* standing for *Motorola*. None of these variations change the basic functions described in this book, nor do they affect the key parameters or applications of the particular ICs.

D.1.0 Arithmetic Functions

DESCRIPTION

This section deals with digital ICs that perform the basic arithmetic functions, using the binary system. Human beings are accustomed to the decimal system, but digital logic is based on the binary system, which recognizes only two values, *0* and *1*. The position of each *0* or *1* in a series determines its value, just as we have tens, hundreds, thousands, etc., in the decimal system. The least significant bit in the binary system is 2°, which can be either *0* or *1*. The next significant bit has the value of 2^1 (*0* or *2*) and after that, the powers of 2 increase (2^2, 2^3, etc.). The decimal number 9 is represented by 1001, which adds 2^3 (8) to the least significant bit, 2^0 (1).

The following fundamentals of binary arithmetic apply to all of the arithmetic functions described in this section. Figure D.1.0 illustrates how the addition and subtraction functions are implemented in digital logic. The half adder consists of an exclusive OR and an AND circuit. As illustrated by the truth table, the rules for addition are carefully followed. The full adder is comprised of two half adders and an OR circuit for the carry-out. The truth table illustrates the operation of full addition of two binary values, A and B.

1) *Rules for addition*; 0+0 = 0; 1+0 = 1; 1+1 = 0 (carry 1).

```
Examples:      101      5          111      7
              +001     +1         +011     +3
              ----     --         ----     --
               110      6         1010     10
                ↑                 ↖
              carry              carry
```

2) *Rules for subtraction:* 0–0 = 0; 1–0 = 1; 0–1 = 1 (borrow 1); 1–1 = 0.

Examples:

```
 0 borrow                          01←borrow
  110       6  (minuend)            1001     9  (minuend)
 - 001    - 1  (subtrahend)        -0011   - 3  (subtrahend)
 -----    ---                      -----   ---
  101       5                       0110     6
```

An alternate method of binary subtraction involves either the *1's* complement or the *2's* complement. In *1's* complement all *0* and *1* in the subtrahend are interchanged and this complemented number is then added to the minuend. The last *1* carry is added as least significant bit to the total sum.

```
1's complement example:   5      101       101  (minuend)
                         -3     -011      +100  (subtrahend)
                         --               ----
                          2               1001
                                         →+  1
                                          ----
                                           010
```

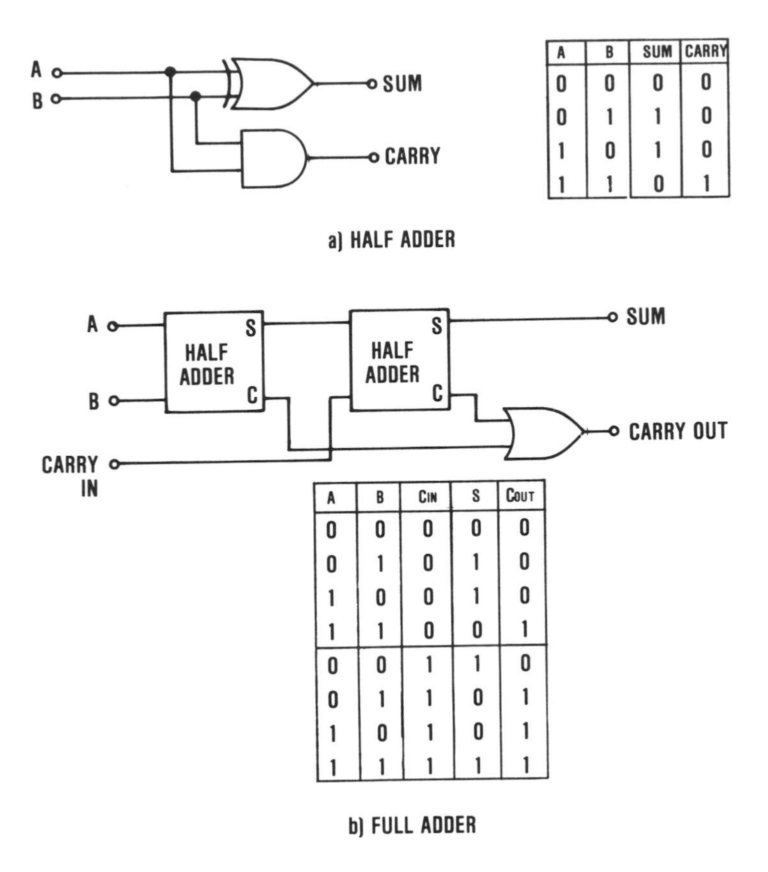

A	B	SUM	CARRY
0	0	0	0
0	1	1	0
1	0	1	0
1	1	0	1

A	B	CIN	S	COUT
0	0	0	0	0
0	1	0	1	0
1	0	0	1	0
1	1	0	0	1
0	0	1	1	0
0	1	1	0	1
1	0	1	0	1
1	1	1	1	1

FIGURE D.1.0 (a) Half Adder
(b) Full Adder

In the *2's* complement method the subtrahend is complemented by changing *0* into *1* and *1* into *0* in the same way, but an extra *1* is added to the subtrahend before it is added to the minuend. The last carry is omitted.

2's complement example:

```
  5    101                    101
- 3   -011 →100+1 =          +101
---                      drop1____
  2                           010
```

3) *Rules for multiplication:* 1×1 = 1; 1×0 = 0.

Examples:

```
   101     5×3          100    4×5
 × 011                × 101
 -----                -----
   101                  100
  101                  000
 -----                100
  1111     15         -----
                      10100    20
```

In many arithmetic operations multiplication is performed by repeated addition.

4) *Rules for division:* 1/1 = 1; 0/1 = 0.

Examples:

```
    101          5            101              5
11)1111       3)15      100)10100           4)20
   11                       100
   ----                     -----
    11                       100
    11                       100
   ----                     -----
    00                       000
```

In many arithmetic operations division is performed by repeated subtraction, and the result is given by the number of times the subtraction can be carried out.

D.1.1 4-Bit Full Adder

DESCRIPTION

This IC accepts two 4-bit binary numbers and adds them in a parallel operation. As illustrated in Figure D.1.1, one set of binary numbers is connected to inputs A1 through A4, and the second number is connected to inputs B1 and B4. Carry information from previous arithmetic operations can be connected to the C_{IN} terminal.

Four standard full adders are used, with a carry signal progressing from adder 1 through adder 4. At the same time, special high-speed parallel carry operation is available for use with other adders in a larger arithmetic system.

The truth table of Figure D.1.1 describes the operation of any one of the full adders. A carry output occurs only when both A and B are in the logic **1**

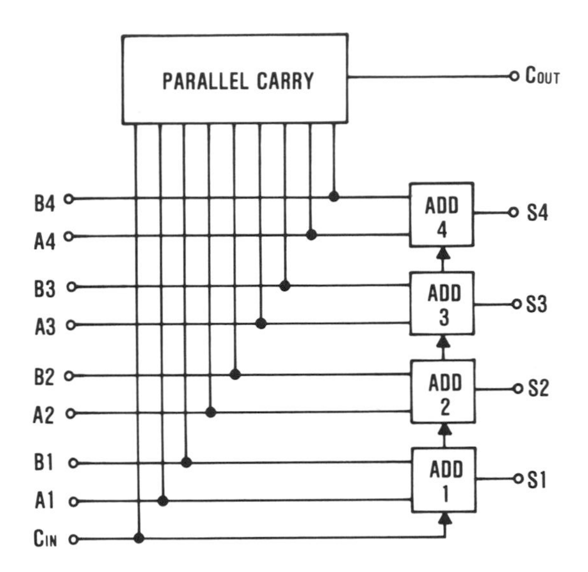

a) BLOCK DIAGRAM

C_{IN}	B	A	C_{OUT}	S
0	0	0	0	0
0	0	1	0	1
0	1	0	0	1
0	1	1	1	0
1	0	0	0	1
1	0	1	1	0
1	1	0	1	0
1	1	1	1	1

b) TRUTH TABLE (ONE STAGE)

FIGURE D.1.1 (a) Block Diagram
(b) Truth Table (One Stage)

state. The lower half of the truth table illustrates the operation when there is a carry-in.

KEY PARAMETERS

The electrical characteristics are those of the particular digital IC family.

a) *Operating speed.* 160 ns is the typical time required from the input of the two binary numbers to the output of the sum and the carry. (These values are typical for low-power CMOS devices.)

b) *Quiescent current.* The total current required by the IC when no operations are performed. 5.0 nA at 5 V is typical for low-power CMOS devices.

c) *Output load.* Each output of this IC can drive two low-power TTL loads. (Typical for low-power CMOS.)

APPLICATIONS

This IC is used in the arithmetic logic unit (ALU) of some computers to perform basic arithmetic functions.

D.1.2 Triple Serial Adder

DESCRIPTION

This IC consists of three identical serial adders. The clock and carry-in input are common to all three adders and there is a separate invert control input for each adder. The detailed logic diagram of the single adder, as well as common logic required to control all of them, is shown in Figure D.1.2. All operations are controlled by the clock which has two inverters to provide sufficient drive to all three adders. Each adder's timing depends on the clocked flip-flop, while the addition itself is performed by the exclusive NOR gates.

A positive logic adder is shown in Figure D.1.2, but, with minor variations, the same type of IC is also available for negative logic.

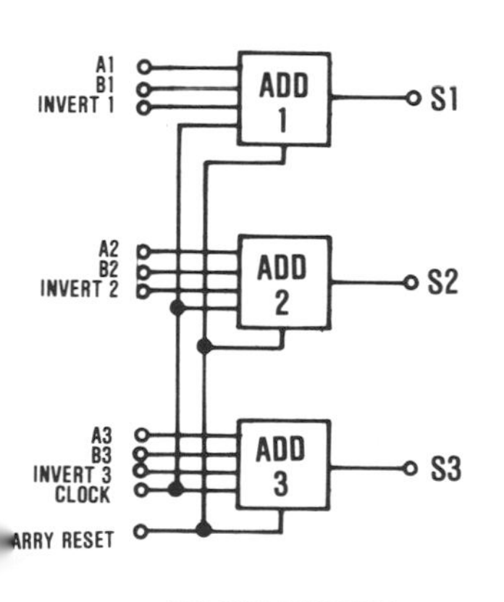

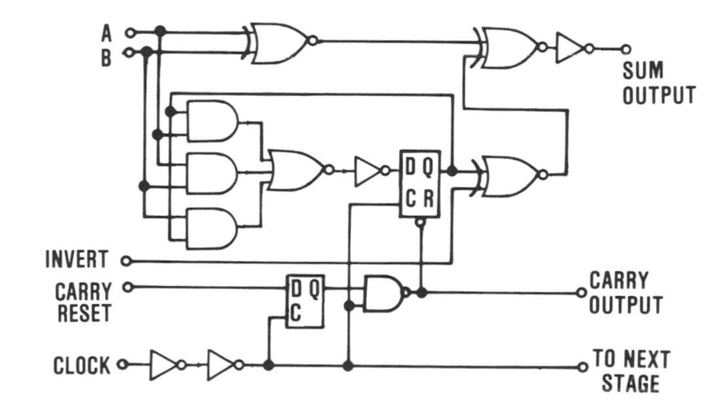

FIGURE D.1.2 (a) Block Diagram
(b) One Adder and Common Logic

KEY PARAMETERS

The electrical characteristics are those of the digital IC family.

a) *Operating speed*. Up to 5.0 MHz. (Typical value for low-power CMOS.)
b) *Clock requirements*. Single-phase clock is required. Clock pulse characteristics depend on digital IC family used.
c) *Quiescent current*. Current required when no operation is being performed. 5.0 nA at 5 V are typical for low-power CMOS.
d) *Output load*. Each output of this IC can drive two low-power TTL loads. (Typical for low-power CMOS.)

APPLICATIONS

This IC is used in the arithmetic logic unit (ALU) of some computers. It can also be used in digital correlators, digital servo control systems, data link systems, flight control computers, and wherever serial arithmetic units are required.

D.1.3 NBCD Adder

DESCRIPTION

Specifically designed for NBCD (natural binary coded decimal) four-bit addition and subtraction, this IC is similar to the four-bit full adder described in D.1.1. The main difference is that in the NBCD adder only the first ten values of the four-bit binary number can be used. The natural binary code assigns the values of 1, 2, 4 and 8 to the four-bits, with the highest binary number consisting of 9.

As illustrated in Figure D.1.3, there are four input lines for the NBCD value for A and four input lines for the NBCD value of B. A carry-in input is also available. The output consists of the four sum lines and the carry-out. The partial truth table shown in Figure D.1.3 illustrates how some of the possible values of A and B are added with a carry-in of *0* or *1*.

When NBCD arithmetic involves subtraction as well as addition, a special "9's complementer" IC can be used together with the NBCD adder so that one of the two NBCD numbers is correctly complemented for subtraction. The description of the "9's complementer" is contained in Section D.1.5.

KEY PARAMETERS

The electrical characteristics are those of the particular family of digital ICs.

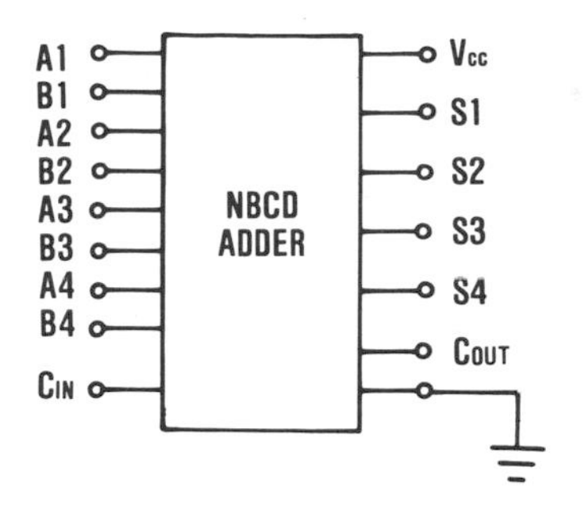

a) FUNCTION BLOCK

A				B				C	C	S			
4	3	2	1	4	3	2	1	IN	OUT	4	3	2	1
0	0	0	0	0	0	0	0	0	0	0	0	0	0
0	0	0	0	0	0	0	0	1	0	0	0	0	1
0	1	0	0	0	0	1	1	0	0	0	1	1	1
0	1	0	0	0	0	1	1	1	0	1	0	0	0
0	1	1	1	0	1	0	0	0	1	0	0	0	1
0	1	1	1	0	1	0	0	1	1	0	0	1	0
1	0	0	0	0	1	0	1	0	1	0	0	1	1
0	1	1	0	1	0	0	0	0	1	0	1	0	0
1	0	0	1	1	0	0	1	1	1	1	0	0	1

b) PARTIAL TRUTH TABLE

FIGURE D.1.3 (a) Function Block
(b) Partial Truth Table

a) *Noise immunity.* The ability of the input logic to distinguish between logic levels and noise. A typical value is 45% of the power-supply voltage. (This applies to low-power CMOS devices.)
b) *Quiescent current.* The current required by the IC when no operations are performed. 5.0 nA at 5 V is typical for low-power CMOS devices.
c) *Output load.* Low-power CMOS devices can drive up to two low-power TTL levels from each output terminal.

APPLICATIONS

This IC is specifically designed for digital systems that operate on the NBCD number system. It finds application in industrial controls, special purpose calculators, and other systems that require NBCD addition and subtraction functions.

D.1.4 Look-Ahead Carry Block

DESCRIPTION

The look-ahead carry generator can be connected across four binary adders to determine when carry signals will occur. As illustrated in the function block of Figure D.1.4, there is a carry input, four carry generating inputs, and four carry propagating inputs. The internal logic generates three carry outputs and two logic signals to indicate that a "carry generate"

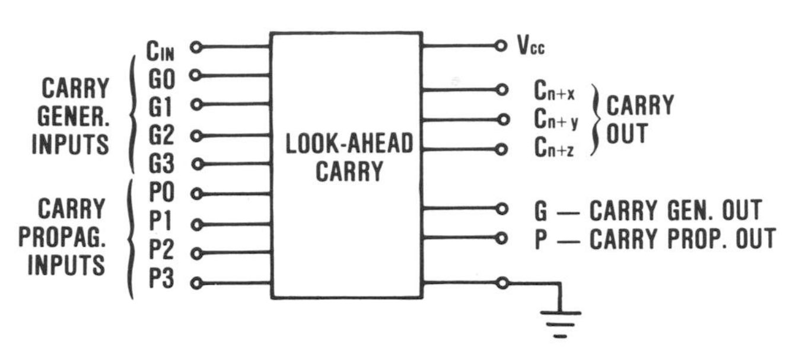

FIGURE D.1.4

or a "carry propagate" output has occurred. Each of the three carry outputs is determined by logic equations involving a combination of "carry generate" and "carry propagate" inputs.

The look-ahead carry function block can be cascaded to provide a full look-ahead across N-bit adders and can perform other essential functions in high-speed arithmetic logic units.

KEY PARAMETERS

The electrical characteristics are those of the particular family of digital ICs.

a) *Operating speed.* The time required from data-in to carry-out. 140 ns is typical for low-power CMOS devices.
b) *Quiescent current.* The current drawn by the IC when no operations are performed. 5.0 nA at 5 V is typical for low-power CMOS devices.
c) *Noise immunity.* Relative levels of input noise that will be rejected as compared to logic. 45% of power-supply voltage is typical for low-power CMOS.
d) *Output load.* In a typical low-power CMOS each output is capable of driving two low-power TTL loads.

APPLICATIONS

Used in conjunction with adders in arithmetic logic units (ALU) of digital computers and other numerical digital systems.

D.1.5 9's Complementer

DESCRIPTION

Binary subtraction consists of adding the complement of the number to be subtracted. The 9's complementer IC provides the subtraction capability

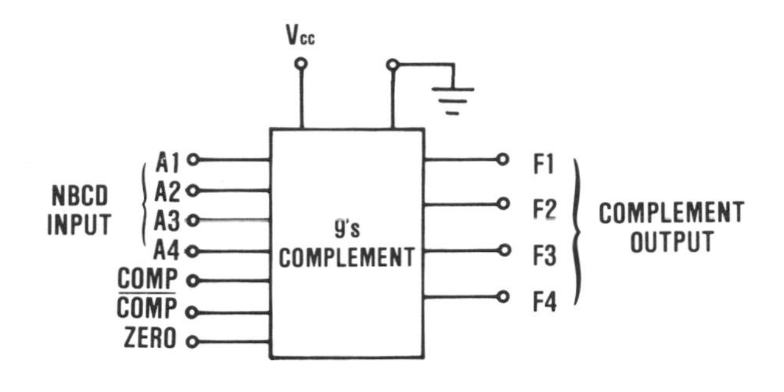

a) FUNCTION BLOCK

MODE	F1	F2	F3	F4	COMP	Z
THROUGH	A1	A2	A3	A4	0	0
COMP	$\overline{A1}$	A2	$A2\overline{A3}+\overline{A2}A3$	$\overline{A2}\,\overline{A3}\,\overline{A4}$	1	0
ZERO	0	0	0	0	X	1

b) TRUTH TABLE

FIGURE D.1.5 (a) Function Block
(b) Truth Table

for the NBCD adder described in Section D.1.3. As illustrated in the function block of Figure D.1.5, the NBCD number is applied to terminals A1 through A4 and the complement output appears automatically at terminals F1 through F4. Three control signals can be used to make this IC more versatile. When the COMP terminal is **1**, the IC generates the 9's complement. When the complement of the COMP terminal $\overline{\text{COMP}}$ is **1**, the output will be the same as the input, and when the ZERO terminal contains a **1**, the output will be all zeroes. This is clearly indicated in the truth table of Figure D.1.5, which also illustrates how the 9's complement is defined. In effect, when the decimal equivalent of the input is zero, the decimal equivalent of the output will be nine, and so on.

When used with the proper gating logic and an NBCD adder, the 9's complement IC becomes an effective add and subtract control.

KEY PARAMETERS

The electrical characteristics of this IC are those of the particular digital IC family.

a) *Propagation delay*. The time from input of an NBCD number until its complement appears at the output. 160 ns is typical at power-supply voltages of 10 V for low-power CMOS.
b) *Quiescent current*. The current drawn when no operations are performed. 5.0 nA at 5 V supply voltage is typical for low-power CMOS.
c) *Output load*. When implemented with low-power CMOS, this IC can drive two low-power TTL loads from each terminal.

APPLICATIONS

Used in arithmetic logic units, ALU of digital computers, particularly those using BCD calculations.

D.1.6 4-Bit Magnitude Comparator

DESCRIPTION

This IC compares two 4-bit binary numbers and indicates, by means of a **1** output, which one is greater than the other, or if they are equal. To accommodate binary numbers greater than four-bits, several comparator ICs can be cascaded. The output of the comparator dealing with the four least significant bits is connected to the cascade input of the four next significant bits. As illustrated in the function block of Figure D.1.6, one 4-bit binary number is applied to terminals A0 through A3 and the second number to terminals B0 through B3. The three possible results, A larger than B, A equal to B, and A smaller than B, constitute the output. For cascading, these outputs are connected to the corresponding cascade inputs at the next more significant 4-bit magnitude comparator.

To appreciate the truth table of Figure D.1.6, we should understand that the A and B in the comparing columns actually correspond to the A and B binary digits in the column headings. In the left comparing column, all of the A's and B's should be considered A2 and B2, etc. Note that in the first row, first column, A3 is indicated as larger than B3, and in the cascading column, A is also larger than B. The output therefore will be a logic **1** only in the A-greater-than-B column. In the bottom row, the situation is reversed and A3 is smaller than B3, with no cascading inputs at all. The result is that any output A is smaller than B.

KEY PARAMETERS

The electrical characteristics are those of the particular digital IC family.

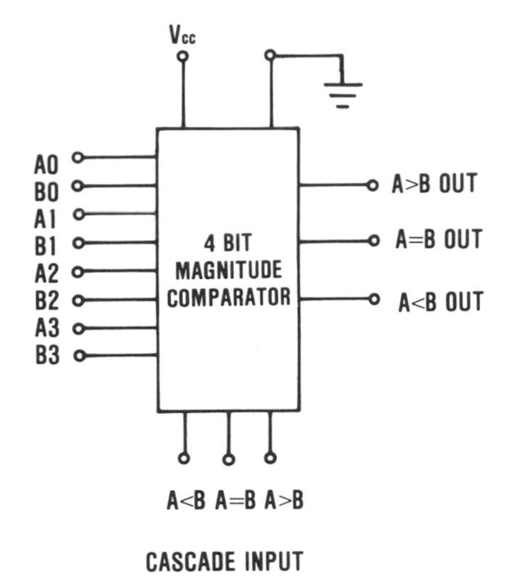

a) FUNCTION BLOCK

COMPARING				CASCADING			OUTPUTS		
A3,B3	A2,B2	A1,B1	A0,B0	A<B	A=B	A>B	A<B	A=B	A>B
A>B	X	X	X	X	X	1	0	0	1
A=B	A>B	X	X	X	X	1	0	0	1
A=B	A=B	A>B	X	X	X	1	0	0	1
A=B	A=B	A=B	A>B	X	X	1	0	0	1
A=B	A=B	A=B	A=B	0	0	1	0	0	1
A=B	A=B	A=B	A=B	0	1	1	0	1	0
A=B	A=B	A=B	A=B	1	0	1	1	0	0
A=B	A=B	A=B	A<B	X	X	X	1	0	0
A=B	A=B	A<B	X	X	X	X	1	0	0
A=B	A<B	X	X	X	X	X	1	0	0
A<B	X	X	X	X	X	X	1	0	0

X = INDIFFERENT

b) TRUTH TABLE

FIGURE D.1.6 (a) Function Block
(b) Truth Table

a) *Quiescent current.* The total current required when no operations take place. 5.0 nA at 5 V supply voltage is typical for low-power CMOS.
b) *Noise immunity.* The relation of the logic input signal to noise, referred to the supply voltage. 45% of the supply voltage is typical noise immunity for low-power CMOS devices.
c) *Output loading.* Low-power CMOS devices can typically drive two low-power TTL loads from each terminal.
d) *Code compatibility.* This comparator can operate on straight binary or the NBCD data.

APPLICATIONS

Comparators are part of the central processing unit (CPU) logic and are also used in correlation and detection circuits of automatic instrumentation, industrial controls, testers, and special purpose digital control devices.

D.1.7 Parallel Binary Multiplier

DESCRIPTION

This IC provides 2-bit by 2-bit parallel multiplication as well as simultaneous addition of two other binary numbers to the product. As illustrated in the logic diagram of figure D.1.7, there are two multiplicand inputs, X0 and X1, and two multiplier inputs, Y0 and Y1. A total of five cascading or adding inputs, K0 and K1, and M0, M1 and M2, are used when the 2-bit by 2-bit multiplier is connected in larger arrays to deal with larger digital numbers. There are five sum and carry outputs—S0, S1, S2, and C1 (S3) and C0. In a typical application, a number of these ICs is connected in an array to perform an m-bit by n-bit parallel binary multiplication function.

Figure D.1.7 illustrates the basic equations of the arithmetic performed by this IC and the logic diagram of a single multiplier cell illustrates the principle of its operation. The two binary bits are applied at terminals X and Y, while the cascading input is terminal K and the adding input is terminal M. The sum of the product, plus the addition, is available at output S and, if there is a carry bit, it is available at output C. A total of three AND circuits, two exclusive OR circuits and an OR circuit comprise the entire multiplier cell.

KEY PARAMETERS

The electrical characteristics are those of the particular digital IC family.

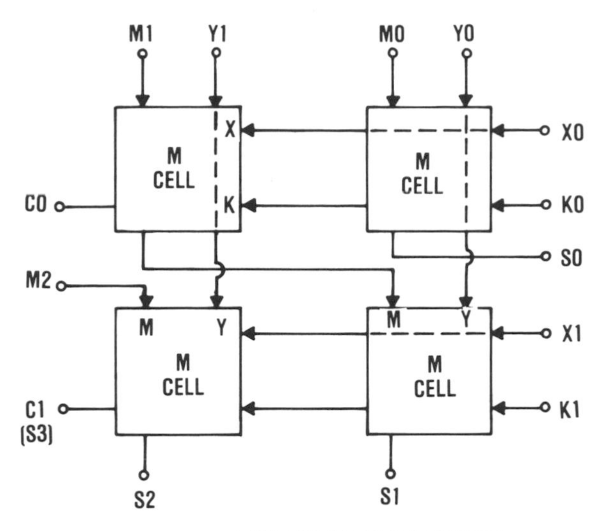

a) MULTIPLIER BLOCK DIAGRAM

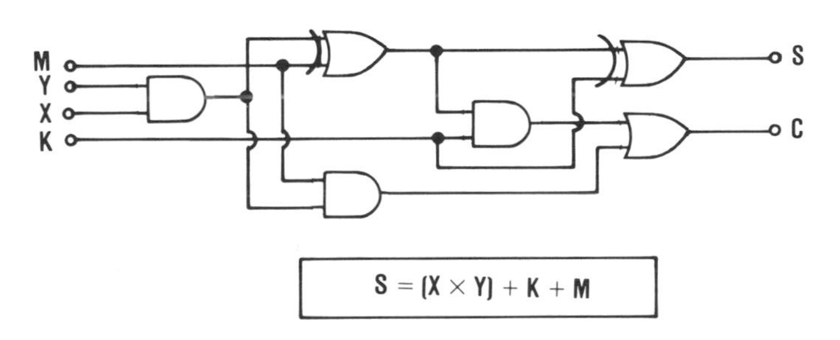

b) MULTIPLIER CELL LOGIC

FIGURE D.1.7 (a) Multiplier Block Diagram
(b) Multiplier Cell Logic

a) *Quiescent current.* The current required when no operations are performed. 5.0 nA at 5 V power-supply voltage is typical for low-power CMOS.
b) *Noise immunity.* The noise level with relation to the logic level at specified power-supply voltage. 20% of the power-supply voltage is a typical noise level for low-power CMOS devices.
c) *Output loading.* Low-power CMOS devices are typically able to drive two low-power TTL loads from each terminal.

APPLICATIONS

This IC is used in arithmetic logic units (ALU) of computers to perform such functions as multiplication, division, square root, polynomial evaluation, reciprocals, and division. This IC is also used in special mathematical systems for fast Fourier transform processing, digital filtering, convolution and correlation functions, and process and machine controls.

D.1.8 Binary Rate Multiplier

DESCRIPTION

This IC provides a rate that is the clock input pulse rate multiplied by 1/16 times the binary input. If, for example, the binary input number is 12, there will be 12 output pulses for every 16 input pulses. As illustrated in Figure D.1.8(a), a 4-bit binary counter supplies four parallel output lines to the rate select logic section. An external clock signal drives the counter. If there was a previous carry-in signal, it is supplied to the counter. It is also possible to clear the counter at any time or to set it to all *1*, representing the number 15. The rate select logic section receives four lines of binary rate select input, a strobe signal and, if desired, a cascade input from another rate multiplier of the same type. Both a complement and a regular output terminal are available, as well as a carry-out and a "15" output.

The partial truth table of Figure D.1.8(b) illustrates the operation of this IC. Note that when the binary input is all "0", the output terminals will be low and high respectively. When the binary input is the equivalent to a decimal *1*, there will be one output pulse for every 16 input pulses. Similarly, if the binary input is the equivalent of the decimal 16, there will be 16 output pulses. In all events, both the carry-out and the "15" output will be a logic **1**.

When more than four bits of binary information are involved, binary rate multiplier ICs can be cascaded either in the add or in the multiply mode. In the add mode, the first multiplier will have the selected number of pulses for every 16 clock input pulses, and the second unit will have the number of pulses selected from it for every 256 input pulses. The outputs are then added together. In the multiply mode, the clock rate of each IC remains the same but the results of the two outputs are multiplied together.

KEY PARAMETERS

The electrical characteristics are those of the particular digital IC family.

a) *Power dissipation*. The maximum power that one IC can dissipate for a given temperature range. For low-power, CMOS devices, 500 mW is typical over a temperature range of -40 to +60°C.
b) *Noise margin*. The percentage of noise voltage the device can

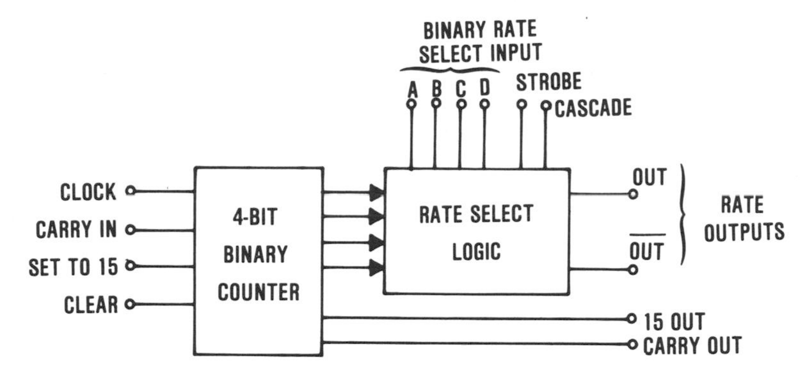

a) RATE MULTIPLIER BLOCK DIAGRAM

BINARY INPUT				NO. OF PULSES			
A	B	C	D	OUT	$\overline{OUT}$	C_{OUT}	15
0	0	0	0	L	H	1	1
1	0	0	0	1	1	1	1
0	1	0	0	2	2	1	1
1	1	0	0	3	3	1	1
0	0	1	0	4	4	1	1
1	0	1	0	5	5	1	1
0	1	1	0	6	6	1	1
1	1	1	0	7	7	1	1
0	0	0	1	8	8	1	1
1	0	0	1	9	9	1	1
0	1	0	1	10	10	1	1
1	1	1	1	15	15	1	1

b) PARTIAL TRUTH TABLE

FIGURE D.1.8 (a) Rate Multiplier Block Diagram
(b) Partial Truth Table

tolerate with relation to the power-supply voltage. A typical value is 20% for low-power CMOS.

c) *Maximum clock frequency*. Depends, to some extent, on the power-supply voltage. 3.5 MHz is a typical value at 15 V for low-power CMOS devices.

APPLICATIONS

A binary rate multiplier can be used to add, subtract, divide, raise to a power, and solve algebra and differential equations. It can also generate natural logarithms and trigonometric functions, and can perform frequency division. It is most frequently used in numerical control devices, frequency synthesizers, digital filtering, and some types of measuring instrumentation.

D.1.9 BCD Rate Multiplier

DESCRIPTION

This IC operates basically the same as the binary rate multiplier described in Section D.1.8. The main difference is that the binary rate select input terminals A, B, C and D, in Figure F.1.9(a), can accept values only from 0 to 9. The selection is limited to ten clock pulses, as opposed to 16 in the binary rate multiplier of Section D.1.8. This particular IC also has an enable control signal both at input and output.

The partial truth table of figure D.1.9(b) illustrates that when the BCD input is all zero, there will be no pulses out but one output will be low and its complement will be high. When the BCD code indicates the number 2, there will be two pulses at both outputs. Note that when the BCD input goes beyond the binary number 9, the output will be either 8 or 9 pulses. These, however, represent illegal BCD codes.

Like the binary rate multiplier of Section D.1.8, the BCD rate multiplier IC can also be cascaded to deal with binary inputs of more than four bits. Cascading is limited to binary code decimal numbers.

KEY PARAMETERS

The electrical characteristics are those of the particular digital IC family.

a) *Quiescent current.* The current used by the IC when no operations are performed. 5.0 nA are typical at 5 V power supply for low-power CMOS.
b) *Maximum clock frequency.* This depends, to some extent, on the power-supply voltage. 3.5 MHz is typical at 15 V for low-power CMOS.
c) *Noise margin.* The amount of noise tolerated for a given power-supply voltage. 20% is typical for low-power CMOS.

APPLICATIONS

This IC can perform arithmetic operations for binary coded decimal

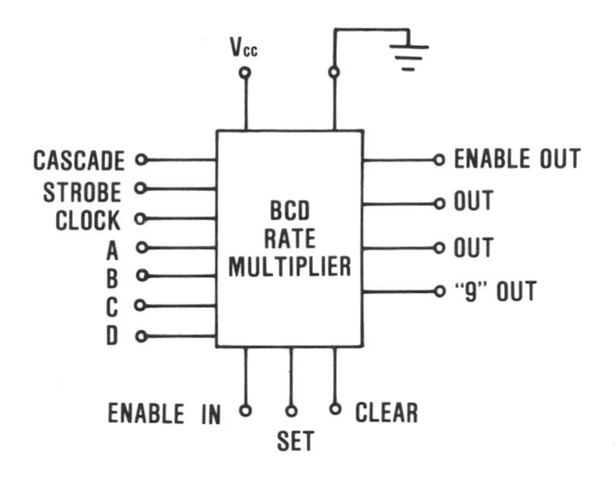

a) FUNCTION BLOCK

	BCD INPUT				NO. OF PULSES			
	A	B	C	D	OUT	OUT	E	"9"
	0	0	0	0	L	H	1	1
	1	0	0	0	1	1	1	1
	0	1	0	0	2	2	1	1
	1	1	0	0	3	3	1	1
	0	0	1	0	4	4	1	1
	1	0	1	0	5	5	1	1
	0	1	1	0	6	6	1	1
	1	1	1	0	7	7	1	1
	0	0	0	1	8	8	1	1
	1	0	0	1	9	9	1	1
*	0	1	0	1	8	8	1	1
*	1	1	0	1	9	9	1	1

* ILLEGAL BCD CODE

b) PARTIAL TRUTH TABLE

FIGURE D.1.9 (a) Function Block
(b) Partial Truth Table

(BCD) codes and is frequently used in motor speed controls, frequency synthesizers, digital filters, and industrial numerical control systems.

D.1.10 8-Bit Serial/Parallel Multiplier

DESCRIPTION

This IC accepts an 8-bit multiplicand in the parallel X input terminals as illustrated in the logic diagram of Figure D.1.10, and stores this data in eight internal FFs. The 8-bit multiplier word is entered serially in the Y input, with the least significant bit first. The actual multiplication is performed in the adder/subtractor and register section and the result appears at the S, sum, terminal. A clock signal times the operation and controls both the input of the Y signal and the output of the S signal. A clear, CLR, signal clears the FFs and registers for each new operation.

When used with multiplicands longer than eight bits, several of these multipliers can be connected in cascade. In this arrangement the sum (S) output of one device is connected to the K input of the next device. The M (mode) input is used to indicate which of the devices in the cascaded arrangement contains the least significant bit.

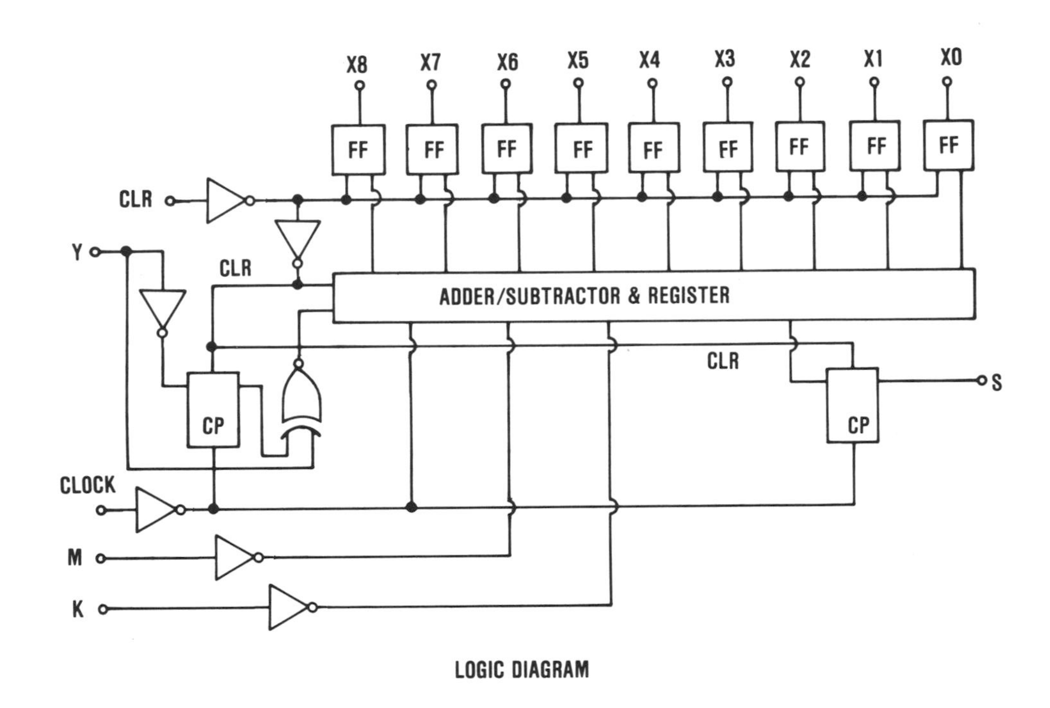

FIGURE D.1.10 Logic Diagram

KEY PARAMETERS

This device is generally available in TTL ICs, and its electrical characteristics are those of the TTL IC digital family.

a) *Clock frequency*. In this TTL family the minimum clock frequency is 25 MHz and a typical operating frequency is 40 MHz.
b) *Clock pulse width* 15 ns is a typical pulse width for this clock frequency.

c) *Maximum power dissipation.* 100 mW is the typical maximum power dissipation for this TTL IC.

APPLICATIONS

Because of the relatively high clock rate, it is possible to use this IC in mini and maxi computer arithmetic logic units (ALU) for multiplication including the 2's complement without correction.

COMMENTS

This type of IC is not available in CMOS, and the clock frequency used with it (40 MHz) is higher than most CMOS devices are capable of.

D.1.11 4-Bit Arithmetic Logic Unit (ALU)

DESCRIPTION

This IC can perform 16 separate logic and 16 separate arithmetic functions on two 4-bit inputs. It generates a 4-bit output function as well as a comparison of the two inputs, a look-ahead and carry signal, and a carry-out signal. The mode control, as illustrated in Figure D.1.11(a), determines whether logic or arithmetic operations are to be performed. The four function selection inputs, S0-S3, determine which of 16 possible functions are to be performed. A list of selectable functions appears in Figure D.1.11(b). In the case of the 16 logic functions, the + signs indicate the logic OR and exclusive OR functions. The dot between the two letters A and B indicates logic AND. In the columns describing the 16 arithmetic functions, the + and - signs indicate addition and subtraction respectively. When two functions A B appear together, multiplication is indicated.

KEY PARAMETERS

The electrical characteristics are those of the particular digital IC family.

a) *Quiescent current.* The current required by the IC when no operations are performed. 5.0 nA at 5 V is typical for low-power CMOS devices.
b) *Noise immunity.* The ability of the input logic to distinguish between logic levels and noise. A typical value is 45% of the power-supply voltage. (For low-power CMOS devices.)
c) *Output load.* Low-power CMOS devices can drive up to two low-power TTL levels from each output terminal.

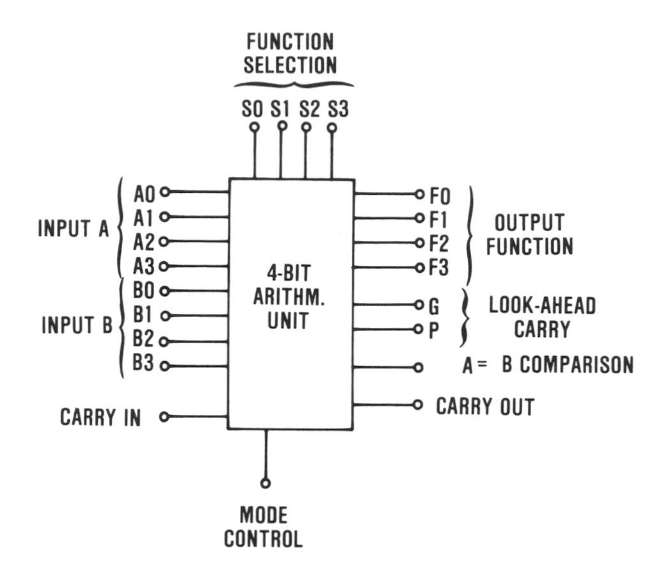

a) FUNCTION BLOCK

16 LOGIC FUNCTIONS		16 ARITHMETIC FUNCTIONS	
$\overline{A}$	$\overline{A}\cdot B$	A — 1	A + (A + B)
$\overline{AB}$	A + B	AB — 1	A + B
$\overline{A} + B$	B	$A\overline{B}$ — 1	$A\overline{B}$ + (A + B)
LOGIC "1"	A + B	— 1	A + B
$\overline{A + B}$	LOGIC "0"	A + (A + $\overline{B}$)	A + A
$\overline{B}$	$A \cdot \overline{B}$	AB + (A + $\overline{B}$)	AB + A
$\overline{A + B}$	A • B	A — B — 1	$A\overline{B}$ + A
$A + \overline{B}$	A	$A + \overline{B}$	A

b) LIST OF SELECTABLE FUNCTIONS

FIGURE D.1.11 (a) Function Block
(b) List of Selectable Functions

APPLICATIONS

This IC is particularly useful where both Boolean variables and straight arithmetic operations are required. It is used in industrial process control, on-line and batch type, numerical machine controls, robot-like devices, and other digital systems.

COMMENTS

Because of the relatively complex function selection, manufacturer's application notes should be consulted before using this IC in a specific equipment.

D.2.0 Buffer/Inverter

DESCRIPTION

The buffer/inverter is one of the key functions used in all digital logic assemblies. As the name implies, this circuit provides a stage of buffering, separation in impedance, between other logic elements and, when it acts as inverter, changes the input signal into its complement. While buffer/inverters are most frequently used in combination with other key logic elements, they are also available as separate ICs, with six individual stages on a single IC, as shown in Figure D.2.0.

To illustrate the critical function of the buffer/inverter, Figure D.2.0 includes the actual circuit used to implement a buffer/inverter in TTL and CMOS logic. In both circuits the input impedance is higher than the output impedance and the output signal is the inverse, complement, of the input signal.

In the TTL version a buffer/inverter is also available in which R3, Q3 and D2 have been omitted. This allows the user to connect an external collector resistor between the output terminal and Vcc or some other voltage. These "open collector" buffer/inverters are particularly useful for level-shifting applications.

Buffer/inverters are available for every logic family and are combined with other circuits to provide special buffering and inverter functions. Some buffers do not invert the input signal; this is indicated by the omission of the circle at the apex of triangular buffer/inverter symbol.

KEY PARAMETERS

The electrical characteristics are the same as those of the particular IC digital family.

a) *Input load factor*. The number of input loads represented by each buffer/inverter. A typical load factor for TTL and CMOS devices is 1.
b) *Output load factor (fan-out)*. The number of output loads of the particular digital IC family that the buffer/inverter can drive. A typical TTL buffer/inverter can drive ten TTL inputs. CMOS buffer/inverters can drive a much higher number of CMOS inputs but usually only two TTL loads.

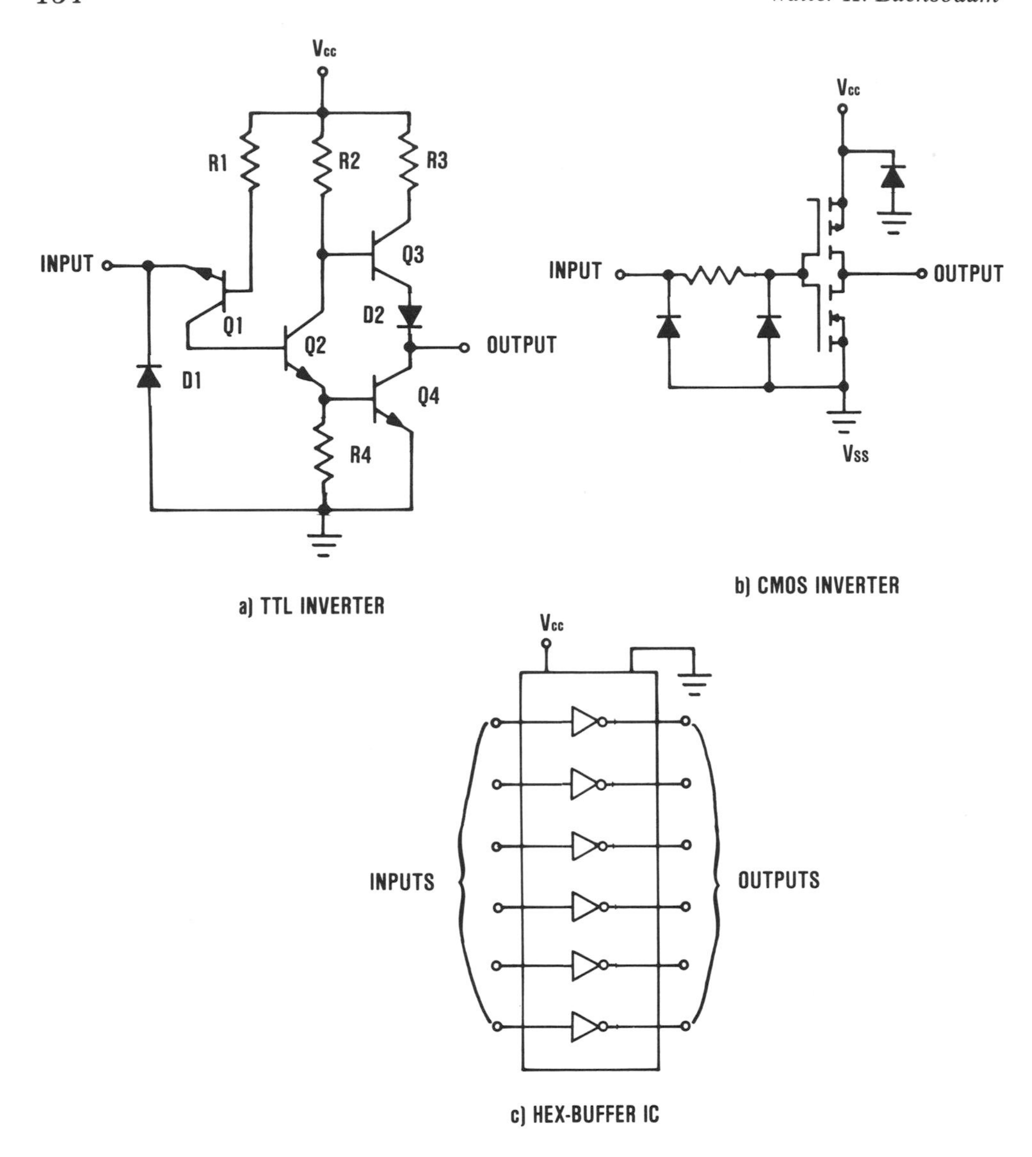

FIGURE D.2.0 (a) TTL Inverter
(b) CMOS Inverter
(c) Hex Buffer IC

c) *Propagation delay time*. The minimum time a signal is delayed between input and output. For a TTL buffer/inverter 13 ns is typical. For a CMOS device 30 to 80 ns is typical.
d) *Quiescent current*. The current drawn by the entire IC when no operations are performed. For TTL devices 2.0 mA is typical, while 2.0 nA is typical for CMOS.

APPLICATIONS

Buffers/inverters are used in all types of digital logic combinations to provide a separation between logic circuits and to increase the fan-out or drive capability of a particular output.

COMMENTS

The description of a hex buffer/inverter indicates that six identical circuits are contained on one IC. Most manufacturers produce the hex buffer ICs for inverting or noninverting applications. Be sure to check the manufacturer's identification number to determine which is which.

D.2.1 Strobed Buffer/Inverter

DESCRIPTION

This IC provides the basic buffer/inverter function as well as two additional control functions. As illustrated in Figure D.2.1(a), all six inputs are controlled by a single "inhibit" signal, and all six outputs are controlled by a single "disable" signal. This permits each of the six inputs and outputs to have a total of three possible states, as indicated by the truth table of Figure D.2.1(b). Because this is an inverting buffer, when the input is **0** and both the inhibit and disable signals are **0**, the output will be **1**. When the input is **1**, the output will be **0** if both the inhibit and disable signals remain **0**. In short, with inhibit and disable signals at **0**, the circuit acts as a simple hex inverter/buffer. When the inhibit signal is **1**, the output will be **0** regardless of what the input signal will be. If the disable signal is **1**, the output will respresent a high impedance regardless of what the input or the inhibit signals will be.

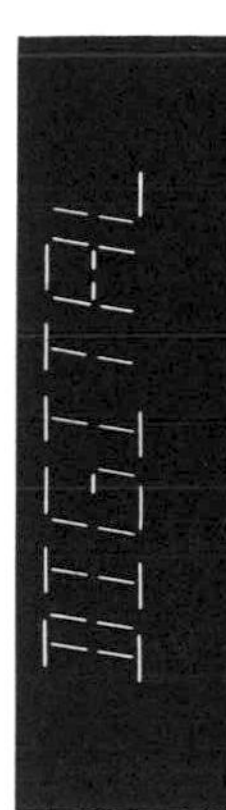

KEY PARAMETERS

The electrical characteristics of this IC will be the same as those of the particular digital IC family.

a) *Propagation delay time.* The time delay between input and output of any terminal. 200 ns is typical for CMOS devices.
b) *Minimum disable setup time.* The minimum time it takes from the application from the disable signal until all six buffer/inverters are disabled. 20 to 50 ns is typical for CMOS.

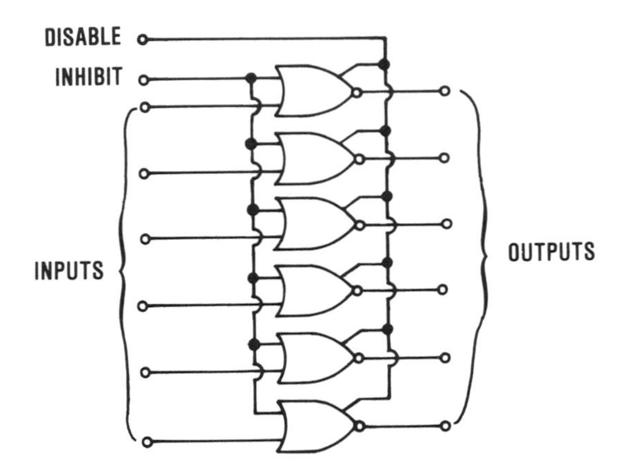

a) STROBED HEX BUFFER BLOCK DIAGRAM

INPUT	INHIB.	DISAB.	OUTPUT
0	0	0	1
1	0	0	0
X	1	0	0
X	X	1	HiZ

X — VALUE IMMATERIAL

b) TRUTH TABLE

FIGURE D.2.1 (a) Strobed Hex Buffer Block Diagram
(b) Truth Table

c) *Minimum hold time for disable signal.* The minimum time required for the disable signal to remain on in order for data to be disabled. 25 to 75 ns is typical for CMOS devices.

APPLICATIONS

This IC is useful as three-state hex inverter for interfacing other ICs with data buses. Other controlled interface applications are also possible.

D.2.2 Schmitt-Trigger Buffer/Inverter

DESCRIPTION

This IC, containing six identical circuits, provides the Schmitt-Trigger function and acts as buffer/inverter, as illustrated in the logic diagram of Figure D.2.2. It is particularly useful where slowly changing waveforms, such as sinewaves, must be squared up, or where noise immunity must be improved. The Schmitt-Trigger circuit conducts over only a narrow "window," a portion of the total supply voltage. When the input signal is below this hysteresis voltage, nothing passes between input and output. When the input signal reaches the bottom portion of the hysteresis voltage, the Schmitt-Trigger circuit goes into saturation, and when the upper level of the hysteresis voltage is reached, the circuit is cut off. The input signal is changed into a pulse that has the voltage level of the particular logic family. In all other respects the Schmitt-Trigger inverter/buffer functions like all other inverter/buffer circuits.

KEY PARAMETERS

The electrical characteristics are the same as those for the particular digital IC family.

a) *Quiescent current.* The current drawn by the entire IC when no operations are performed. For low-power CMOS devices, 0.5 nA and 5 V is typical. For high-voltage CMOS devices, 0.02 mA is typical up to 15 V.
b) *Hysteresis voltage.* The voltage range over which the Schmitt-Trigger fires. 0.55 V is typical for low-power CMOS with a 5 V power supply. For high-voltage CMOS devices 0.9 V is typical for a 5 V power supply.
c) *Threshold voltage.* The actual positive and negative voltage that causes the Schmitt-Trigger circuit to fire and to turn off. For low-power CMOS devices, 2.7 V is a typical positive going and 2.1 V is a typical negative going threshold voltage.
d) *Propagation delay time.* The time required between the arrival of the input signal and the availability of the output signal. For low-power CMOS 125 ns is typical and for high-voltage CMOS devices 140 ns is typical, both at 5 V power supply voltage.

APPLICATIONS

Schmitt-Trigger buffer/inverters are used in wave and pulse shapers, as monostable and as astable multivibrators. They are also used widely as

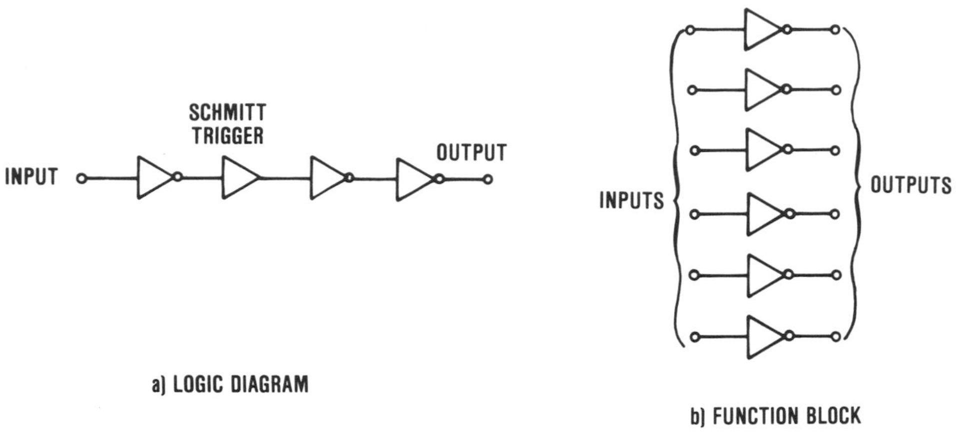

SCHMITT TRIGGER
INPUT
OUTPUT

a) LOGIC DIAGRAM

INPUTS
OUTPUTS

b) FUNCTION BLOCK

FIGURE D.2.2 (a) Logic Diagram
(b) Function Block

receiving circuits in digital transmission systems because the Schmitt-Trigger action greatly improves noise immunity.

D.2.3 True/Complement Buffer

DESCRIPTION

Usually available with four identical circuits on a single IC, this buffer provides both the true output and its complement. As illustrated in Figure D.2.3, the complement of the input is provided by a single inverter stage while the true output is provided by two stages. The circle at the input of the second stage indicates that, while this circuit is an inverter, its output will be the true or original signal and not the inverted one.

In all other respects the buffer/inverter circuits of Figure D.2.3 are the same as the standard buffer/inverter described in Section D.2.0.

KEY PARAMETERS

The electrical characteristics are the same as those of the particular digital IC family.

a) *Quiescent current.* The total current drawn by the IC when it is not operating. 0.02 μA is typical for high-voltage CMOS devices at power-supply voltages up to 15 V.
b) *Propagation delay time.* The time required for the input signal to reach the output. In a true/complement buffer it is important that

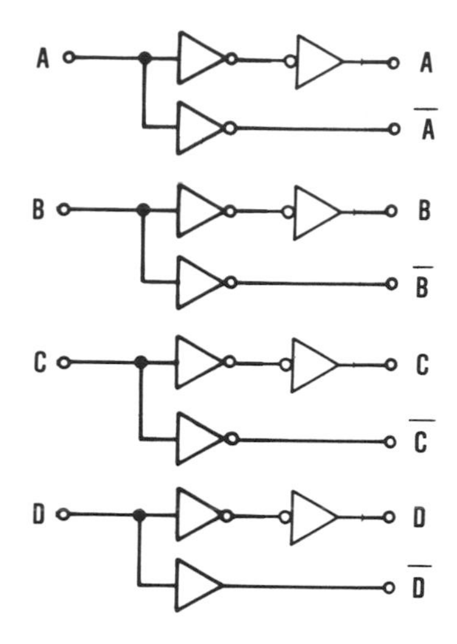

FIGURE D.2.3. True/Complement Logic Diagram

the propagation delay for both outputs be exactly the same. 60 ns is a typical value for high-voltage CMOS devices.

APPLICATIONS

These buffers are particularly useful for MOS clock drivers, providing both clock phases. They are also used to drive ladder or "weighted R" resistor networks, transmission lines, and digital displays. The fact that both the true and the complement of the input is available makes this IC very versatile in digital data link equipment.

D.2.4 Non-Inverting 3-State Buffer

DESCRIPTION

This IC contains six identical non-inverting buffers, arranged in two groups, as illustrated in Figure D.2.4. The first group of two buffers is controlled by the "disable A" signal. The second group of four buffers is controlled by the "disable B" signal. This provides three possible output states. When the respective disable signal is **0**, the output will be the same as the input, with the usual buffer effect. When either disable signal is **1**, the

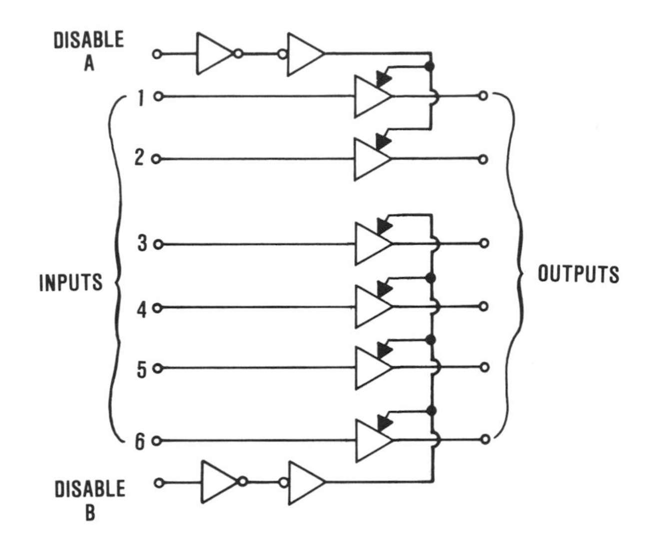

FIGURE D.2.4. 3-State Buffer Logic Diagram

buffers it controls will be turned off and, regardless of input signal, the output will represent a high impedance.

KEY PARAMETERS

The electrical characteristics will be the same as for the particular digital IC family.

a) *Turn-on delay time.* The time required from the input of either A or B disable signal until the respective outputs reach the high impedance stage. 75 ns is typical for low-power CMOS devices at 5 V power supply.
b) *Turn-off delay time.* The time required after either the A or B disable signal has disappeared until the outputs return to their normal state. 75 ns is a typical value for low-power CMOS devices at 5 V power-supply voltage.
c) *3-State propagation delay time.* The time required from input to output signal at different conditions. Typical low-power CMOS values are:

   ***1** to high impedance.* 75 ns
   ***0** to high impedance.* 80 ns
   *High impedance to **1**.* 65 ns
   *High impedance to **0**.* 100 ns

APPLICATIONS

3-state buffers are frequently used at the input/output ports or bus interfaces of computers and other digital devices. This type of IC is found wherever the flow of parallel digital data is controlled.

COMMENTS

If the disable A and disable B signals are combined, all six buffers can be controlled by a single signal. The particular arrangement of two and four permits the user to provide separate disable signals for control and data signals on a typical digital data bus.

D.3.0 Counters/Dividers

DESCRIPTION

The essential function in all counters/dividers is performed by the basic flip-flop (FF) which changes state only on either a positive or a negative going transaction. If a pulse or square-wave signal is applied to the input of an FF with positive logic, it will change state only on the rising edge of the signal. For every two pulses or square waves at the input, only one square wave will be available at the output. Every FF has a Q and a $\overline{Q}$ output, one being the complement of the other. A detailed discussion of FFs is presented in Section D.7.0.

The term *counter* implies that pulses or square waves are counted, and this function is provided by adding logic gates to the basic FF configuration. The term *divider* more accurately describes the function of the FFs themselves since each FF stage divides the input frequency by two. In some applications the input frequency is divided by the series of FFs into another frequency that is a predetermined fraction of the input. In other applications the pulses applied to the input are counted and a logic output signal is generated when a previously specified number of pulses has passed through the counter. Because both counting and division can be performed by all devices described in this section, we will, from now on, use only the term *counter*.

A basic "ripple" counter is illustrated in the logic diagram of Figure D.3.0(a). A square-wave signal is supplied at input A and A/2 appears at the Q output terminal of FF1. As illustrated, each stage divides its input frequency by two. The FFs are numbered according to the binary system. Note that FF1 operates at half of the input of frequency, while FF8 will operate at 1/16 of the input frequency. If the output of the fourth stage (FF8)

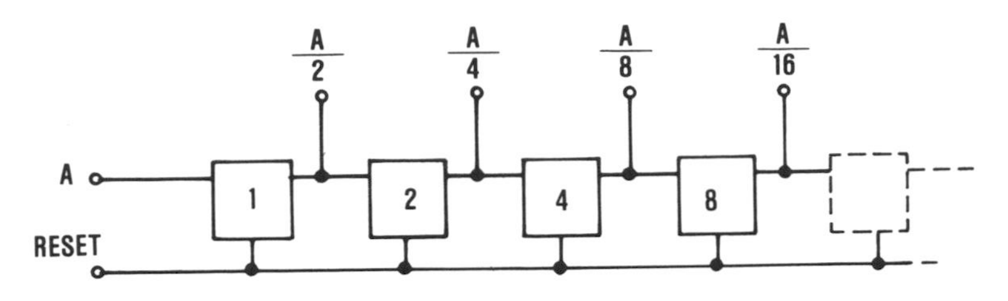

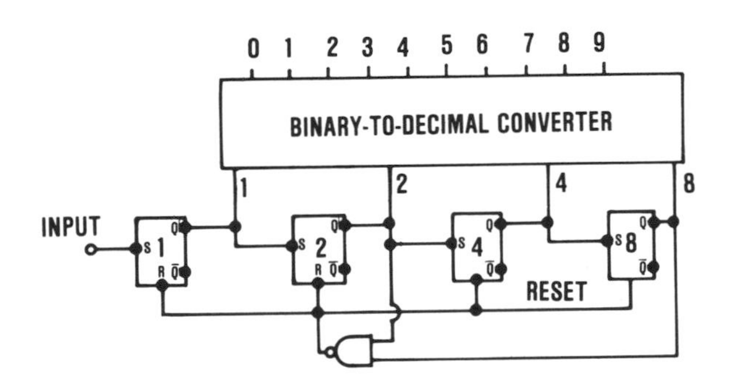

FIGURE D.3.0 (a) Logic Diagram of Basic Ripple Counter
(b) Logic Diagram of Basic Decade Counter

were connected to the reset line, this counter would count only the first 16 pulses. Once the 16th pulse has set FF8, the reset signal would prevent any further counting.

The logic diagram of D.3.0(b) illustrates the minimum circuitry for a decade counter. Here, the NAND circuit senses the Q signal from FF2 and the Q signal from FF8. When both occur at the same time, the counter has counted a total of ten pulses of square waves, and the reset line is activated. The four Q outputs will then represent the binary equivalent of the decimal number ten. During the first nine pulses which appear at the input, the binary equivalents of the corresponding decimal numbers will appear at the Q outputs, and these are converted to the decimal numbers 0 through 9.

Almost all of the counters described in this section are available with either straight binary or binary-coded decimal (BCD) outputs. Counters are also available in two main categories, depending on the manner of operation. The basic ripple counter shown in Figure D.3.0 is considered asynchronous because it operates independently of a clock. It could be connected to a photocell which counts people passing through a gate. In this application it would count asynchronous events. In many digital logic applications, however, the counter operates under the control of a clock signal. This greatly reduces the chances of triggering by noise or other unwanted signals, but it also requires that the input signal be synchronized with the clock signal. In many synchronous counters the input and the clock are the same signal. Special features, such as programmable division by any number or up and down counting, are available in both synchronous and asynchronous, binary and BCD counters. A wide variety of counters and special features is available in each of the different digital IC families.

KEY PARAMETERS

The electrical characteristics are essentially those of the particular digital IC family.

a) *Maximum operating frequency*. This is the highest input frequency that a ripple (asynchronous) counter can accept. For synchronous counters this also determines the highest clock frequency. The maximum operating frequency depends, to some extent, on the digital IC family used. TTL counters have a higher operating frequency than CMOS types.
b) *Minimum input signal rise time*. Unless a Schmitt-Trigger circuit is used in the input, slowly changing signals will not activate the counter. Sine waves and varying DC voltages are acceptable only when a Schmitt-Trigger input is used.
c) *Cascading*. When several counter ICs are used in cascade, a carry-out signal from the last stage must be available as input for the next counter IC.
d) *Input polarity*. Indicates whether the FFs in the counter will change on positive (rising) or negative going (falling) edges.
e) *Available connections*. On some counters the Q terminal of each FF is available, and in others some of the FF input circuits are also available. This permits the user to connect stages of the counter for a variety of applications. In some counters decoding circuitry is included, and only the fully decoded outputs are available.
f) *Noise characteristics*. Indicate the relation of logic levels to noise signals. To some extent, this characteristic is dependent on the particular digital IC family.
g) *Output load*. In some counters, output driver circuits are included, while in others only a minimum output load can be connected. To some extent, output load capability depends on the particular digital IC family.
h) *Propagation delay*. Normally stated as the time required from the input to the first Q terminal output. Total delays for the entire

counter must be calculated, considering the number of stages and the frequency division performed. The basic propagation delay time depends on the particular digital IC family.

i) *Control signal delay.* The time required from the application of a particular control signal, such as reset, preset, up/down, until the control signal takes effect.

APPLICATIONS

Counters are used for frequency division, as control timers, to drive time displays, as part of phase-locked loops, and in a host of other applications where either time or a number of events are being counted.

COMMENTS

The various types of counters described in the following pages do not exhaust the full gamut of available counters. Special purpose counters, such as those used for digital clock radios, for color TV circuitry and other special applications, are also available.

D.3.1 Counter, Synchronous

DESCRIPTION

Synchronous counters are available in a large variety of sizes and with a number of standard output decoders. The essential features of this type of counter can be seen from the illustration of Figure D.3.1(a). Each of the flip-flops (FF) is controlled by the clock signal. Whatever data are entered must be entered in synchronism with the clock, because only when the clock enables each FF can the counting or dividing operation take place. As indicated in the waveform diagram of Figure D.3.1(b), all the output signals, Q1 through Q4, have the same symmetrical pulse shape. If we assume that the data input is the same as the clock signal, each output pulse will be the width of one complete square-wave cycle of the clock. Note that each output occurs one complete clock cycle after the previous one. The reset signal can be enabled at any time and, in the illustration of Figure D.3.1(b), it is kept at logic **0** after the start. The clock inhibit signal, however, has been set to logic **1** after seven clock cycles and all output signals remain at **0** thereafter. As indicated, the counter is advanced one count during each positive clock signal transaction, as long as the clock inhibit signal is at **0**.

KEY PARAMETERS

The electrical characteristics are essentially the same as those of a particular digital IC family.

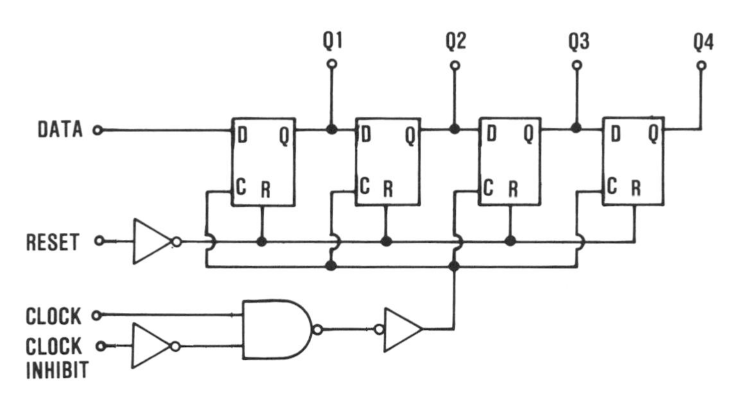

a) LOGIC DIAGRAM OF SYNCHR. COUNTER

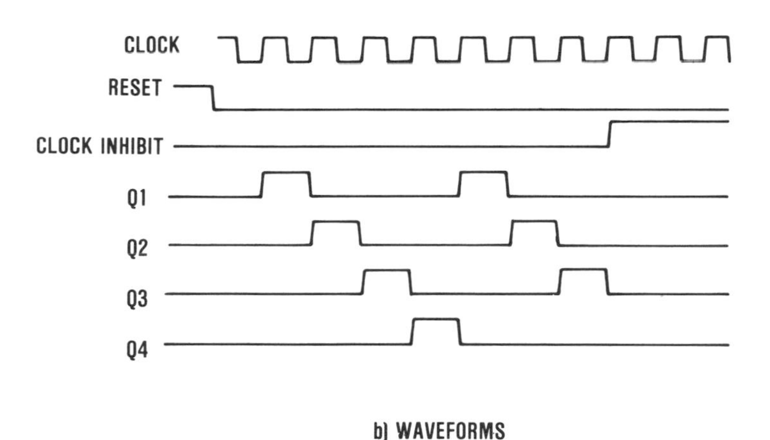

b) WAVEFORMS

FIGURE D.3.1 (a) Logic Diagram of Synchronous Counter (b) Wave Forms

a) *Maximum clock input frequency.* Up to about 5.5 MHz for CMOS and 35 MHz for TTL devices.
b) *Clock pulse width.* The maximum width of the clock input signal. This ranges from 60 to 200 ns for CMOS ICs and as low as 20 ns for TTL devices.
c) *Reset pulse width.* The minimum width of the reset pulse. CMOS devices range from 60 to 300 ns, while TTL devices reset pulse widths are approximately 20 ns.
d) *Reset removal time.* The minimum time required until the reset signal is removed and the counter operation is again enabled. A typical value for CMOS devices is 150 to 400 ns.
e) *Maximum power dissipation.* In some instances stated for the total package, in others per output stage. For CMOS a typical value is 500

mW from -40 to +60°C. TTL ICs are available with power dissipations of 100 mW for up to six output stages.

APPLICATIONS

Synchronous counters are used in decade and binary computer control and timing circuits, for decade counters in clock displays, divide-by-N counting, and for frequency division.

COMMENTS

Most commercially available synchronous counters include additional gating and decoding circuits, making their logic diagram more complex than that presented in Figure D.3.1.

D.3.2 Counter, Asynchronous (Ripple)

DESCRIPTION

This type of counter can accept data which occur at any time. Whenever a new transition appears at the input, the effect of this change ripples down through the series of flip-flops (FF). Because there is no clock input to control the transition of each FF, an asynchronous counter is more subject to random noise than a synchronous counter. The logic diagram of Figure D.3.2(a) illustrates the simplicity of a straight binary asynchronous counter. Only the data and reset inputs and an output from each FF are required. Assuming that the data is a square wave, as illustrated in the waveform diagram of D.3.2(b), and the reset line is at logic **0**, the output of the four stages will "ripple" through, as illustrated. Note that the frequency of Q1 is half of the frequency of the data input, Q2 is half of the frequency of Q1, Q3 half the frequency of Q2, etc. Because of the inverter connected to the data input, the first FF appears to trigger on the negative going edge. When the reset signal changes from logic **0** to logic **1**, all of the outputs are reset to **0**.

KEY PARAMETERS

The electrical characteristics are essentially those of the particular digital IC family.

a) *Quiescent current.* Total current drawn when counter is not operating. 5.0 nA is typical at 5 V for CMOS ICs.

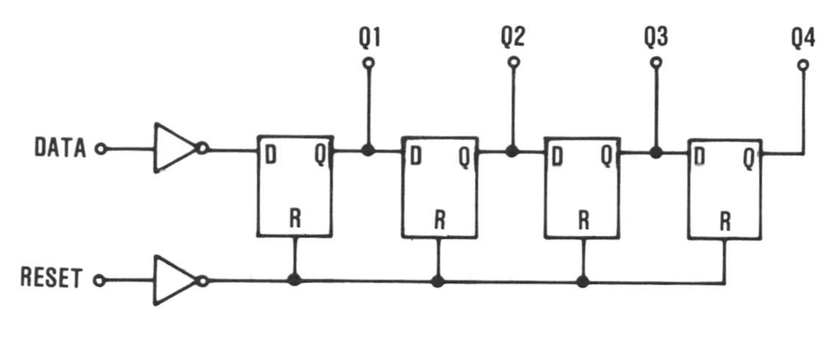

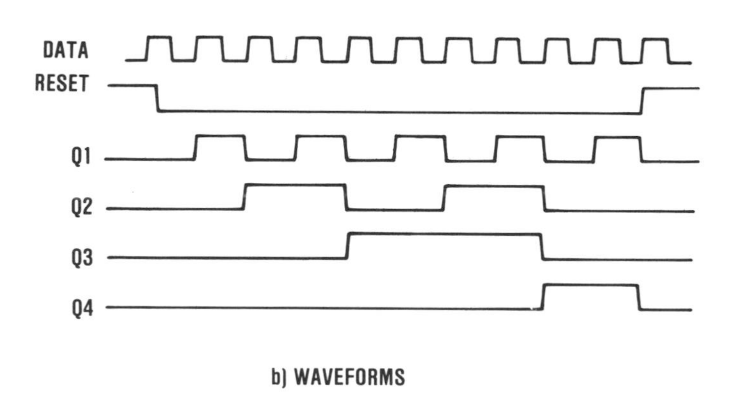

FIGURE D.3.2 (a) Logic Diagram of Asynchronous Counter (b) Wave Forms

b) *Maximum input frequency.* 8 MHz is typical for CMOS, and up to 50 MHz is possible with TTL ICs.
c) *Input transition time.* Very short rise and fall times of the input signal are limited by the maximum input signal frequency. Very long rise and fall times can be accommodated in those devices which have Schmitt-Trigger or similar hysteresis type input circuits. Many CMOS devices have no stated limits for this parameter.
d) *Reset pulse width.* Minimum values for CMOS devices range from 250 to 500 ns. For TTL devices this will range from 50 to 100 ns.
e) *Reset removal time.* The time required after the reset has gone to logic **0** until the counter is operational again. For CMOS ICs, 150 to 250 ns is typical, while 20 to 100 ns is usually specified for TTL ICs.

APPLICATIONS

Asynchronous counters are used for event counting, pulse counting in radio isotope detection applications, timing control, and similar applications. They are particularly useful in applications where the events to be counted or measured occur at random or pseudo-random intervals.

COMMENTS

Most asynchronous counters are supplied with additional control and output logic circuits, making them more complex than indicated in the logic of Figure D.3.2. Some manufacturers list them as ripple counters, while others classify them as asynchronous.

D.3.3 Counter, Programmable, Divide-by-N

DESCRIPTION

Each flip-flop (FF) in this counter can be programmed for either **0** or **1**, allowing the user to determine the desired division of the input (clock) frequency. A set of program inputs and an enabling signal perform this task. In addition, the clock can be inhibited, there is a master reset input and a special cascade input and output terminal. As illustrated in Figure D.3.3, the cascade input terminal is marked CF (carry forward) while the output terminal is labeled 0. The PE terminal enables the programmable divide-by-N inputs P1 through P4 and the output appears on terminals Q1 through Q4. When larger numbers are to be divided down, several of these ICs can be cascaded without any additional logic. The clock inhibit input allows disabling of the pulse counting function.

KEY PARAMETERS

The electrical characteristics are essentially the same as those of the particular digital IC family.

a) *Maximum clock frequency.* Ranges from 3 to 6 MHz for CMOS devices and up to 50 MHz for TTL ICs.
b) *Minimum clock pulse width.* Low-power CMOS devices generally use 80 to 250 ns, while TTL ICs can operate with clock pulse widths of 25 ns.
c) *Minimum program enable pulse width.* For low-power CMOS devices 80 to 250 ns is typical.
d) *Minimum master reset pulse width.* Low-power CMOS devices require from 200 to 350 ns.

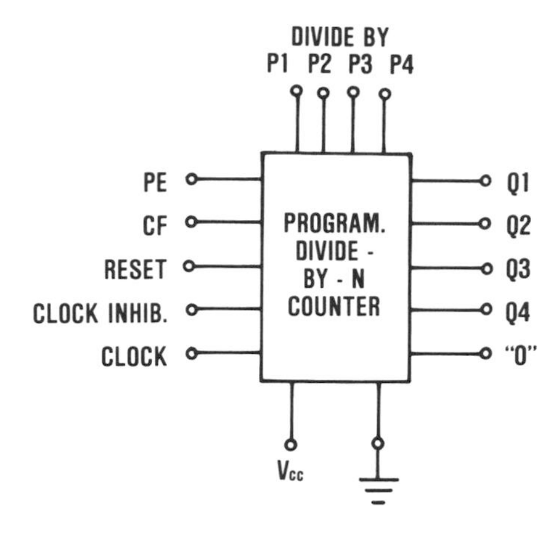

FIGURE D.3.3

APPLICATIONS

Programmable divide-by-N counters are particularly useful in frequency synthesizers, phase-locked loops, and other frequency division applications.

COMMENTS

A variety of programmable or presettable counters is available. By presetting the stages of a counter, one can achieve timing because the counter can be made to reset or "fill up" when the preset count has been reached. Another application is to divide the input frequency by a preset or programmed number. Either function is available in a variety of different configurations.

D.3.4 Counter, Up/Down

DESCRIPTION

The key feature of this IC is the ability of an otherwise ordinary counter to count either down or up. This makes it possible to enter a number of pulses during one period and then to subtract a second number from the first by counting down that number of pulses. While many up/down counters are programmable or presettable, the function block illustrated in Figure D.3.4 omits this feature and shows only the controls necessary for the up and down counting operation and cascading. When the counter is full, the carry-out terminal is used to feed the overflow into the next cascaded IC. The CF terminal serves as input of either the original signal or the overflow from a previous cascaded section. A master reset, a clock inhibit and a clock input are provided, but the key feature is the up/down mode control. When this terminal is at **1**, the counter operates in a normal counting-up mode. When the up/down mode control is **0**, the counter will count down.

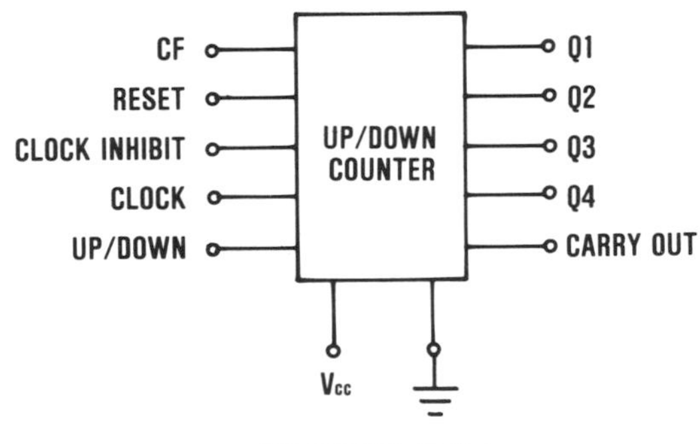

FIGURE D.3.4

KEY PARAMETERS

The electrical characteristics will be the same as for the particular digital IC family.

a) *Up/down setup time.* The time required to change counter operation from up to down or back to up. Typical values for low-power CMOS range from 50 to 170 ns. For TTL devices, 20 to 30 ns is typical.

APPLICATIONS

Whenever up/down counting capability is required. Widely used in difference counting and frequency synthesizer applications. This IC is also found in A/D and D/A converter systems and in computers to generate magnitudes and polarity signs.

COMMENTS

Up/down counters are available with programmable inputs, with divide-by-N features, and with other capabilities.

D.3.5 Counter, Presettable Up/Down

DESCRIPTION

This IC presents a combination of the up/down counting feature described in D.3.4 and the programmable or presettable feature described in D.3.3. Like these two types of counters, this IC can also be cascaded by using the carry-in and carry-out terminal. As illustrated in Figure D.3.5, there are four preset input terminals, P1 through P4, which are enabled by a **1** on terminal PE (preset enable). The up/down mode control (U/D) operates in the same manner as described for the counter of D.3.4. The outputs of the four flip-flop (FF) stages are shown in Q1 through Q4. A master reset terminal is available but there is no clock inhibit.

KEY PARAMETERS

The electrical characteristics are essentially the same as for the particular digital IC family

a) *Clock pulse frequency.* For low-power CMOS this ranges from 2 to 5 MHz, and up to 35 MHz for TTL devices.
b) *Minimum clock pulse width.* 60 to 150 ns for CMOS devices.
c) *Preset enable or reset removal time.* The time required after either the reset or preset inputs before the next clock will trigger the counter. 60 to 150 ns is minimum for CMOS.

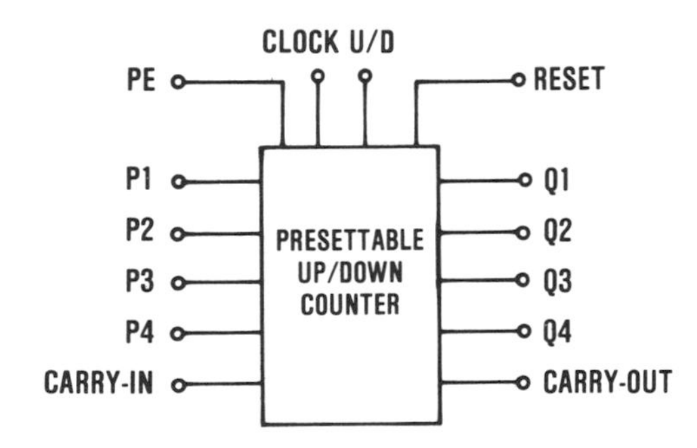

FIGURE D.3.5

DIGITAL

d) *Up/down setup time.* The time required to change operation from up to down, or back to up counting. CMOS devices have a minimum of 100 to 350 ns.
e) *Preset enable or reset pulse width.* The minimum is 75 to 200 ns for CMOS devices.

APPLICATIONS

Up/down difference counting, multistage synchronous or ripple counting in frequency synthesizers as well as synchronous frequency dividers for phase-locked loops.

D.3.6 Counter, 5-Decade, Multiplexed Output

DESCRIPTION

This IC is an example of a specific system function implementation. As illustrated in Figure D.3.6, the IC contains five decade counters in cascade. In effect, then, the first counter, 10^0, will count from 0 through 9, the second counter 10^1, will count from 10 to 90, and so on. The total maximum count of this unit is 99,999. The outputs of the five decade counters go to the multiplex (MUX) and control section which also receives the scan clock and scan reset input. A binary-coded decimal (BCD) output is available at Q1 through Q4, controlled by the tri-state BCD control level. The MUX-OUT section indicates which of the five BCD digits appears at any instant at Q1 through Q4. In other words, if a series of five separate decimal displays were used here, such as the visual display of a multimeter, the digit select would enable the 1s, 10s 100s, etc., in sequence, at Q1 through Q4. A single BCD 7-segment

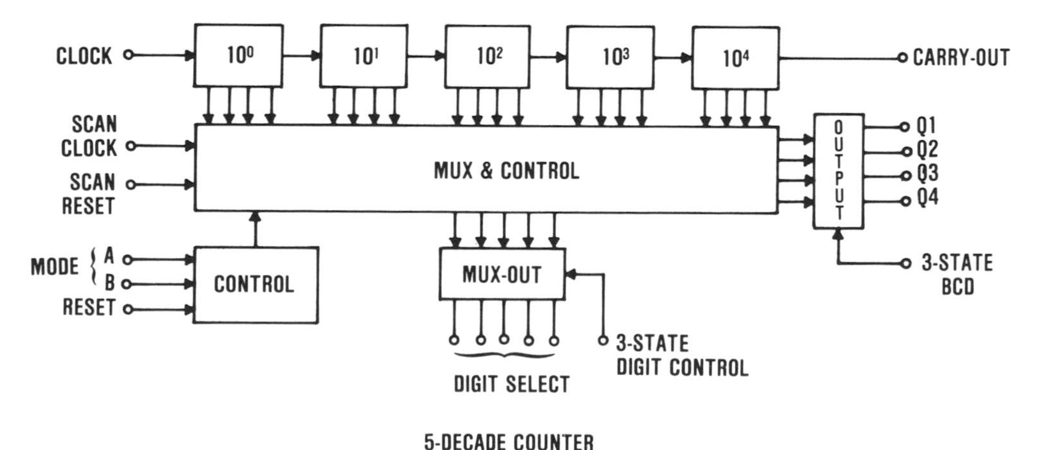

FIGURE D.3.6. 5-Decade Counter

converter and display driver is then connected to all five numerical displays in parallel. The MUX-OUT, controlled by the tri-state digit control signal, then selects, in sequence, each of the five numerical displays to be illuminated.

KEY PARAMETERS

The electrical characteristics are the same as for the particular digital IC family.

a) *Maximum clock pulse frequency*. For CMOS devices, 0.5 to 1.2 MHz.
b) *Minimum clock pulse width*. 375 to 1,000 ns is minimum for CMOS devices.
c) *Master reset pulse width*. 450 to 2,000 ns is minimum for CMOS ICs.

APPLICATIONS

This IC was specifically intended for applications in real time or event counters where frequent updating is required and where multiplexed displays are used.

COMMENTS

The five-decade counter with MUX output is a special purpose IC and illustrates how counters, in general, can be used as small systems.

D.3.7 Counter, Combined with Decoder and Display Driver

DESCRIPTION

A basic decade up/down counter is combined with a latching circuit, a decoder and driver, to connect directly to seven segment displays. As illustrated in Figure D.3.7, the counter receives inputs from the clock signal conditioner, a set of Schmitt-Trigger circuits and drivers which accommodate either the up or the down clock. In this instance it is possible to use two separate clocks instead of a single clock controlled by an up/down mode control signal. The control logic portion also accepts the reset signal and the "toggle" enable (TE). When the counter has reached the end of the count, the state of the counter can be fed directly to the decoder or, depending on the condition of the "latch enable control line," this information can be stored in the decade latching circuit. The decoder changes the BCD information into segment information and, through the driver stages, drives a 7-segment numerical display.

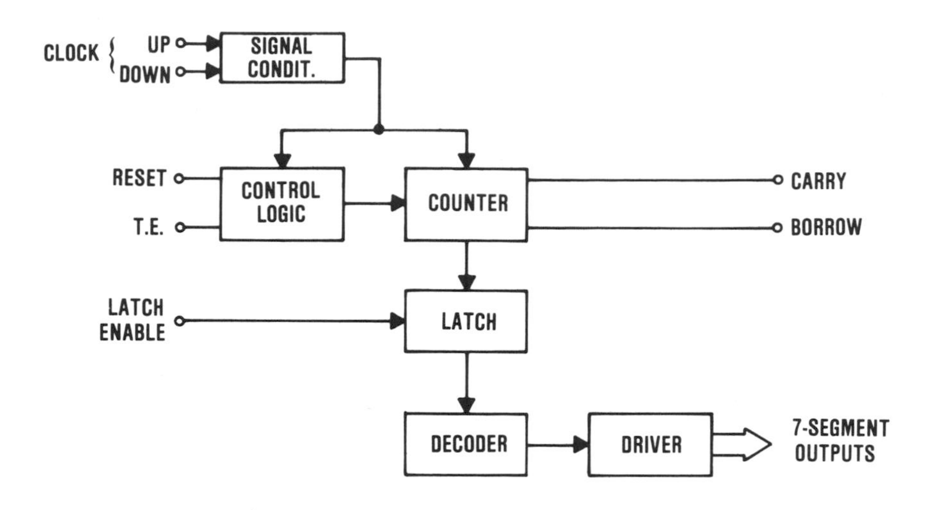

FIGURE D.3.7. Combination Counter and Decoder

KEY PARAMETERS

The electrical characteristics are essentially the same as those of the particular digital IC family.

APPLICATIONS

This IC is particularly useful in rate comparators and other counting applications where a numerical display is required. It can also be used in various radiation counters where both up and down counting of random pulses is required.

D.4.0 Decoders/Encoders

DESCRIPTION

Decoders are essentially an arrangement of logic elements which are combined to change from one digital code to another. The term "decoder" is most frequently used but, depending on the point of view, the term "encoder" is equally correct.

Figure D.4.0 shows the logic diagram of a 3-bit binary-to-1-of-8 decoder. The inputs A, B and C can represent any logic function, and the outputs 1 through 8 will then provide the addition or the logic OR function of these three inputs. In order to produce the complement of the input functions, the OR circuits have inverters at the inputs indicated by the small circle. As example, output 4 is $A + \overline{B} + \overline{C}$. If the input A, B and C were each logic **1**s, output 4 would be **1**. If A=0, B=1, C=1, then output 4 is **0**. A close look at the logic diagram of Figure D.4.0 shows that it represents the conversion from a 3-bit binary to an octal number. Output 1 represents the octal number 7 and output 8 represents the octal number 0.

The most widely used types of decoders are presented in the following pages. In addition to these standard ICs, special purpose decoders are found in some digital logic systems. Decoders are available in all of the commonly used digital IC families, such as CMOS, TTL, ECL, etc.

KEY PARAMETERS

The electrical characteristics are essentially the same as those of the particular digital IC family.

a) *Maximum power dissipation.* The maximum power dissipated by the IC when in operation. This will vary according to the digital IC family and also according to the number of stages or logic elements used.
b) *Quiescent current.* The current used by the entire IC when no operations take place. This will vary according to the digital IC family and according to the number of logic elements used.

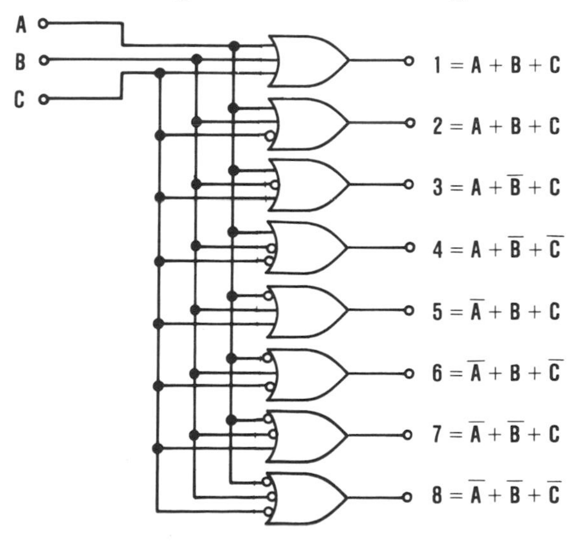

FIGURE D.4.0. Logic Diagram of 3-Bit Binary to 1-of-8 Decoder

c) *Noise immunity*. The relation of the noise amplitude as a percentage of the power supply voltage. This parameter usually depends on the digital IC family.
d) *Propagation delay time*. The time required from input to output. While this depends, to some extent, on the digital IC family, the complexity of the logic gating and the number of stages between input and output are also important factors.
e) *Control signals*. Some decoders have means of enabling all or part of the inputs or outputs. Special input and output terminals to permit cascading decoders are also available in some decoders.

APPLICATIONS

Decoders are used in digital systems whenever it is necessary to change from one code to another. Specific applications are presented for the different decoders described in the following pages.

D.4.1 Decoder, BCD-to-Decimal

DESCRIPTION

This type of decoder is probably the most widely used in all digital systems because it changes the inherent binary codes used within the system to the decimal code used by the human operators. Figure D.4.1 illustrates the function block of a basic BCD-to-decimal decoder. Four input lines, representing the four BCD values, result in ten output lines, representing the decimal numbers 0 through 9. This type of decoder is often used in combination with decade counters and with decimal displays. By using only the three least significant inputs, a 3-bit binary-to-octal decoder is obtained, with outputs only on terminal 0 through 7.

KEY PARAMETERS

The electrical characteristics are essentially the same as those of the particular digital IC family.

a) *Quiescent current*. The total current flowing in the IC when no operations are performed. 5.0 nA at 5 V power supply is typical for low-power CMOS.
b) *Noise immunity*. The relation of noise amplitude to V_{cc}. 45% of the power-supply voltage is typical for low-power CMOS.
c) *Propagation delay time*. The time required from the input to the output of the IC. Typical values are 300 ns at 5 V power supply for low-power CMOS devices.

APPLICATIONS

BCD-to-decimal decoders are used for code conversion, address decoding, memory selection control, read-out decoding, and demultiplexing in digital systems such as mini- and microcomputers, digital voltmeters, etc.

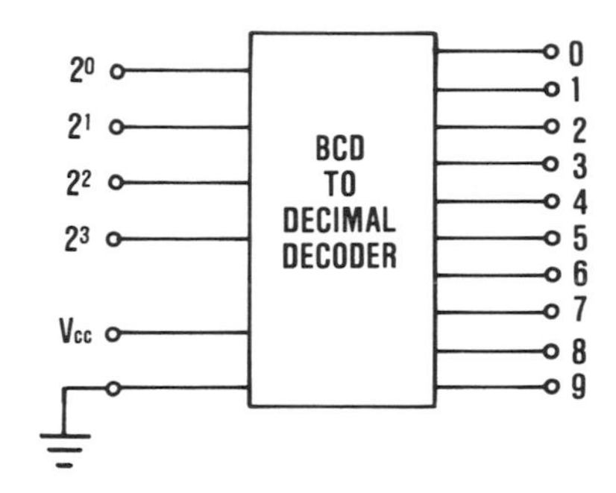

FIGURE D.4.1

D.4.2 Decoder, BCD-to-7-Segment

DESCRIPTION

This type of decoder accepts the BCD information and converts it into seven outputs for connection to a 7-segment display. These numerical displays can be implemented by light-emitting diodes (LED), liquid crystal displays (LCD), gas discharge displays, fluorescent displays, or incandescent displays. The 7-segment outputs may be connected directly only to a low brightness fluorescent display. In all other instances some kind of driving circuit is required.

As illustrated in Figure D.4.2, there are three additional inputs, not part of the code conversion circuit. When the lamp test input terminal is connected to ground, all seven segments will be illuminated. The purpose of the blanking in-and output terminals is to provide so-called "ripple-blanking" operation. In this mode a series of 7-segment numerical indicators are illuminated in sequence in order to reduce the total power requirements. This means that a blanking signal is applied to the "blank" terminal and, after the BCD-to-7-segment decoder has been disabled for a brief period, the output "blank-out" is connected to the next decoder "blank-in."

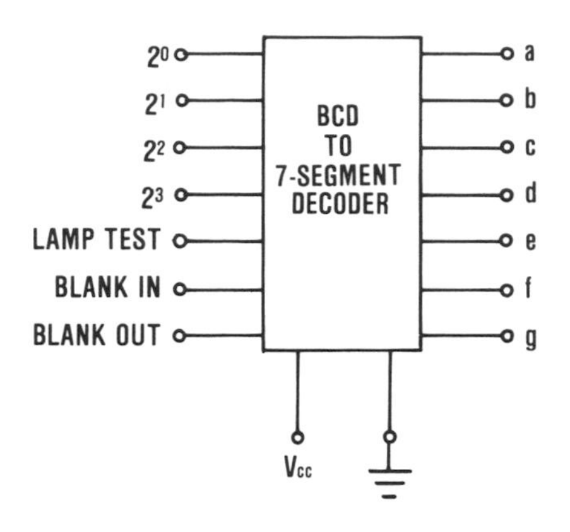

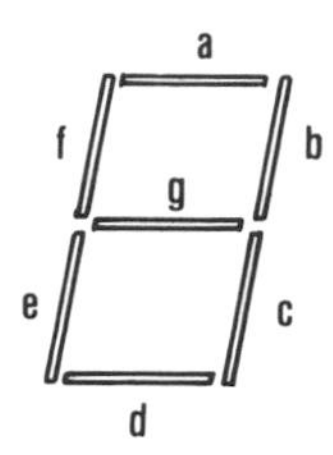

FIGURE D.4.2

KEY PARAMETERS

The electrical characteristics are essentially the same as those of the particular digital IC family.

a) *Maximum power dissipation.* The maximum power dissipated by the entire IC. 500 mW is a typical value for CMOS.
b) *Noise immunity.* The relation between noise and V_{cc}. 45% of the power-supply voltage is typical for CMOS.
c) *Propagation delay time.* The time from input to output. 450 ns is a typical value at 5 V power supply for CMOS.

APPLICATIONS

These ICs are used to display BCD information on 7-segment numerical displays. They are found in multimeters, counters, other test equipment, commercial electronic scales, and wherever a numerical read-out is required.

COMMENTS

The BCD-to-7-segment decoder described here is often found in a single IC, combined with a counter, latching circuit and output drivers suitable for the particular type of display, as described in D.3.7.

D.4.3 Decoder, Excess-3 (Gray)-to-Decimal

DESCRIPTION

As illustrated in Figure D.4.3(a), this type of decoder has a 4-bit input and a straight decimal output. The 4-bit input, however, is not encoded according to the BCD system but rather according to the excess-3 (gray)

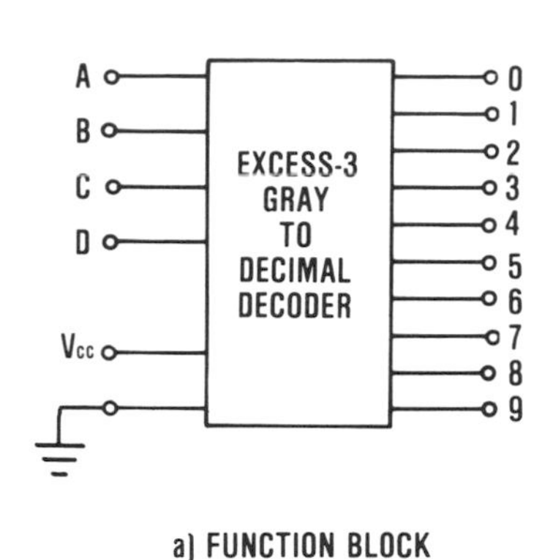

a) FUNCTION BLOCK

A	B	C	D	0	1	2	3	4	5	6	7	8	9
0	1	0	0	0	1	1	1	1	1	1	1	1	1
0	1	1	0	1	0	1	1	1	1	1	1	1	1
1	1	1	0	1	1	0	1	1	1	1	1	1	1
1	0	1	0	1	1	1	0	1	1	1	1	1	1
0	0	1	0	1	1	1	1	0	1	1	1	1	1
0	0	1	1	1	1	1	1	1	0	1	1	1	1
1	0	1	1	1	1	1	1	1	1	0	1	1	1
0	1	1	1	1	1	1	1	1	1	1	0	1	1
0	1	1	1	1	1	1	1	1	1	1	1	0	1
0	1	0	1	1	1	1	1	1	1	1	1	1	0

b) TRUTH TABLE

FIGURE D.4.3 (a) Function Block
(b) Truth Table

code. The difference becomes apparent in the truth table of Figure D.4.3(b). This type of code is usually used in special military or industrial applications and is not frequently found in digital computers. In general, this type of decoder operates in the same manner as the BCD-to-decimal version, but the internal connection of the gates must be different.

One important feature of this decoder is that all output terminals will be "1" whenever an invalid code appears at the input. As apparent from the truth table, it is possible to have as many as six invalid codes since up to 16 possible combinations exist for a 4-bit signal.

KEY PARAMETERS

The electrical characteristics are essentially the same as those of the particular digital IC family.

a) *Maximum power dissipation*. 20 mW is typical for low-power TTL.
b) *Quiescent current*. The current drawn when no operations are being performed. 400 mA is typical for low-power TTL.
c) *Propagation time delay*. The time required from input to output. In some combinations a two-gate delay occurs and this is typically 18 ns, while a three-gate delay is typically 22 ns for low-power TTL devices.

APPLICATIONS

This type of code conversion often takes place in long-distance communications equipment and in specialized industrial and military equipment where rotating components, such as synchro and servo devices, are used.

D.4.4 Decoder, 8-Bit Priority

DESCRIPTION

The main purpose of a priority encoder is to generate a binary address for the active input with the highest priority. As illustrated in the function block of Figure D.4.4(a), the 8-bit priority encoder has eight data inputs, D0 through D7 and one enable input. The output consists of a three-bit binary address, Q0 through Q2, a group select and enable output. To understand the details of the operation of this function block, the reader is referred to the truth table of Figure D.4.4(b). Note that when the enable input is **0**, it does not matter what the data inputs are because all output lines will be at **0**. Only when the enable input is a logic **1** does the priority encoder operate. If all of the data inputs are **0**, the group select and the 3-bit binary address will also be **0**, but the enable output will be **1**. The remainder of the truth table illustrates, in effect, a conversion from an octal to a 3-bit binary code. The only difference between an ordinary octal converter and the 8-bit priority encoder is that an inherent priority has been assigned to the data inputs. When D7 is **1**, it does not matter what the status of the other data input is; the output code will be the equivalent of binary 7. D7 has the highest priority. If D7 is **0**, the next highest priority will go to D6, and so on.

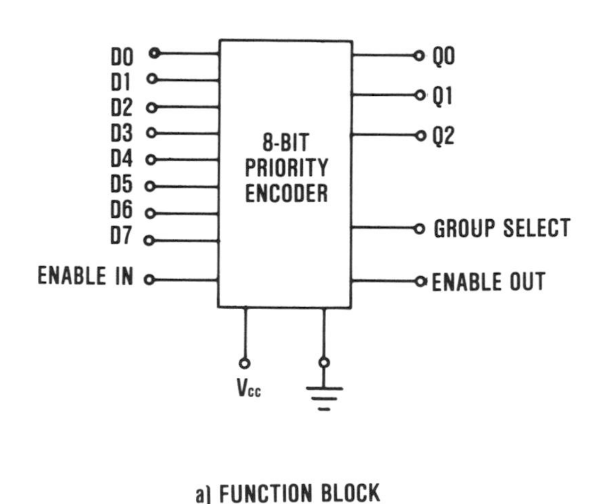

a) FUNCTION BLOCK

INPUT									OUTPUT				
E_{IN}	D7	D6	D5	D4	D3	D2	D1	D0	GS	Qz	Q1	Q0	E_0
0	X	X	X	X	X	X	X	X	0	0	0	0	0
1	0	0	0	0	0	0	0	0	0	0	0	0	1
1	1	X	X	X	X	X	X	X	1	1	1	1	0
1	0	1	X	X	X	X	X	X	1	1	1	0	0
1	0	0	1	X	X	X	X	X	1	1	0	1	0
1	0	0	0	1	X	X	X	X	1	1	0	0	0
1	0	0	0	0	1	X	X	X	1	0	1	1	0
1	0	0	0	0	0	1	X	X	1	0	1	0	0
1	0	0	0	0	0	0	1	X	1	0	0	1	0
1	0	0	0	0	0	0	0	-1	1	0	0	0	0

b) TRUTH TABLE

FIGURE D.4.4 (a) Function Block
(b) Truth Table

KEY PARAMETERS

The electrical characteristics are essentially the same as those of the particular digital IC family.

a) *Quiescent current.* The total current used by the IC when no operations are performed. 5.0 nA at 5 V power suppy is typical for low-power CMOS.
b) *Noise immunity.* The relationship between noise signals and the power-supply voltage. 45% of power-supply voltage is typical for low-power CMOS.
c) *Propagation delay time.* The time it takes from the input to the output signal. Typical delay time for the effect of the enable signal and the effect of the data signals to appear on outputs is approximately 250 to 300 ns at 5 V power supply for low-power CMOS.

APPLICATIONS

Priority encoders are used in digital-to-analog and in analog-to-digital conversion systems. Another application is in communications systems where fixed priorities exist for certain types of messages and the data inputs represent signaling lines.

D.4.5 Decoder, Binary-to-2-of-8

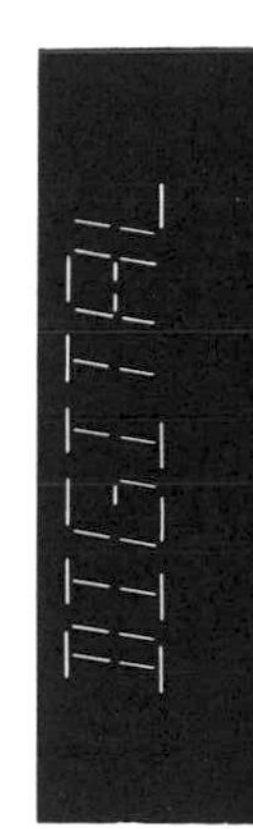

DESCRIPTION

While somewhat similar to the binary-to-decimal decoder, this IC provides more flexibility and operates in a different way. As illustrated in Figure D.4.5(a), the IC contains two identical decoders which receive the same 2-bit address. Each decoder is controlled by two enabling signals through an AND circuit. The detailed operation of this IC is understood by examining the truth table of Figure D.4.5(b). Decoder A and its outputs are enabled only when one of the two input signals is **1** and the other **0**. The output of decoder B is enabled only when both input signals to the AND circuits are **0**. The address A0 and A1 is simply a two-bit binary number which is decoded into four separate individual lines.

KEY PARAMETERS

The electrical characteristics are essentially the same as those for the particular digital IC family.

a) *Maximum power dissipation.* 50 mW is typical for low-power TTL.

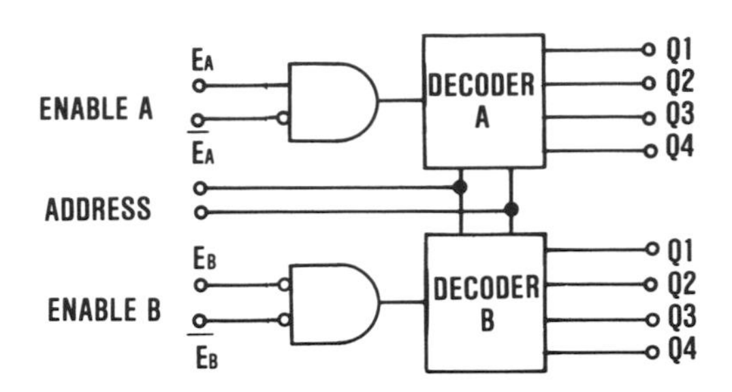

a) LOGIC DIAGRAM, BINARY-TO-2-OF-8 DECODER

ADDRESS		EA		OUTPUT A				EB		OUTPUT B			
A_0	A_1	E	E	Q_1	Q_2	Q_3	Q_4	E	E	Q_1	Q_2	Q_3	Q_4
X	X	0	X	1	1	1	1	1	X	1	1	1	1
X	X	X	1	1	1	1	1	X	1	1	1	1	1
0	0	1	0	0	1	1	1	0	0	0	1	1	1
1	0	1	0	1	0	1	1	0	0	1	0	1	1
0	1	1	0	1	1	0	1	0	0	1	1	0	1
1	1	1	0	1	1	1	0	0	0	1	1	1	0

b) TRUTH TABLE

FIGURE D.4.5 (a) Logic Diagram—Binary to 2-of-8 Decoder
(b) Truth Table

b) *Propagation delay time.* From the appearance of the address to the output. A typical vaue is 20 ns, while the typical value for the delay from the appearance of the enabling signals to an output is between 15 and 20 ns for low-power TTL.

APPLICATIONS

This IC is particularly useful in large logic control systems where binary signals are processed asynchronously. It often replaces combinations of individual AND and OR gates.

D.4.6 Decoder, 16-Key Encoder

DESCRIPTION

This IC provides all the necessary logic to fully encode a keyboard consisting of 4 × 4 single-pole, single-throw (SPST) switches. An internal debounce circuit is provided which requires only one external capacitor. Provision to eliminate the key "rollover" effect is also included. As illustrated in Figure D.4.6, connections are made to the four rows and to the four columns (X and Y) of the keyboard. An internal oscillator, tuned by one capacitor, generates the signals required to scan the X and Y inputs to determine when one of the keys is depressed. The output to the data bus consists of four lines, representing the binary equivalent to the hexadecimal (16) key input. The "data available" line indicates to the remote circuitry that data are available, and the inverting amplifier is provided to connect that signal back to the "output enable" terminal. When a "handshake" interface system is used, an external flip-flop is required to indicate the data

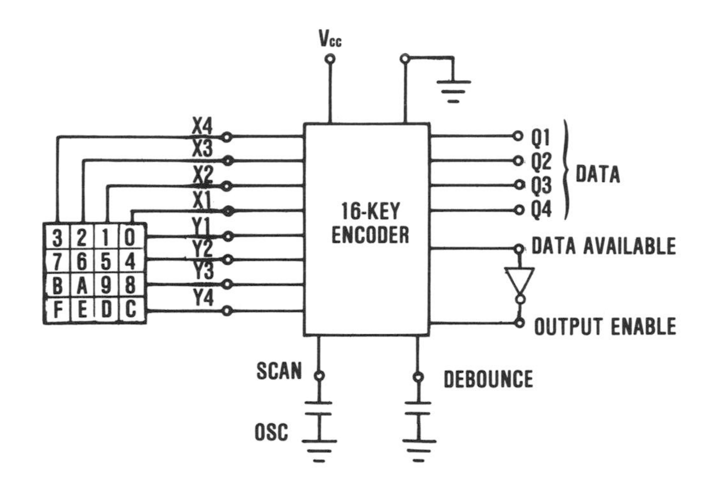

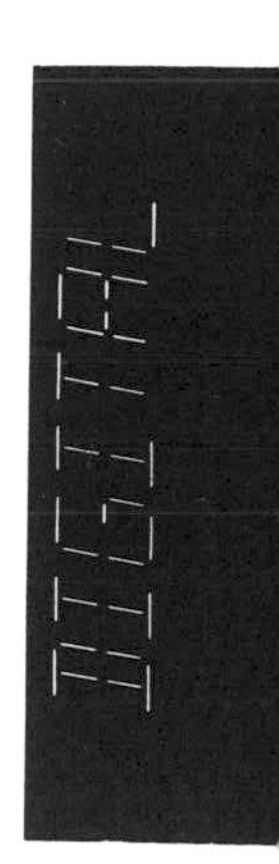

FIGURE D.4.6

available information and then reset when the output enable response is received from the remote terminal.

The 16-key encoder IC can also be used with a system clock, in which case the data available and the output enable can be synchronized. The oscillator capacitor used for scanning oscillator can be eliminated and the system clock can provide the scanning signal as well.

KEY PARAMETERS

The electrical characteristics are essentially the same as those of the particular digital IC family.

a) *Maximum power dissipation.* 500 mW is typical for CMOS devices.
b) *Propagation delay time.* The time required from the detection of an input signal at the encoder until the binary output signal and the data available signal are present. 60 to 80 ns is typical for CMOS devices.

APPLICATIONS

This type of encoder is used whenever numerical input keyboards, particularly 16-key arrays, are used.

COMMENTS

In addition to the 16-key encoder, other ICs are available to accept different numbers of key inputs. Special ICs for converting the ASCII code from a standard keyboard into the equivalent 7-bit binary number are also available. All of these encoders operate in essentially the same manner as described here.

D.4.7 Decoder, BCD-to-Binary and Binary-to-BCD

DESCRIPTION

This IC is really a 128-bit read-only memory (ROM), but it performs the function of converting binary into binary coded decimal (BCD) and BCD into binary codes. A single IC of this type can deal only with 4-bit binary or BCD words, but larger words can be converted by interconnecting a series of these ICs. As illustrated in Figure D.4.7, the input consists of four data lines and two enable lines, and the output provides four lines for the BCD and another four lines for the binary code. Both enable inputs must be in the **1** condition for the decoder to operate. If either of the enable gates is in the **0** position, all eight outputs will be **1**.

When this IC is used as part of a larger digital system, the enable inputs are particularly important.

KEY PARAMETERS

The electrical characteristics are essentially the same as those of the particular digital IC family.

a) *Maximum power dissipation.* The maximum power the entire IC can dissipate at 25°C. 625 mW is typical for TTL ICs.
b) *Propagation delay time.* The time required from input to output. 45 ns is typical for all delay times in TTL ICs of this type.
c) *Output loading.* The number of equivalent output loads each terminal can drive. 2000 ohm pull-up resistors are provided in each output circuit of this IC, permitting up to five TTL loads to be connected.
d) *Quiescent current.* The current drawn by this IC when the inputs are not enabled. 55 mA is typical for TTL ICs.

APPLICATIONS

This IC is used whenever conversion from binary to BCD and back is required. Cascading these ICs is particularly useful when a binary number larger than four bits is used to drive a series of BCD displays in parallel. A number of parallel 4-bit BCD inputs can also be converted into a multibit straight binary output. This IC is generally found in arithmetic logic unit (ALU) of digital computers.

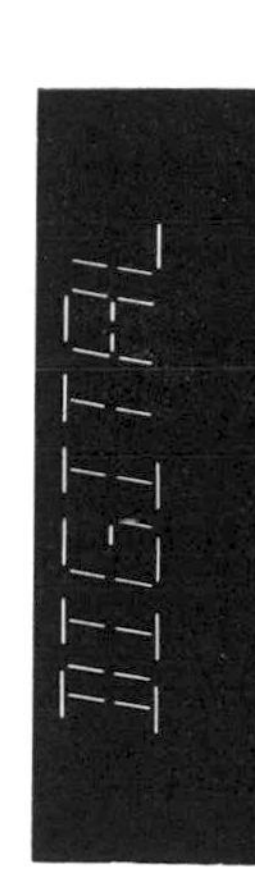

COMMENTS

This IC is available only from a limited number of manufacturers. The binary-to-BCD-to-binary function is often performed by software.

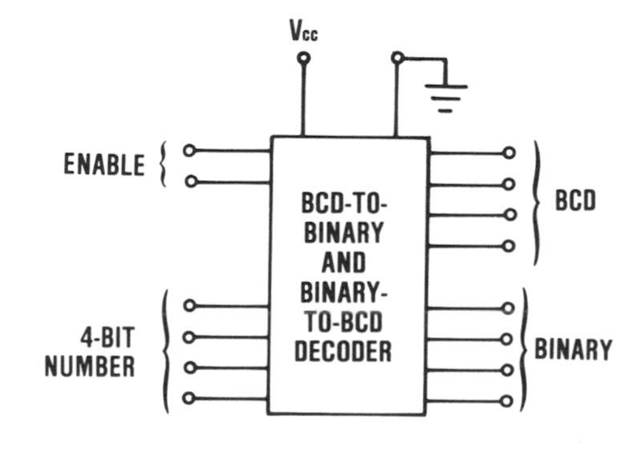

FIGURE D.4.7

D.4.8 Decoder/Encoder, Manchester Code

DESCRIPTION

This IC provides both encoding and decoding functions of the Manchester 2 Code used in military and industrial communications systems. It is divided into two sections, the encoder and the decoder, which operate independently of each other, except for the master reset. While specifically designed to meet the requirements of MIL-STD-1553, this IC also can be used for other time-division multiplexed serial data protocols.

As illustrated in the function block of Figure D.4.8, this 24-pin IC can operate either on bipolar or unipolar data, allowing it to work with a variety of communications systems. The encoder requires two clock signals—the "send" clock, which is twice the desired data rate, and the "encode" clock, which is six times the data rate. Both encoder and decoder functions contain counters and a variety of other digital elements, but their operation is too complex for detailed description in this book. The reader is referred to manufacturer's data.

KEY PARAMETERS

This IC is currently only available as a CMOS device and its electrical characteristics are essentially those of the CMOS digital IC family.

a) *Maximum clock frequency*. The maximum clock frequency for either encoder or decoder is 15 MHz.

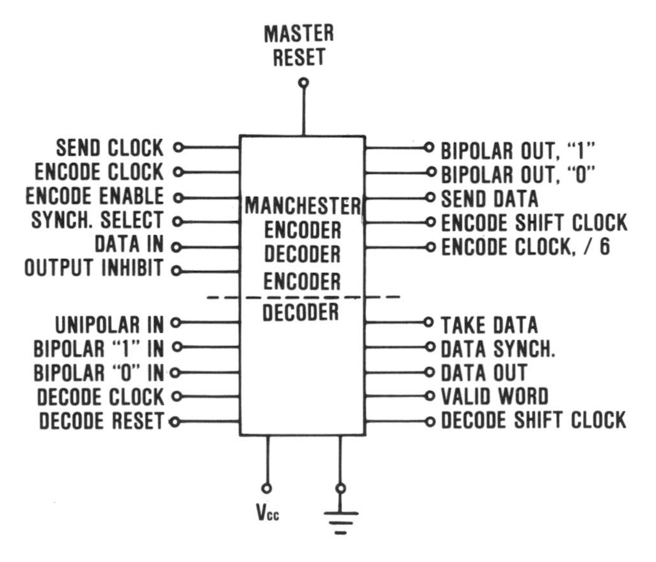

FIGURE D.4.8

b) *Maximum data rate.* 1.25 megabaud. To assure stability over the entire temperature range, the manufacturer recommends a maximum range of 1 megabaud.
c) *Typical operating power.* 50 mW is specified at 5 V power-supply over the standard temperature range.

APPLICATIONS

This IC is specifically designed to provide the encoding and decoding function for a time-division multiplexed data bus using the Manchester 2 code. Aircraft interior communications, both military and commercial, are a typical use for this kind of system.

COMMENTS

Because of the unique and complex nature of this IC, the reader is urged to obtain manufacturer's application data before using this IC for new designs.

D.4.9 Decoder, Binary-to-Serial Output Pulse

DESCRIPTION

The basic function of this IC is to convert binary data into a series of output pulses. Its main application is in digitally controlled dial telephone systems. As illustrated in the function block of Figure D.4.9, four binary input lines result in bursts of pulses equivalent to the digits of the telephone number entered in the binary inputs. To provide all of the needs of a telephone dialing system, a series of other connections is required. An external capacitor is connected to the oscillator and clock terminals and is part of a tuned network which generates the required 10 Hz output pulses. The control inputs, such as "hold," "call request," "redial," "interdigit time," and "make-break ratio" are obtained from other portions of the telephone or the digital control system. The "dial rotating" output provides a signal to the digital control system which indicates that dial pulses are being sent out.

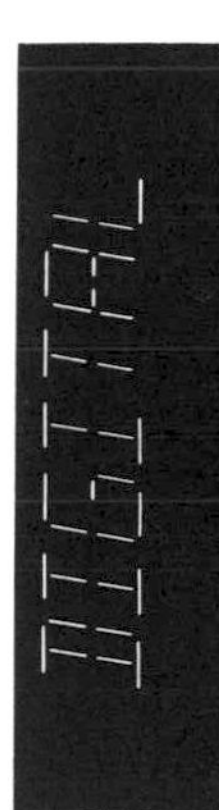

This IC contains a memory which stores the binary input data while the 10 Hz dial pulses are generated and sent out, with the correct time interval between digits. Automatic but selectable interdigit times, 300 or 800 ms, and a dialing rate selectable for either 10 or 20 pulses per second are provided by internal timing circuits. Up to 16 digits can be dialed and the "hold" control can be used for special interdigit delays such as a wait for intermediate dial tones.

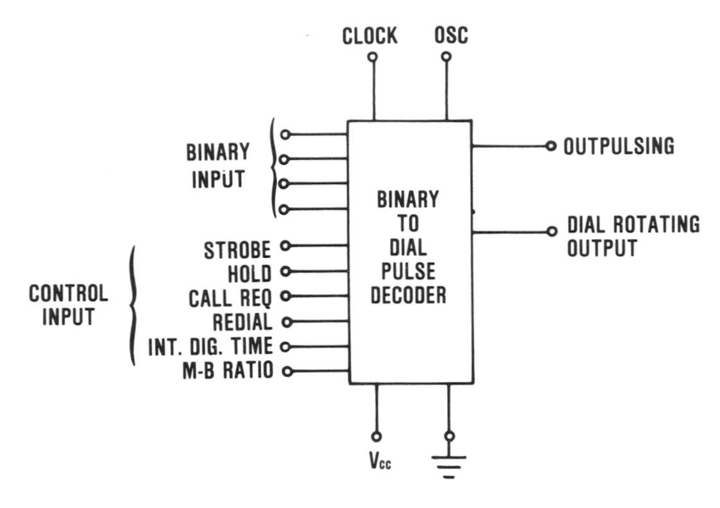

FIGURE D.4.9

KEY PARAMETERS

This device is available in CMOS only, and its electrical characteristics are those of the CMOS digital IC family. The control and output circuits of this device must meet the requirements of the standard telephone dialing and signaling system. The outputs are compatible with discrete transistor driver interfaces. Input circuits are generally CMOS compatible.

APPLICATIONS

This device is specifically designed to interface digital logic circuitry with the standard telephone line dialing system.

COMMENTS

Because of the unique applications of this IC, the reader is advised to consult the manufacturer's application data before designing with this IC.

D.5.0 Display Driver, BCD-to-7-Segment

DESCRIPTION

The universal use of 7-segment numerical displays has resulted in a wide range of ICs capable of driving them. One of the most widely used is illustrated in the block diagram of D.5.0. The BCD input is applied to a set of four flip-flops which are used as a "latch." The "strobe" signal determines when the information presented at the input is stored in the latch and when it is erased. The BCD-to-7-segment decoder is effectively the same as described in Section D.4.2, and there are seven separate driver stages.

As illustrated in D.5.0, a lamp test and a blanking signal are applied to the decoder portion. The purpose of the lamp test signal is to illuminate all seven sections of the numerical display, and the blanking signal can be used to turn off the entire display for zero blanking multidigit displays, or else to control the on-off time when a series of displays operates in a power conserving, multiplexed mode.

The voltage and current capability of the driver stage determines which type of 7-segment numerical displays can be driven. A typical high-voltage CMOS IC can operate up to a maximum of 20 V and can source up to 25 mA at each output. This makes it possible to drive LED displays, low-voltage fluorescent displays, and even incandescent displays. To drive liquid crystal displays (LCDs), some other characteristics are necessary which are provided in the IC described in Section D.5.1.

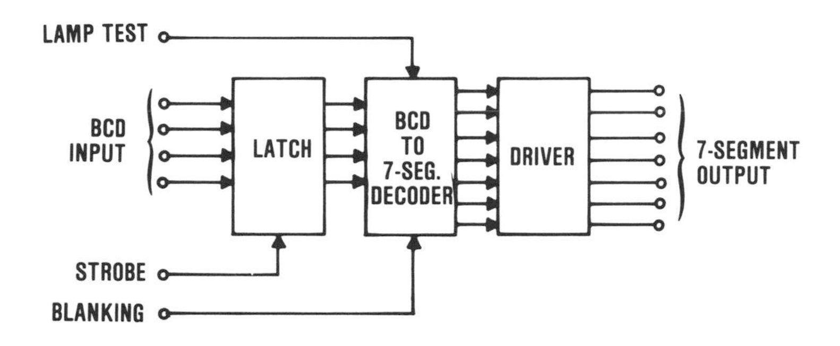

FIGURE D.5.0. Block Diagram—BCD-to-7-Segment LED Driver

KEY PARAMETERS

The electrical characteristics are essentially those of the particular IC family.

a) *Maximum power dissipation.* 500 mW for a high-voltage CMOS IC. 250 mW for TTL ICs.
b) *Quiescent current.* The maximum current drawn when the IC is not operating. 100 mA at 20 V supply and +25°C is typical for high-voltage CMOS.

c) *Propagation delay time.* The time required between data input and output. A maximum of 1,000 ns is typical for high-voltage CMOS devices at 5 V power supply.
d) *Minimum setup time.* The minimum time required from the data input until it is stored in the latches. 150 ns is typical for high-voltage CMOS ICs.
e) *Strobe pulse width.* The minimum width of the strobe signal controlling the storage of the data. 400 ns for high-voltage CMOS at 5 V supply.
f) *Output circuit.* The method of driving the different types of displays determines what type of output circuit should be used. Many high-voltage CMOS devices use an open-collector-emitter circuit, and this requires separate series resistors for each output when driving an LED. Low-voltage fluorescent displays can be driven directly, but incandescent displays require a small pull-up resistor for each output to the power-supply voltage.

APPLICATIONS

This IC is used to drive common-cathode LED displays either as single units, or as multiplexed series of displays. Incandescent and low-voltage fluorescent displays can also be driven with this IC.

D.5.1 Display Driver, BCD-to-7-Segment, LCD

DESCRIPTION

Liquid crystal displays are activated by an AC signal across a selected display segment; this feature is included in the function block of the 7-segment LCD driver illustrated in Figure D.5.1. In addition to the standard BCD input, there is also a display frequency input terminal. Two functions can be performed by this terminal. When this driver IC is used to drive LED or other displays, this terminal, when in the **0** state, causes the selected 7-

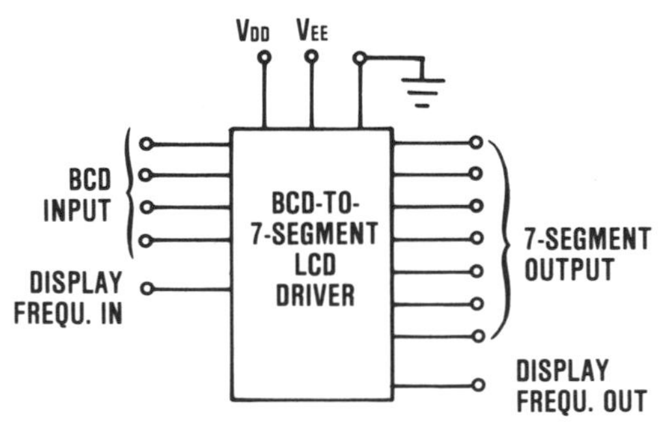

FIGURE D.5.1.

segment outputs to be in the **1** state, and when the display frequency input is **1**, the select outputs will be **0**. In the usual use with LCD, however, the input to the terminal will be a square wave ranging in frequency from 30 to 100 Hz. The selected segments of the display will have a square-wave output across them that is 180° out of phase with the display frequency input.

In addition to the 7-segment output, there is also a display frequency output which provides the level-shifted, high-amplitude display frequency necessary for driving the common electrode in liquid crystal displays (LCD). Note that positive and negative power-supply voltages are required.

KEY PARAMETERS

The electrical characteristics will be essentially the same as those of the particular digital IC family.

a) *Maximum power dissipation.* 500 mW at -40 +60°C is typical for high-voltage CMOS ICs.
b) *Quiescent current.* Maximum curent drawn when IC is not operating. 5 mA for high-voltage CMOS with ±5 V power supply.
c) *Propagation delay time.* The time required from input to output. Maximum is 1300 ns for ±5 V power supply and high-voltage CMOS.
d) *Noise margin.* The noise voltage with regard to power-supply voltage. 1 V noise margin at 5 V power supply is typical for high-voltage CMOS ICs.

APPLICATIONS

This IC can drive most liquid crystal displays. It is used in auto dashboard displays, panel meters, wall and table clocks, calculators, etc.

COMMENTS

A variety of LCD drivers is available, including those containing a storage latch function, those able to accept any 4-bit number and displaying the letters L, H, B, A and the minus sign. In LCD wristwatches, the decoder and display driver function is part of the watch IC.

D.5.2 Display Driver, BCD-to-7-Segment, Hexadecimal

DESCRIPTION

This IC contains internal driver circuits, similar to those described in D.5.0, but also contains individual 290 ohm resistors in series with each output driver so that external resistors are not needed when this IC is

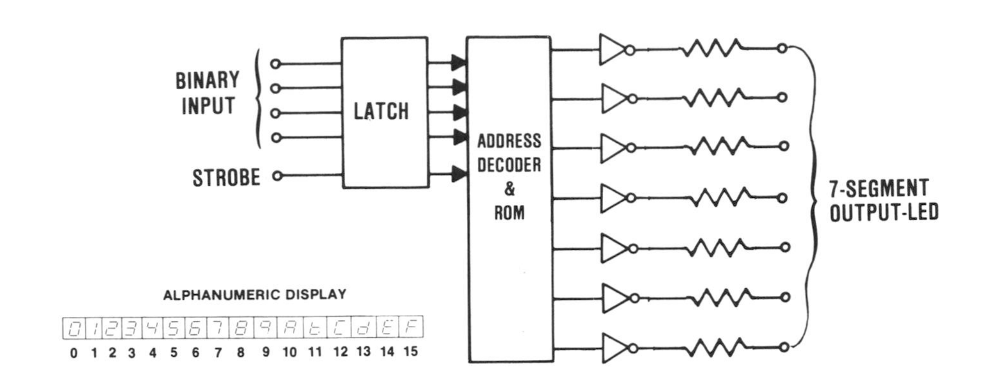

FIGURE D.5.2. Block Diagram—Hexadecimal 7-Segment Driver

connected to an LED display operating at 5 V. Another major difference between this IC and the one described in D.5.0 is its ability to accept a full binary input. The decoder section uses a read-only memory (ROM), with the binary input acting as address. The word stored in memory is then the equivalent of the 7-segment display. As illustrated in Figure D.5.2, the display consists of the standard numerals 0 through 9, and an ingenious arrangement of the 7-segment to display the letters A through F. In most other respects this IC compares with most of the display driver ICs.

KEY PARAMETERS

The electrical characteristics are essentially the same as those of a particular digital IC family.

a) *Maximum power dissipation.* 450 mW is typical for low-power CMOS.
b) *Quiescent current.* Maximum current drawn by the IC when not in operation. 10 mA is typical for low-power CMOS.
c) *Maximum output current per segment.* 11.5 mA is maximum for each of the outputs.

APPLICATIONS

The key feature of this IC is its ability to display the entire hexadecimal number set. For that reason it is used in various computer development and display control systems where hexadecimal is important.

D.5.3 Display Driver, 4-Digit, Expandable

DESCRIPTION

This single IC can drive four 7-segment digits with a decimal point (DP) for each. As illustrated in Figure D.5.3, there are seven data lines and the DP as the input, corresponding to the 7-segment output lines and the output decimal point. A 2-bit binary digit address selects, in sequence, which of the four digits, indicated by the 4-digit output lines, will be activated. Four separate control lines are provided to allow maximum flexibility in the application of this IC. The "chip enable" turns the entire IC on and off. The "write enable" controls entering of the digit address into the temporary memory, the latch. The "digit enable" line allows one of the four digit outputs to appear, and the "segment enable" line controls the appearance of the 7-segment output and the respective DP.

One of the key features of this IC is that it contains an internal oscillator which sequentially scans and presents the stored data to the seven high-power, tri-state output driver circuits which can drive the LED display directly. The four separate digit outputs are connected to the common cathode of the four digit 7-segment display. The output driver stages are active when the "segment enable" is **0** and go into the tri-state condition when this control lead is **1**. This permits a variation of the duty cycle and therefore the brightness of the display, and also permits turning the display off to conserve power.

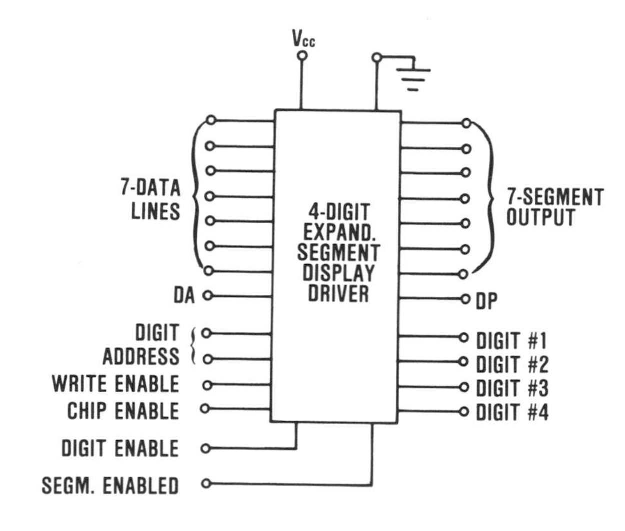

FIGURE D.5.3

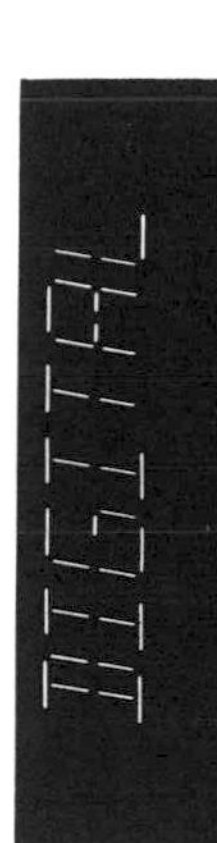

The design of this IC permits individual control of any segment in the display, and the number of segments per digit can be expanded without external components. It is possible, for example, to cascade two of these ICs to drive a 16-segment alphanumerical display. By further cascading of these ICs it is possible to control any array of discrete LEDs in any desired manner. Another feature of this IC is that when both the segment enable and digit enable lines are **0**, the entire IC is inoperative, drawing the minimum standby current.

KEY PARAMETERS

The electrical characteristics are essentially those of the CMOS digital IC family

a) *Maximum power dissipation.* 1200 mW at 25°C.
b) *Quiescent power dissipation.* The power drawn by the IC when in the special power saver mode. 5 mW is typical.
c) *Output load current.* 100 mA is typical.
d) *Propagation delay time from input to output.* Maximum of 1,000 ns at 5 V power supply
e) *Multiplex scan frequency.* Nominal value is 525 Hz for 5 V power supply.

APPLICATIONS

As single IC it is used to drive a 7-segment, 4-digit LED display directly. By adding series resistors and four transistors, connected at the common cathode, this IC can drive a gas discharge type display. When two of these ICs are used in cascade, they can drive a set of four 16-segment alphanumerical displays. In larger arrays these ICs can be used to control any number of discrete LED elements.

COMMENTS

The timing relations of the data input lines, the digit address, and the four enable lines are essential and may be critical in providing the desired display. The manufacturer's application data should be consulted before a final design using this IC is complete.

D.5.4 Display Driver, 7-Segment Character Generator

DESCRIPTION

The basic functions performed by this IC are essentially the same as those of the BCD-to-7-segment LED driver described in D.5.0. The method of

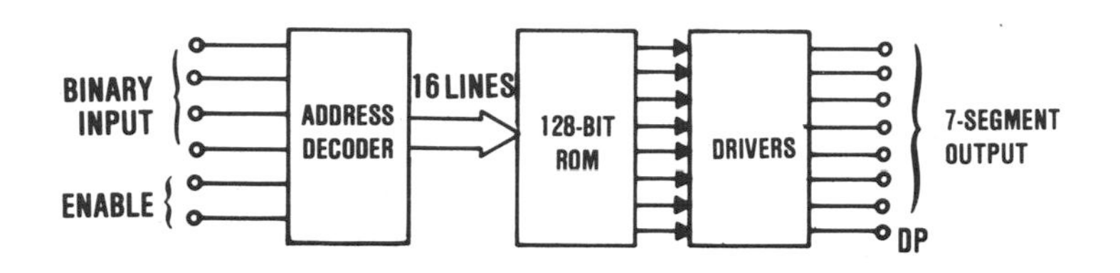

FIGURE D.5.4. Block Diagram—7-Segment Character Generator

performing these functions, however, is quite different, as illustrated in the block diagram of Figure D.5.4. The binary input signal, under the control of the two enable inputs, is decoded in the address decoder into one of 16 lines which address the 128-bit read-only memory (ROM). Depending on the selected address line, the ROM readout will be the selected 7-segment outputs plus the decimal point (DP). A set of eight driver amplifiers is provided for direct connection to a 7-segment LED or incandescent lamp numerical display. In this character generator there is no need to store the BCD input in a latch, nor is there a set of BCD-to-7-segment decoders. The address decoder is a straightforward binary-to-16 line logic gate network. One interesting feature is the availability of fusable links in the 128-bit ROM. While these links are ordinarily destroyed in such a pattern as to translate from one of the 16 lines to the selected 7-segment output, it is possible to obtain other output patterns if the fusable link destruction pattern is changed during the manufacturing process.

KEY PARAMETERS

This IC is generally available as TTL device, and its electrical characteristics are those of this particular digital IC family.

a) *Maximum power dissipation.* 250 mW is typical for this TTL IC.
b) *Quiescent current.* Current required by this IC when no operations are performed. 55 mA is the maximum quiescent current at 5 V power supply for this TTL IC.
c) *Propagation delay time.* The time required from input to output. 45 ns is the maximum time delay for any operation in this IC.
d) *Output current.* All outputs are open collector, requiring pull-up resistors with a maximum loading of 20 mA.

APPLICATIONS

This IC can drive low-voltage lamp indicators directly as well as LEDs through pull-up resistors. Like most other 7-segment drivers, this IC will find application in connection with 7-segment numerical displays in various appliances, test instruments, clocks, etc.

D.5.5 Display Driver—High-Voltage, High-Current

DESCRIPTION

The dual high-power driver circuit whose logic diagram is illustrated in Figure D.5.5 is typical of the devices used to interconnect various types of decoders or character generator devices to interface the normal CMOS voltage and current levels with higher voltage and current indicators. Each IC contains two identical sets of a NAND gate, followed by an inverter driver amplifier which drives an emitter follower Darlington amplifier to provide the high-current and high-voltage capabilities.

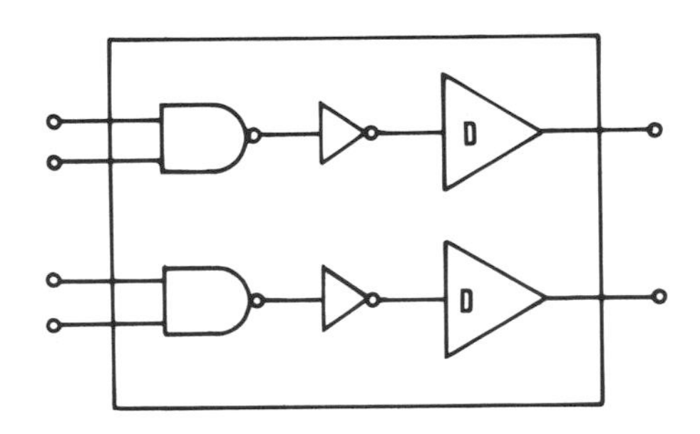

FIGURE D.5.5. Logic Diagram—Dual High Power Driver

KEY PARAMETERS

The electrical characteristics are essentially those of the particular digital IC family.

a) *Maximum voltage.* 30 V is typical maximum for CMOS.
b) *Maximum output current.* 250 mA is typical for this CMOS IC.

APPLICATIONS

Wherever high voltage and high current is required to drive a display, lamp, relay, etc.

D.6.0 Error Detector—Parity Generator/Checker

DESCRIPTION

The parity bit method of error detection depends on a fixed word length in which all of the **1**s are added together. For even parity, the sum of these logic **1**s must be even; for odd parity, it must be odd. To assure that the sum is respectively odd or even, a separate parity bit is transmitted which is set either to **1** or to **0**, whichever is required to make the sum of the **1**s even or odd.

The logic diagram of Figure D.6.0 shows a simple 9-bit parity generator/checker. The nine inputs are added through a tree arrangement of exclusive OR circuits. When the sum of the **1**s is even, the "even" output will be a logic **1**. When the sum is odd, the "odd" output will be a logic **1**. The inhibit signal is used to prevent operation of the parity generator/checker, and when its input is **1**, both the even and the odd output will be at **0**.

When used as a parity generator, the parity bit is supplied along with the data to generate an even or odd parity output. When this IC is used as parity checker, the received data bits and parity bits are compared for correct parity. The even or odd outputs then indicate an error in the received data. It is possible to cascade a series of these ICs to check parity or generate parity for longer words.

KEY PARAMETERS

a) *Maximum power dissipation.* 500 mW is typical for high-voltage CMOS devices.

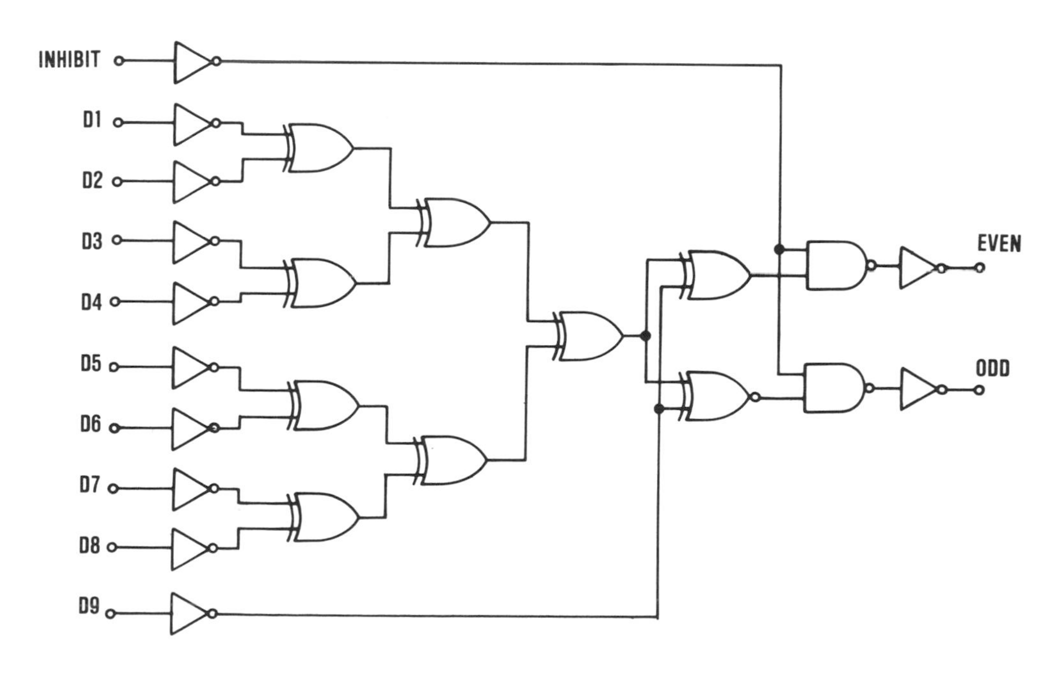

FIGURE D.6.0. Logic Diagram—Parity Generator/Checker

b) *Quiescent current*. The total current drawn by the IC when it is not in operation. 0.04 μA is typical at a 5 V power supply for high-voltage CMOS.
c) *Propagation delay time*. The time between data input and data output. A typical value is 350 ns for high-voltage CMOS devices at 5 V power supply.
d) *Inhibit-to-output delay time*. The time required from the application of the inhibit signal until the output of both even and odd terminals is **0**. 150 ns is a typical value for high-voltage CMOS.

APPLICATIONS

This type of IC is used whenever error correction and detection of digital data must be performed. It is frequently used in digital communications systems, at the I/O ports of computer systems, and in specific portions of digital computers themselves.

COMMENTS

Although the illustration of Figure D.6.0 shows a 9-bit parity generator/checker, ICs are also available for 8-bit, 12-bit, 16-bit, etc., word length operations.

D.6.1 Error Detection/Correction Circuit

DESCRIPTION

In more sophisticated error detection and correction circuits, the relatively simple parity generator/checker described in D.6.0 is insufficient and more complex systems are required. The logic diagram of the error detection and correction circuit shown in Figure D.6.1 is typical of the building block type of ICs used in large computer systems for parity checking, single bit error detection/correction, and double bit error detection. Note that the logic diagram of Figure D.6.1 consists of a tree of exclusive OR circuits, similar to that of Figure D.6.0. In a typical application, this type of circuit would automatically generate parity for every byte in a sequence of digital data. To obtain high-speed operation, most computer error detection/correction ICs are from the ECL digital IC family.

KEY PARAMETERS

a) *Maximum power dissipation*. 520 mW is a typical value for ECL ICs.

b) *Propagation delay.* The time required between data input and output. 5 ns is a typical value for ECL ICs.

APPLICATIONS

This type of IC is used as a building block, together with other ICs of the same or similar function, to detect and correct errors in large data streams in medium and large size computers and digital communications systems.

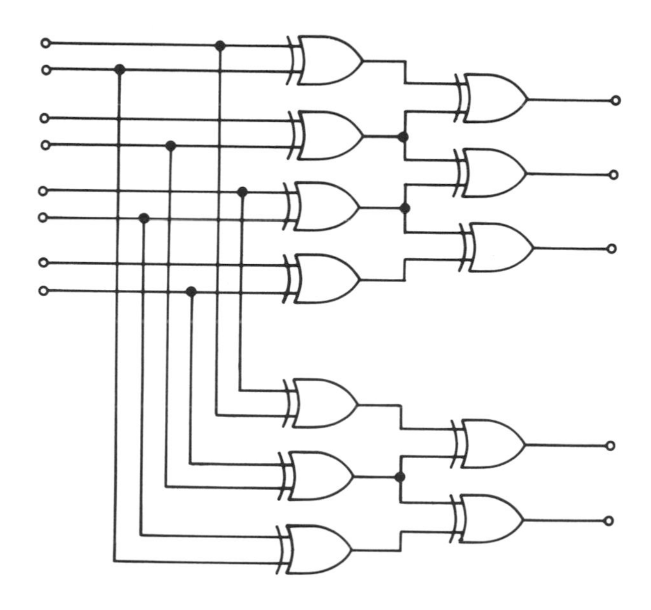

FIGURE D.6.1. Logic Diagram of Error Detection-Correction Unit

D.6.2 Error Detector/Cycle Redundancy Check (CRC)

DESCRIPTION

This IC performs error detection by means of a fairly sophisticated method—the cyclic encoding and decoding schemes which are based on the manipulation of polynomials in modulo arithmetic. During the encoding process, the data stream is divided by a selected polynomial, and this division results in a remainder which is attached to the message as check bits. To detect errors, the bit stream containing both data and check bits is divided again by the same selected polynomial. If there are no detectable errors, this division should result in a zero remainder. In usual computer and communications practice, a standard set of polynomials is most widely used. One of these is based on a series starting with X^{16}, another is used on a similar 16th power series with a reverse notation, a third one is based on a X^{12}, and a fourth one is based on X^8. For international data transmission, there are CCITT codes, generally based on X^{16}.

The polynomials themselves are stored in a read-only memory (ROM) as illustrated in Figure D.6.2. Three control lines are capable of selecting any one of eight different polynomials stored in the ROM and presetting the 16-bit register with that polynomial code. The data to be encoded or checked is clocked into the 16-bit register and, depending on the function and whether or not a remainder occurs, the error detector indicates that an error has been found.

Among the controls for the 16-bit register is an enable signal to allow a check word to be entered from the ROM, a reset signal, and a master reset signal.

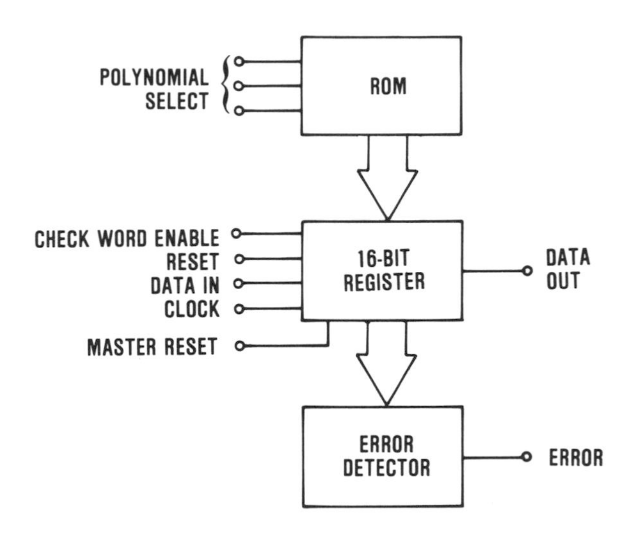

FIGURE D.6.2. Block Diagram—CRC Generator/Checker

KEY PARAMETERS

The electrical characteristics of this IC are essentially the same as those of the TTL digital IC family.

APPLICATIONS

This IC is used in data communications systems and terminals and in digital computers. It is particularly useful in large memory systems. Digital cassette and cartridge systems, floppy and other disc storage systems are locations where the CRC generator/checker will be found. The cyclic redundancy method of error detection is also used in some military and space communications systems.

COMMENTS

This method of error detection requires additional data bits at the end of the actual message train to carry along the remainder which is the result of the polynomial division. The size of the remainder, and therefore the number of bits to be added to the actual message, depends on the polynomial selected and on the nature of the data. Some understanding of digital communications theory is required for the optimum design with this IC.

D.7.0 Flip-Flops (FF)

DESCRIPTION

One of the key building blocks of all digital logic systems, the flip-flop (FF) is available in a variety of different FF circuits with a host of different features.

As illustated in Figure D.7.0(a), the basic flip-flop consists of two gates with the appropriate cross connection. The actual circuitry of a D-type, positive edge triggered FF is shown in Figure D.7.0(b), and illustrates the application of different gating circuits to the basic FF function. Note the presence of the "clock," "reset" (clear), and "set" inputs in addition to the actual data input, terminal D. The logic level at input D is transferred to the Q output only during the positive going edge of the clock pulse. When the clock pulse is at either the **0** or the **1** level, the D input signal has no effect. The "set" input signal can set the condition of the FF any time, regardless of the presence or absence of a clock pulse. Similarly, the "reset" signal can restore the condition of the FF any time.

The detailed operation of the FF is summarized by the truth table of Figure D.7.0(c). The set and reset (clear) operations can occur regardless of the presence of the clock pulse, but data inputs at terminal D can occur only during the positive edge of the clock pulse when the set and reset signals are **0**. If both set and clear signals are **1**, simultaneously, it is possible for both Q and Q signals to be **1** at the same time. The actual condition of the outputs is undetermined and unpredictable in this situation.

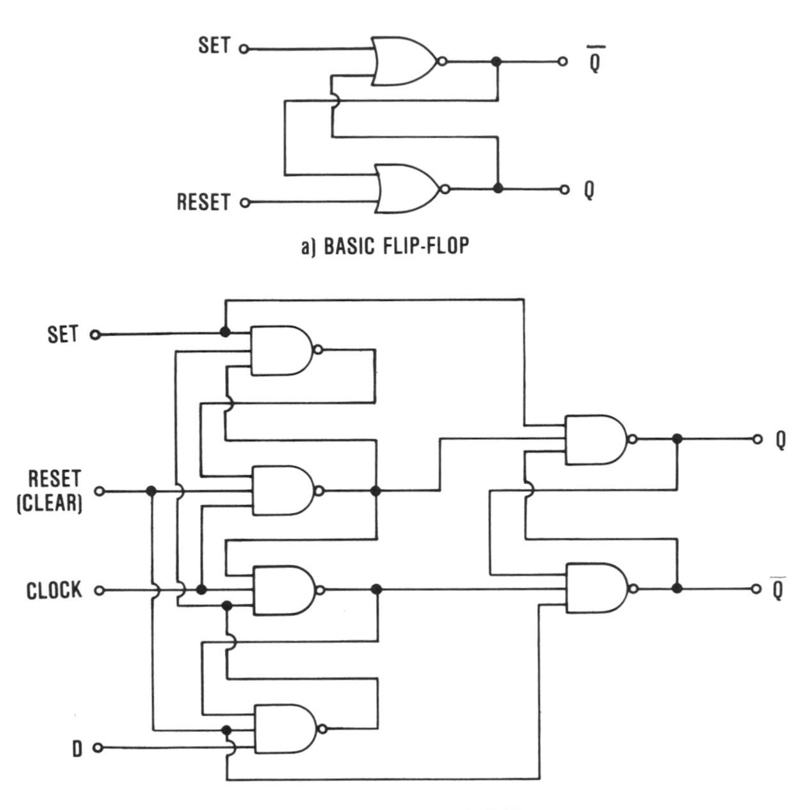

CLOCK	DATA	RESET	SET	Q	$\overline{Q}$
rising edge	0	0	0	0	1
rising edge	1	0	0	1	0
falling edge	X	0	0	NC	NC
X	X	1	0	0	1
X	X	0	1	1	0
X	X	1	1	1	1

X—DON'T CARE
NC—NO CHANGE

c) TRUTH TABLE

FIGURE D.7.0 (a) Basic Flip-Flop
(b) D-Type Flip-Flop
(c) Truth Table

KEY PARAMETERS

a) *Maximum power dissipation.* The maximum power that the IC can dissipate over the specified temperature range, usually from -40 to +85°C.

b) *Quiescent current.* The current drawn when the FFs on the entire IC are not operating.
c) *Toggle rate.* The maximum frequency at which the FF can change state.
d) *Propagation delay, clock to output.* The time required from the clock pulse transition until a change takes place in the FF output terminal.
e) *Propagation delay, set or reset to output*. The time delay from the application of the set or reset signal until the output state changes.
f) *Clock pulse width.* The minimum width of the clock pulse required for reliable operation.
g) *Set or reset pulse width.* The minimum width of the set or reset pulse.

APPLICATIONS

FFs of every type are found in IC applications such as counters, shift registers, microprocessors, memories, etc. ICs containing a small number of FFs on which all terminals are available are used as auxiliary circuits in digital systems. FFs are used as short-time delays, set-reset switches, and in a host of other control applications.

COMMENTS

FFs are available in every digital logic IC family and can be obtained in a wide variety of characteristics and operational features. The following pages highlight the most frequently used specific FF ICs.

D.7.1 Flip-Flop, Dual-Type D

DESCRIPTION

This universally used flip-flop (FF) contains two separate type D FFs, as illustrated in the functional diagram of Figure D.7.1. Each of the two FFs is identical and can be used separately or interconnected with each other for whatever purpose may be required. They can be used as shift register elements or as type D FFs for counter or toggle applications.

KEY PARAMETERS

a) *Quiescent current.* The current drawn by the IC when no operations take place. 2.0 nA are typical at 5 V for low-power CMOS.
b) *Toggle rate.* The frequency with which the FF can change state. 4.0 MHz is typical for low-power CMOS and up to 50 MHz is typical for low-power TTL ICs.

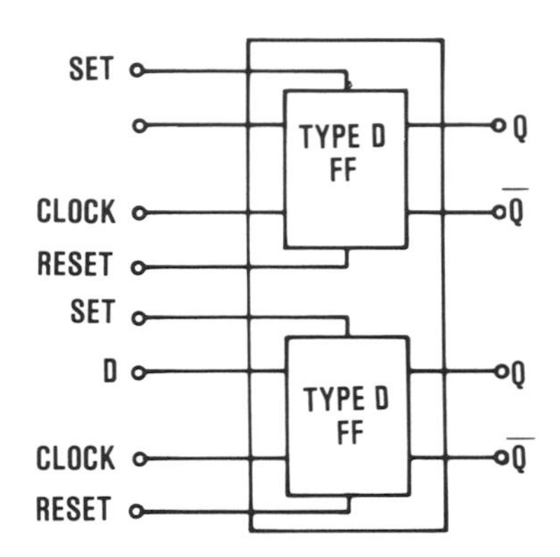

FIGURE D.7.1. Dual Type D FF Function Block

c) *Propagation delay, clock to output.* The time required from the clock pulse transition until a change occurs in the output terminal. 175 ns is typical for low-power CMOS and 10 ns is typical for low-power TTL ICs.
d) *Propagation delay, set or reset to output.* The time required for either the set or reset signal to change the output. 175 ns is typical for low-power CMOS and 13 ns is typical for low-power TTL ICs.
e) *Clock pulse width.* 250 ns is typical for low-power CMOS and 12 ns is typical for low-power TTL ICs.

D.7.2 Flip-Flop, Quad-Type D

DESCRIPTION

In this IC four identical FFs are controlled by a common clock and reset signal, as illustrated in the logic diagram of Figure D.7.2. Note that the data input to each FF and both Q and $\overline{Q}$ outputs is buffered by separate amplifiers. Only during the positive going clock signal can the data present at the D-inputs be transferred to the Q outputs. The reset operation is possible at any time and is independent of the clock.

KEY PARAMETERS

a) *Quiescent current.* The current drawn by the IC when no operations take place. 5 nA is typical at 5 V for low-power CMOS.
b) *Toggle rate.*The highest frequency at which the FFs can change state. 4.5 MHz is typical for low-power CMOS and 45 MHz is typical for TTL ICs.
c) *Propagation delay, clock to output.* The time from the clock pulse

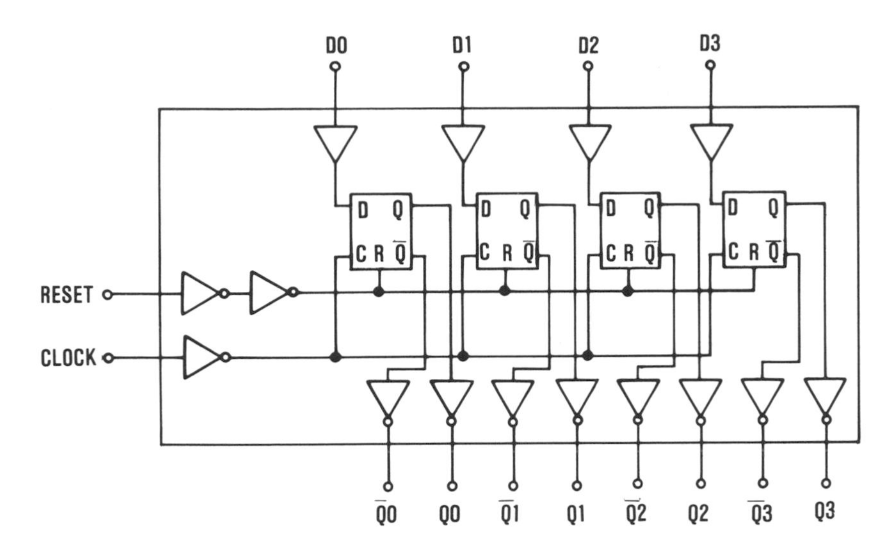

FIGURE D.7.2. Quad Type D FF Function Block

transition until a change occurs in the output. 220 ns is typical for low-power CMOS and 15 ns is typical for low-power TTL ICs.

d) *Propagation delay, reset to output.* The time required after the application of the reset signal until the output is reset. 325 ns is typical for low-power CMOS and 40 ns is typical for low-power TTL ICs.

e) *Clock pulse width.* 110 ns is typical for low-power CMOS and 20 ns is typical for low-power TTL ICs.

APPLICATIONS

These ICs can be connected as shift registers or counter elements as well as for specialized switching and timing control circuits.

COMMENTS

While the IC described here contains four FFs with common clock and reset, other ICs containing up to 16 FFs are available.

D.7.3 Flip-Flop, Dual J-K

DESCRIPTION

The J-K FF, and the type D FF described in the previous pages, are the most popular and widely used types of FFs. The main difference between them is that the J-K terminals replace the D-input.

As illustrated in the logic diagram of Figure D.7.3(a), the J-K FF requires more logic elements but also can perform different functions. The detailed functions available are summarized in the truth table of Figure D.7.3(b). Note that where there were two possible data inputs in the type D-FF, there are four possible data inputs in the J-K type because either or both of these terminals can assume the **0** or the **1** state. In the truth table of the J-K FF the change from the present condition of Q to the next condition of the outputs is indicated, because the change rather than the static operation is the key function of the J-K FF.

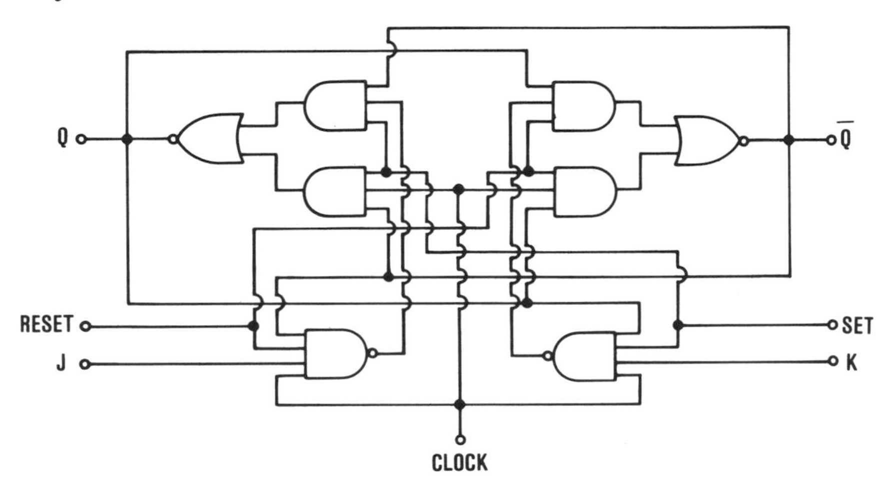

a) LOGIC DIAGRAM — J-K FLIP-FLOP

					PRESENT	NEXT	
CLOCK	J	K	S	R	Q_1	Q	$\overline{Q}$
_/¯	1	X	0	0	0	1	0
_/¯	X	0	0	0	1	1	0
_/¯	0	X	0	0	0	0	1
_/¯	X	1	0	0	1	0	1
¯_	X	X	0	0	X	NC	NC
X	X	X	1	0	X	1	0
X	X	X	0	1	X	0	1
X	X	X	1	1	X	1	1

X—DON'T CARE
NC—NO CHANGE

b) TRUTH TABLE

FIGURE D.7.3 (a) Logic Diagram—J-K Flip-Flop
(b) Truth Table

KEY PARAMETERS

a) *Quiescent current.* The current drawn by the IC when no operations take place. 2.0 nA at 5 V is typical for low-power CMOS.
b) *Toggle rate.* The maximum frequency at which the FFs change state. 3 MHz is typical for low-power CMOS and 45 MHz is typical for low-power TTL ICs.
c) *Propagation delay, clock to output.* The time delay from the clock pulse transition to a change in the output. 175 ns is typical for low-power CMOS and 11 ns is typical for low-power TTL ICs.
d) *Propagation delay, set to output.* The time from the application of the set pulse until a change occurs at the output. 175 ns is typical for low-power CMOS and 16 ns is typical for low-power TTL ICs.
e) *Propagation delay, reset to output.* The time required from the application of the reset signal until a change occurs at the output. 350 ns is typical for low-power CMOS and 16 ns is typical for low-power TTL ICs.
f) *Clock pulse width.* 165 ns is typical for low-power CMOS and 12 ns is typical for low-power TTL ICs.

APPLICATIONS

J-K FFs are found in all digital equipment for control, register, or toggle functions.

COMMENTS

Although a dual J-K FF IC is described here, the number of J-K FFs that are available in a single IC depends only on the pin connections available. When common clock, set and reset controls are used, more than two J-K FFs can be made available on a 16-pin standard IC.

D.7.4 Flip-Flop, Gated J-K, Master-Slave

DESCRIPTION

This IC illustrates the many functions that a single FF can perform. As illustrated in the function block of Figure D.7.4(a), the J-K FF has a 3-input AND gate connected to the J and to the K terminal. The use of the multiple J and K inputs controls the transfer of information into the master section during clock operation at the positive edge of the clock pulse. The truth tables in Figure D.7.4(b) illustrate both synchronous and asynchronous operations of this IC and provide some insight into the particular functions that it can perform.

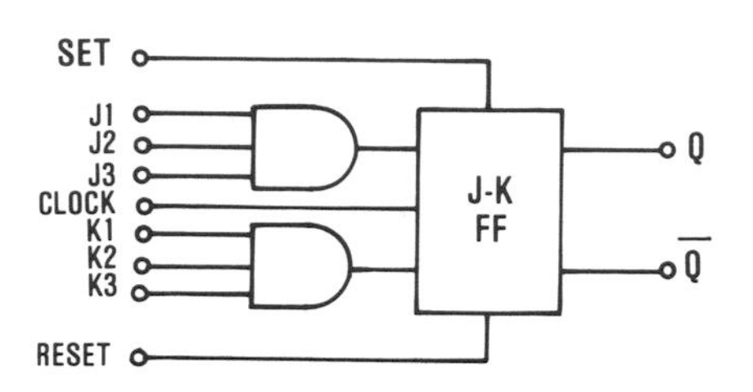

a) LOGIC DIAGRAM — GATED J-K MASTER-SLAVE FF

INPUT BEFORE CLOCK		OUTPUT AFTER CLOCK	
J	K	Q	$\overline{Q}$
0	0	NC	Nc
0	1	0	1
1	0	1	0
1	1	TOGGLES	

J = J1 • J2 • J3
K = K1 • K2 • K3
b) SYNCHR. OPERATION

J & K = DON'T CARE

S	R	Q	$\overline{Q}$
0	0	NC	NC
0	1	0	1
1	0	1	0
1	1	0	0

c) ASYNCHR. OPERATION

FIGURE D.7.4 (a) Logic Diagram—Gated J-K Master-Slave FF
(b) Synchronous Operation
(c) Asynchronous Operation

KEY PARAMETERS

a) *Maximum power dissipation.* 500 mW is typical for high-voltage CMOS devices.
b) *Quiescent current.* The current drawn when the IC is not operating. 0.02 μA is typical for high-voltage CMOS devices.
c) *Toggle rate.* The highest frequency at which the FF can change state. 7 MHz is typical for high-voltage CMOS at 5 V.
d) *Propagation delay, clock to output.* 250 ns is typical for high-voltage CMOS.
e) *Propagation delay, set or reset to output.* 150 ns is typical for high-voltage CMOS.
f) *Clock pulse width.* 70 is typical for high-voltage CMOS ICs.

APPLICATIONS

This type of FF is particularly useful in registers, counters, and control circuits where multiple data inputs must be combined.

COMMENTS

The IC described above is merely an example of the combinations of FF and gate arrangements that are available as standard ICs.

D.8.0 Gates, Logic

DESCRIPTION

The six basic logic gates illustrated in Figure D.8.0 form the building blocks of almost all digital functions provided in ICs. Practically all digital medium and large-scale integrated circuits contain these fundamental logic gates, and a large variety of specific gate function arrays are available in every digital IC family.

Figure D.8.0(a) illustrates the two-input OR circuit, its Boolean function and its truth table. Figure (c) shows the same symbol with a circle at the output, indicating the complementing or inverting function, the NOR circuit. Figures D.8.0(b) and (d) illustrate the AND and the NAND circuits, respectively. The Exclusive OR and the Exclusive NOR circuits, as shown in Figures D.8.0(e) and (f), together with their Boolean notation and truth table, comprise special applications of the OR and NOR function.

While the Boolean equations of each of the six logic gates illustrated in Figure D.8.0 are different, the truth tables of some of them are identical. Note, for example, that the truth table of the NAND circuit is identical to the truth table of the exclusive OR. The truth table of the AND circuit is identical to the exclusive NOR circuit. This implies that in many cases these circuits can be used interchangeably.

KEY PARAMETERS

a) *Propagation delay time*. The time required between input and output. This parameter depends on the particular digital IC family.
b) *Input loading factor*. The voltage and/or current required for each input. This depends largely on the particular digital IC family.
c) *Output loading (fan-out)*. The output capability of each terminal, stated as either current source or sink, or as number of input loads. This parameter depends on the particular digital IC family.

APPLICATIONS

Logic gates are the building blocks of all digital ICs and are used in all digital systems.

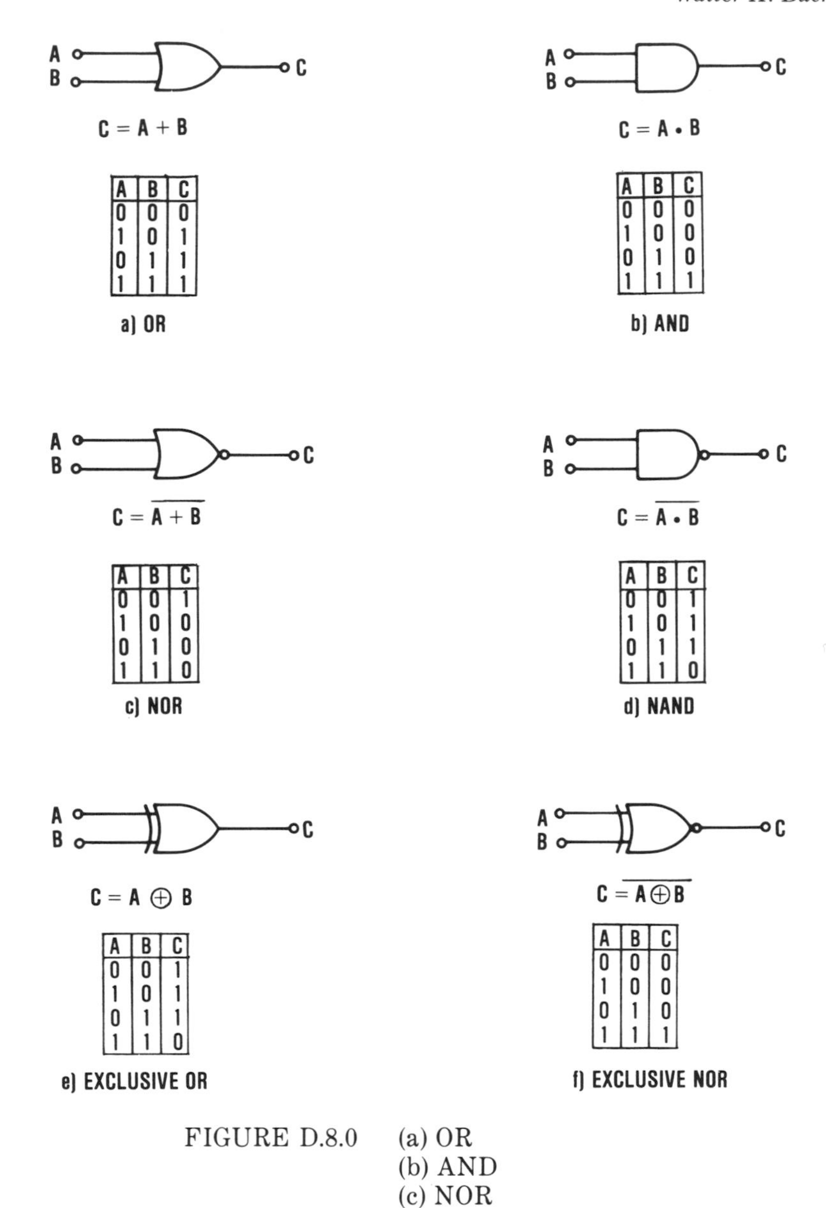

a) OR

A	B	C
0	0	0
1	0	1
0	1	1
1	1	1

b) AND

A	B	C
0	0	0
1	0	0
0	1	0
1	1	1

c) NOR

A	B	C
0	0	1
1	0	0
0	1	0
1	1	0

d) NAND

A	B	C
0	0	1
1	0	1
0	1	1
1	1	0

e) EXCLUSIVE OR

A	B	C
0	0	1
1	0	1
0	1	1
1	1	0

f) EXCLUSIVE NOR

A	B	C
0	0	0
1	0	0
0	1	0
1	1	1

FIGURE D.8.0 (a) OR
(b) AND
(c) NOR
(d) NAND
(e) Exclusive OR
(f) Exclusive NOR

COMMENTS

Because logic gates are so important, a large assortment of gate ICs are available in each digital IC family. Only a few representative examples are included here.

D.8.1 Gates, Multiple

DESCRIPTION

The most widely used package in all digital IC families is the DIP configuration with 16 terminals. The three multiple gate arrangements shown in Figure D.8.1 illustrate the most popular multiple gate configurations. Every digital IC family provides at least these three configurations as well as a number of specialized ones. There are quad-2-input OR, AND, NAND, NOR, and exclusive ORs available, as well as triple 3-input and dual-4-input gates of all types. There is also a single 8-input gate type, usually as AND or NAND configurations. Variations of these standard configurations include the use of external pull-up resistors to provide interfacing with the desired voltage and current levels, or else output driver amplifiers are included to increase the output loading capability (fan-out).

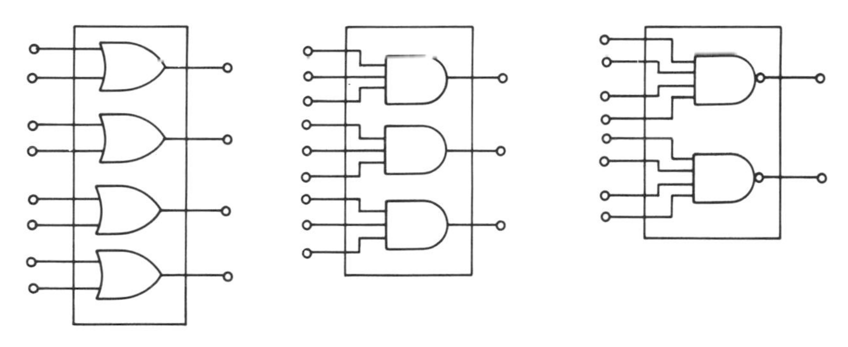

FIGURE D.8.1 (a) Quad 2-Input OR
(b) Triple 3-Input AND
(c) Dual 4-Input NAND

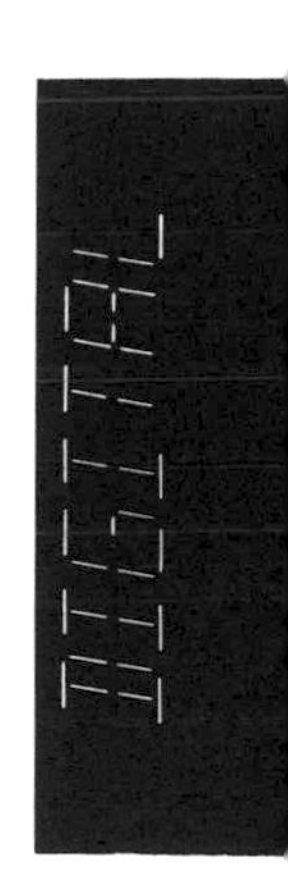

KEY PARAMETERS

The key parameters described in Section D.8.0 apply equally to the multiple gate ICs and must be referred to the particular digital IC families to obtain particular values.

APPLICATION

Multiple logic gate ICs are used as supplementary functions for large-scale integrated circuits on which the gates are part of the entire configuration. For small digital systems, the multiple gate ICs may be used to perform the required logic functions.

D.8.2 Gates, Array

DESCRIPTION

A wide variety of logic gate arrays are available in ICs, some of which perform such frequently used functions as binary-to-BCD or BCD-to-7-segment display code conversions. The type of logic gate array described here can perform several different functions and therefore offers a very flexible logic assembly.

The gate array illustrated in Figure D.8.2 can serve as a 4-bit AND/OR selector, and a quad-2-channel data selector, or as a quad exclusive NOR gate. There are eight data inputs, which are combined in NOR circuits with two control inputs, A and B. The output of each two NOR circuits is fed into an exclusive NOR circuit which is then controlled by the NAND circuits and the combination of inputs A and B. Note that each of the four data output lines contains an inverter driver to provide maximum output loading capability (fan-out).

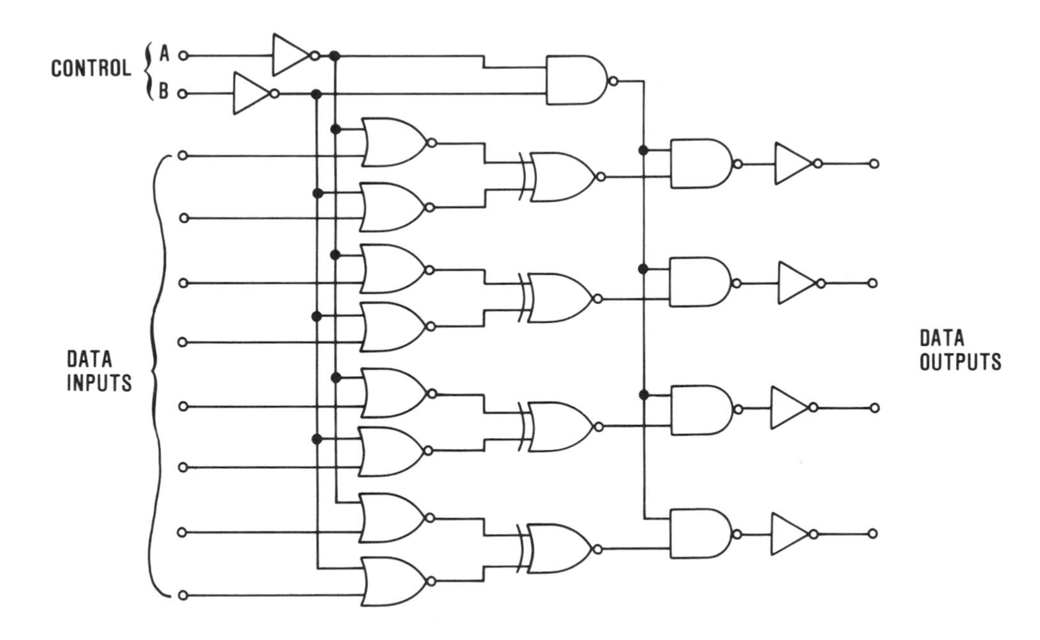

FIGURE D.8.2. Logic Diagram—Typical Gate Array

KEY PARAMETERS

a) *Propagation delay time.* The time between input and output. 250 ns is typical at 5 V for low-power CMOS.
b) *Input loading.* For low-power CMOS, the input voltage at 5 V power supply is typically 2.5 V for **0** and at least 2.75 V for **1** logic level.
c) *Output loading capability (fan-out).* For low-power CMOS, the fan-out is in the order of 50. This means that each output of Figure D.8.2 can drive up to 50 inputs, equivalent to each of the data inputs of that same IC.

APPLICATIONS

This IC is used in a variety of digital systems to perform one of the following functions: 4-bit AND/OR selector, quad-2-channel data selector, or quad exclusive NOR gate array.

COMMENTS

The IC described on this page is only one of a number of different logic gate arrays that are available for flexible interconnections in complex digital systems.

D.8.3 Gates, Expandable/Expander

DESCRIPTION

In some applications a relatively great number of inputs must be combined in the AND, OR, or other logic function. One method of increasing the number of inputs to a particular logic gate function is to use an expandable system, combined with an expander. Figure D.8.3 illustrates two separate ICs, an expandable gate input, an AND circuit, and an 8-input AND expander which can be used with it. The expandable 8-input NAND circuit illustrated in Figure D.8.3(a) consists of two 4-input AND circuits which drive a NOR circuit. The NOR has two special terminals, one for introducing the expansion signal X and the other for its complement, $\overline{X}$. The 8-input expander shown in Figure D.8.3(b) consists of two 4-input circuits with a special output for the complement of the AND gate. To obtain a 16-input NAND function, the expander outputs are simply connected to the corresponding inputs of the expandable 8-input AND circuit. The terminals marked X are connected together, as are those marked X.

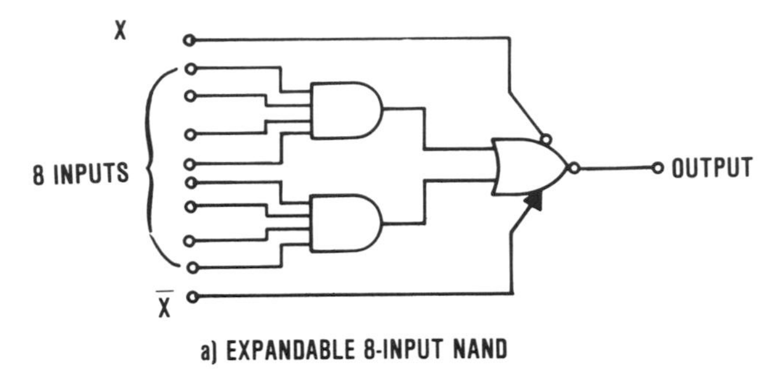

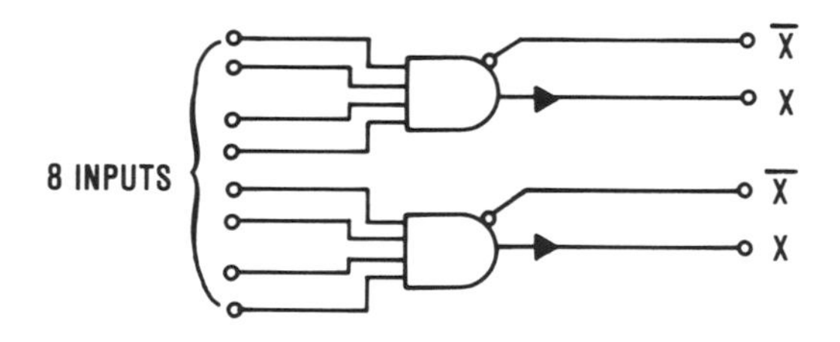

FIGURE D.8.3 (a) Expandable 8-Input NAND
(b) 8-Input AND Expander

KEY PARAMETERS

a) *Propagation delay time.* The time required between input and output. 18 ns is typical for TTL ICs, and 300 ns is typical for high-voltage CMOS ICs.
b) *Input loading.* The input voltage or current required for each pin on the IC. This depends on the particular digital IC family.
c) *Output loading capability (fan-out).* The number of equivalent input circuits that can be connected to each output terminal. This depends largely on the particular digital IC family.

APPLICATIONS

In all digital systems where more than eight inputs must be combined in one of the logic gate functions.

D.9.0 Latch, Basic

DESCRIPTION

The basic latch function can be appreciated from the logic diagram of Figure D.9.0, which shows a simple quad R-S latch. The set and reset input of each of the four flip-flops (FF) determines the setting of the Q output terminal. Each Q output goes through a gated inverter driver and all four drivers are controlled by the enable signal. When the enable signal is **0**, the four outputs are all at the high impedance state, regardless of the R and S inputs. Only when the enable signal goes to **1** can the set and reset inputs control the Q outputs.

This IC is available in almost all digital IC families and the usual configuration consists of four FFs per IC.

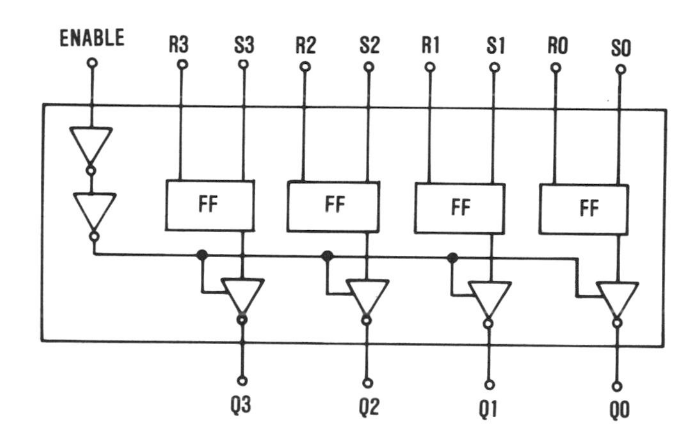

FIGURE D.9.0. Logic Diagram—Quad R-S Latch

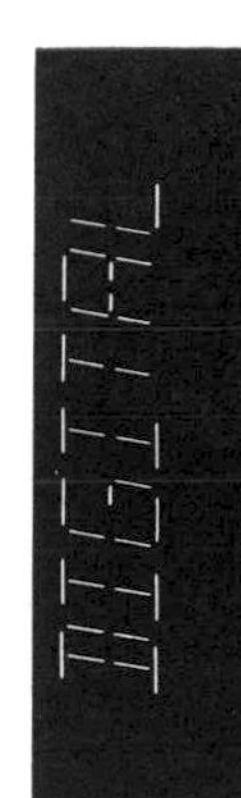

KEY PARAMETERS

a) *Quiescent current.* The current required by the IC when no operations take place. 4.0 nA is typical for low-power CMOS ICs.
b) *Maximum power dissipation.* Depends on the digital IC family. 310 mW is typical for ECL ICs.
c) *Propagation delay time.* The time required from input to output when the enable signal is **1**. 175 ns is typical for low-power CMOS. 16 ns is typical for TTL, and 2.0 ns is typical for ECL ICs.
d) *Enable pulse width.* 80 ns is typical for CMOS, 20 ns is typical for TTL, and 5.0 ns is typical for ECL ICs.

APPLICATIONS

Latches are used whenever parallel digital data must be stored temporarily. Typical applications include D/A converters, BCD-to-7-segment display drivers, and computer input/output (I/O) sections.

D.9.1 Latch, Strobed

DESCRIPTION

This IC provides the same latching function as described in D.9.0 but contains some additional refinements. As illustrated in Figure D.9.1, the function block of the IC contains four inputs and four outputs as well as three control lines. Each of the four inputs corresponds to the D (data) input of a regular R-S flip-flop (FF). The output corresponds to the Q terminal of the FF, but there is usually a buffer driver or control gate between the FF and the output terminal. Control signals, such as the master reset (MR), the strobe (ST) and the disable (DIS) signals, determine the detailed operation of all four latching FFs. As illustrated in the truth table of Figure D.9.1, any D-input can only affect its Q output when the MR and the DIS signals are **0** and the strobe (ST) is **1**. When all three control signals are **0**, the input signal does not matter and the output Q terminal will not change, indicating that the circuit is latched. The master reset (MR) changes all Q terminals to **0** whenever its control input is **1**. Similarly, when the disable signal is **1**, the output Q terminals will all be in the high impedance condition. This makes it possible to connect this type of latching IC to a bus or I/O port in a digital computer system and avoid loading down these circuits.

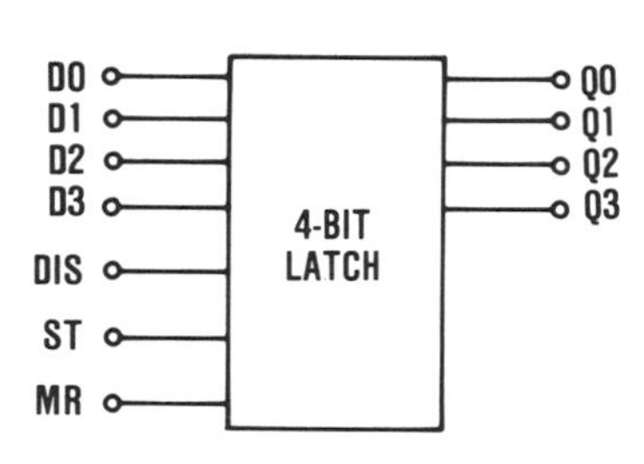

MR	ST	DIS	D	Q
0	1	0	0	0
0	1	0	1	1
0	0	0	X	N.C.
1	X	0	X	0
X	X	1	X	H.I.

X — DON'T CARE
N.C. — NO CHANGE (LATCHED)
H.I. — HIGH IMPEDANCE

FIGURE D.9.1 (a) Function Block
(b) Truth Table

KEY PARAMETERS

a) *Quiescent current.* The total current drawn by the IC when no operations take place. 5.0 nA is typical for low-power CMOS ICs.
b) *Propagation delay time.* The time required from input to output when the strobe signal is **1**. 220 ns is typical for CMOS ICs.
c) *Master reset pulse width.* 200 ns is typical for low-power CMOS.
d) *Strobe pulse width.* 140 ns is typical for low-power CMOS ICs.

APPLICATIONS

This IC is used whenever parallel digital data are stored temporarily. Specific applications include digital computers and communications systems at the input/output (I/O) interfaces and in the bus drive and control sections. This type of latch is also used for numerical displays, in connection with a 7-segment decoder and display driver circuit.

COMMENTS

Although a single 4-bit latch is illustrated in Figure D.9.1, most standard ICs contain two such functions on a single IC.

D.9.2 Latch, Addressable

DESCRIPTION

The block diagram of Figure D.9.2 illustrates how this IC accepts data in serial form and produces a parallel, 8-bit, latched output. The serial data stream is entered into the eight latches, basically eight separate flip-flops (FF), as the address of each data bit is entered on three lines in simple binary notation. In the decoder section the address is decoded into one of eight lines which control each of the eight latch FFs. This enters the serial data, in sequence, into the respective FFs. The write disable signal allows the serial entry into the eight parallel FFs. The reset signal resets all eight latches to **0** whenever the output information is no longer needed.

KEY PARAMETERS

a) *Quiescent current.* The total current drawn by the IC when no operations take place. 5.0 nA is typical for low-power CMOS ICs.
b) *Propagation delay time.* The time required from input to output with the write disable at **0**. 200 ns is typical at 5 V for low-power CMOS.
c) *Data pulse width.* 100 ns at 5 V is typical for low-power CMOS.
d) *Address pulse width.* 200 ns at 5 V is typical for low-power CMOS.
e) *Reset pulse width.* 75 ns is typical for low-power CMOS.
f) *Write disable pulse width.* 160 ns is typical for low-power CMOS.

APPLICATIONS

In digital computers this IC can accept serial data, together with a 3-bit binary address, to convert it to 8-bit parallel bytes. It is particularly useful in the input/output (I/O) section of digital communications systems.

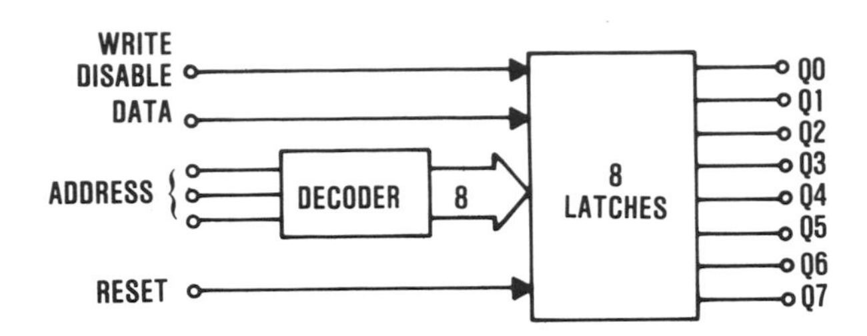

FIGURE D.9.2. Block Diagram—8-Bit Addressable Latches

D.9.3 Latch, Addressable, with Counter

DESCRIPTION

This IC is a more sophisticated version of the addressable 8-bit latch described in D.9.2. As illustrated in the block diagram of Figure D.9.3, this IC contains its own address counter, reset logic and full logic. The reset signal resets not only the eight latches but also the address counter. A simple clock signal can be used to increment the address counter while serial data is applied to the eight latch FFs. The output of the address counter is decoded, providing eight signals to the latches, as described in Section D.9.2. When the eighth bit has been reached, or if the reset pulse occurs, the full logic section provides an output signal to the digital interface that the eight latches have been loaded. This IC is specifically designed for bus interfaces because the enable output signal can go to the bus to indicate that the data is available at the eight data output lines.

KEY PARAMETERS

a) *Quiescent current.* The current drawn by the IC when no operations take place. 5.0 nA at 5 V is typical for low-power CMOS.
b) *Propagation delay time, enable to output.* Time delay between appearance of enable pulse and data output. 160 ns is typical at 5 V for low-power CMOS.
c) *Propagation delay time, strobe to output.* Time delay from appearance of strobe signal to data output. 200 ns is typical at 5 V for low-power CMOS.
d) *Propagation delay time, strobe to full.* Time delay from strobe appearance to appearance of full signal. 200 ns is typical at 5 V for low-power CMOS.

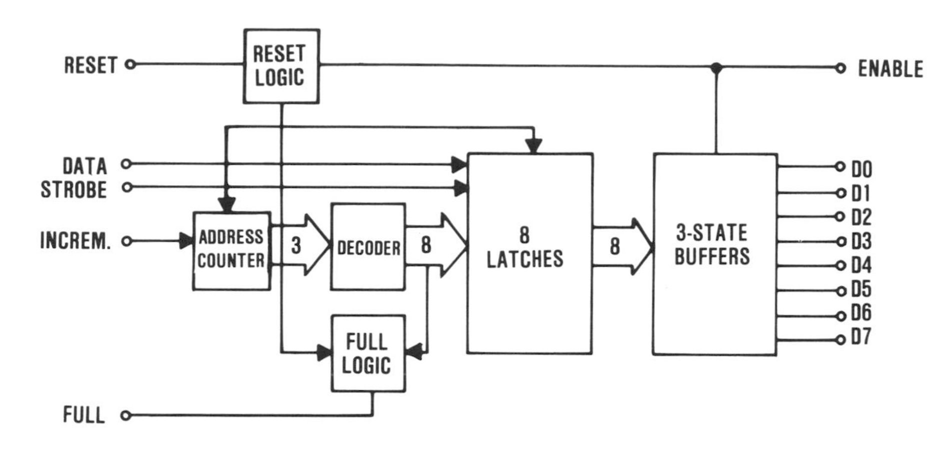

FIGURE D.9.3. Block Diagram—8-Bit Latches with Address Counter

e) *Propagation delay time, reset to output.* Time delay from appearance to reset pulse till all outputs are **0**. 175 ns is typical at 5 V for low-power CMOS.
f) *Pulse width.* Pulse width of the enable, strobe, reset and increment signal range from 200 to 300 ns at 5 V for low-power CMOS.

APPLICATIONS

This IC is particularly useful for converting a serial data stream into eight-bit bytes under control of a clock which would be applied to the increment input. Digital computer interface and bus circuits are the main areas of application.

COMMENTS

This IC is typical of the types that combine latching circuits with other digital control circuits on a single IC.

D.10.0 Line Drivers

DESCRIPTION

The line driver function is available in a variety of IC configurations, and the logic diagram of Figure D.10.0(a) illustrates the essential features of one popular configuration. Two sets of four inverting amplifiers are controlled by an enable signal which is passed through a noninverting amplifier. The operation of each individual amplifier is defined by the truth table of Figure D.10.0(b). When the enable input is **1**, the amplifier output will be a high impedance, regardless of what the input is. Only when the enable input is **0** can the output be the inverse of the input.

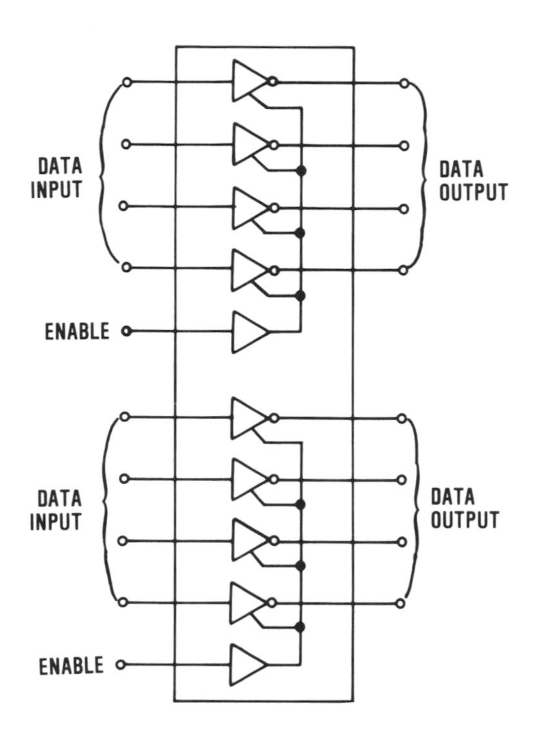

a) LOGIC DIAGRAM — TYPICAL LINE DRIVER IC

ENABLE	INPUT	OUTPUT
1	X	HIGH Z
0	0	1
0	1	0

b) TRUTH TABLE

FIGURE D.10.0 (a) Logic Diagram—Typical Line Driver IC
(b) Truth Table

As the name implies, this type of circuit is used to transmit digital data over a transmission line of some sort. In most instances this is simply one line on a multi-wire bus connecting digital devices to each other. 660 ohms is the nominal impedance used in most digital bus systems, and this should be matched by the output of the line driver when it is enabled. When the driver is not enabled, its output impedance will be much higher. This is important because otherwise the effect of a number of line drivers would load the entire bus system. This type of digital amplifier is frequently called "tri-state" because the output of each line driver can be either **1**, **0**, or high impedance.

KEY PARAMETERS

a) *Power dissipation.* The total power dissipated by the IC during operation. 325 mW is the typical power dissipation for a standard ECL and 20 mW is typical for low-power TTL.
b) *Quiescent current.* The minimum current through the IC when all amplifiers are in the high impedance state. 100 μA is typical for TTL and 20 μA is typical for low-power TTL ICs.
c) *Propagation delay time, data input to output.* 100 ns is typical for TTL ICs and 30 ns is typical for low-power TTL types.
d) *Propagation delay time, output disable.* The time required for the enable signal to change the output to a high impedance state. 6 to 10 ns is typical for standard TTL and 25 to 30 ns is typical for low-power TTL.
e) *Output current capability.* The amount of current that each line driver stage can provide. 30 mA is typical for standard TTL and 3 to 6 mA is typical for low-power TTL.

APPLICATIONS

Line drivers are used to amplify digital signals sufficiently so that they can be transmitted over a bus line, coax cable, or some other transmission medium. Most line driver ICs are part of a multi-wire bus system with transmission distances generally less than 50 ft.

COMMENTS

In some literature, line drivers are referred to as tri-state buffers. They differ from the buffer amplifiers described in Section 2.0 because of the three states possible in their output and because they are specifically intended to match the impedance of a digital data bus. Many line drivers are also capable of level shifting to provide compatibility with different IC families. Level shifters are described in Section I, Interface ICs.

D.10.1 Line Receivers

DESCRIPTION

This IC receives digital signals from a two-wire transmission line, although a multi-wire bus with a common ground can also be received by them. As illustrated in Figure D.10.1(a), each amplifier has two inputs, one inverting and one noninverting, and a single output. It is possible to connect one of the two inputs to the reference voltage and thereby set the logic level required for "above noise" reception. Line receivers are generally differential amplifiers with a hysteresis characteristic, similar to or including the Schmitt-Trigger principle.

A balanced input and balanced output line receiver is illustrated in Figure D.10.1(b). This type of differential input and differential output amplifier is particularly useful where the amplifier IC acts as a repeater. Note that both of the receiver ICs illustrated in Figure D.10.1 are continuously open and not controlled by any gating signal.

KEY PARAMETERS

a) *Power dissipation.* The power required by the entire IC. 100-120 mW is typical under no-load conditions for ECL ICs.
b) *Propagation delay time.* The time required from data input to output. 1.5 to 2.0 ns is typical for ECL ICs.

APPLICATIONS

Line receivers are used in the wire transmission portion of digital communications systems.

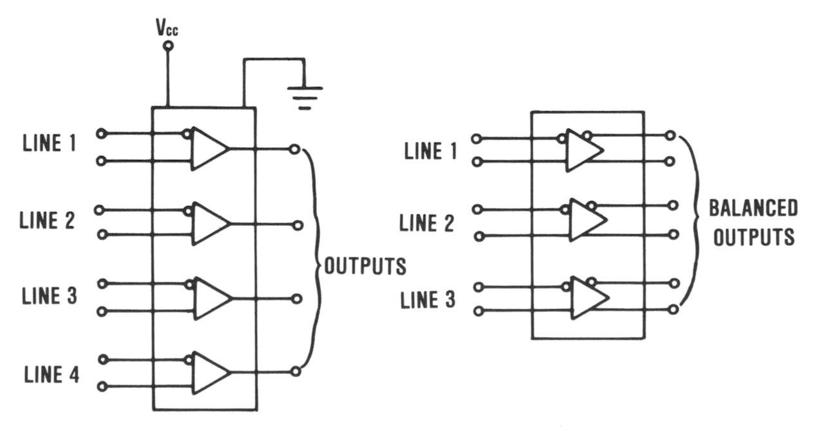

FIGURE D.10.1 (a) Single Output Line Receiver
(b) Balanced Output Line Receiver

COMMENTS

In many instances the analog type of differential amplifier described in Section A, Analog ICs, is used as line receiver. These ICs provide specific characteristics useful for communications circuits, such as balance, common-mode rejection, and transmission line impedance matching capabilities.

D.10.2 Line Transceivers

DESCRIPTION

In this IC the line driver and receiver functions are combined in a special two-stage amplifier network. The logic diagram of Figure D.10.2 shows four line transceivers, controlled by a single enable gate for the receive and another enable gate for the transmitter function. When the enable A signal is **1** and the enable B signal is **0**, the amplifier going from A1 to B1 will be operating while the other amplifier going from B1 to A1 will be in the high output impedance state. This means that data can travel from A1 to B1. When enable A is **0** and enable B is **1**, the operation is reversed and

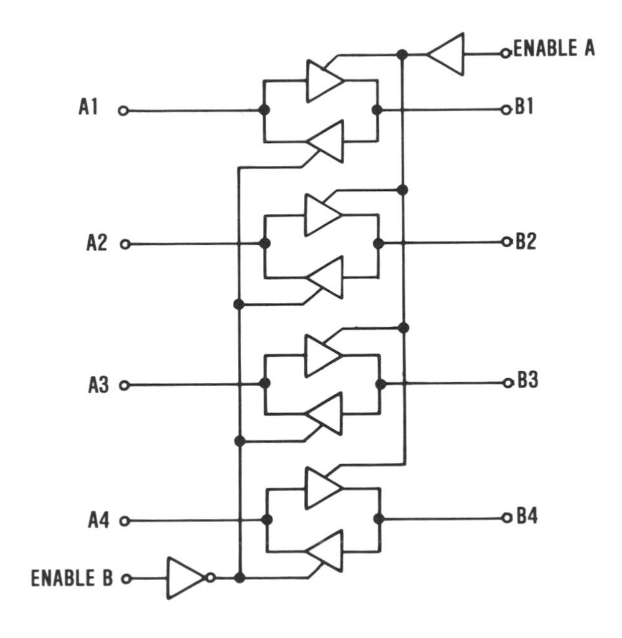

FIGURE D.10.2. Logic Diagram—Quad Line Transceiver

data can travel only from B1 to A1. If all the A terminals are connected to an exterior bus and all the B terminals to an interior bus, then the receipt on transmission of information can be completely controlled by changing the logic state of the enable A and enable B terminals.

While Figure D.10.2 shows only four transceivers, ICs are available with as many as eight transceivers, controlled by a single enable gate for transmission and another one for receiving.

KEY PARAMETERS

a) *Power dissipation.* 100 mW is typical for TTL and 20 mW is typical for low-power TTL ICs.
b) *Propagation delay time.* The time from data input to output. 12 ns is typical for low-power TTL ICs.
c) *Propagation delay time, enable to output.* The time from the enable pulse until an output appears. 20 ns is typical for low-power TTL ICs.

APPLICATIONS

Line transceivers are widely used in the input/output (I/O) sections of digital computers and their peripherals.

COMMENTS

Line transceivers are sometimes found as part of a complete I/O computer section. The key feature is that both directional amplifiers are of the tri-state kind.

D.11.0 Memory

This section presents the functional operation of the various memory ICs on the market. Detailed technical information pertaining to the IC's construction, such as NMOS, MOS, TTL, as well as exact timing characteristics, etc., is not presented here but is available in the manufacturer's data.

Memory ICs store binary bits which represent data. Instructions of a control program used in computer systems, data values, parameters, etc., are a few examples. This binary information is generally formatted as a group of binary bits, defined as a word.

The operation of the memory IC can be understood from Figure D.11.0. When the memory array is updated with new data words, a write signal is generated. This causes the data word on the data bus-in lines to be stored in

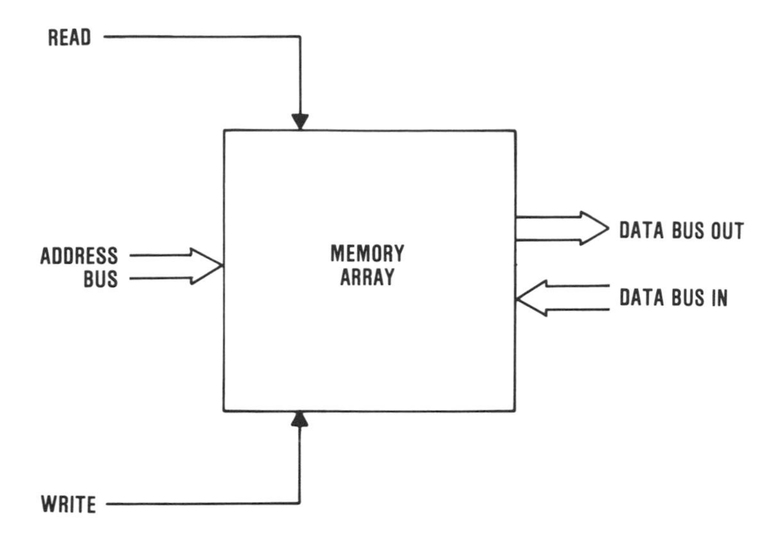

FIGURE D.11.0. Basic Memory Functions

the memory array. The location in the array is determined by the binary value on the address bus. If the data stored in the memory array is to be accessed, a read signal is activated. This transfers the stored data to the data-out lines. The detailed operation varies from memory IC to IC. Some ICs have only a read operation because the binary bit patterns are always stored in the IC. Other ICs do not have the input address bus.

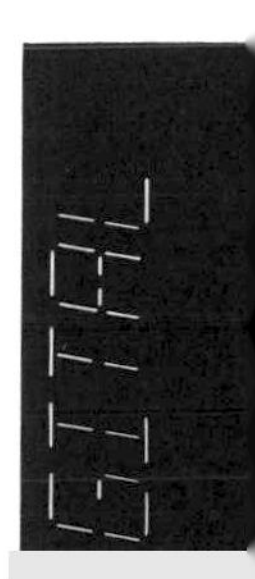

Data Access: There are two basic types of memory ICs by this classification:

a) *Random Access.* The memory array is constructed as a block of words with each word consisting of "n" number of bits. Access to any word in the memory array is random. Address decode logic (ADL) is included in the IC to perform this function and the inputs to the ADL are connected to the address bus. The binary value on the address bus is decoded for the ADL, causing the activation of the word select line. Each select line is connected to one of the words in the memory array. A read or write operaton will result in a data transfer in or out of the memory IC.

b) *Serial Access.* The memory array is constructed to store blocks of data bits in series. There is only one input and one output data line and no address decode logic is required. (The exception to this is the charge coupled device (CCD) where the address decode logic is used to access the blocks of data.) If the data block is to be read out of the IC, all of the data must first be moved until the desired block is positioned at the output of the IC. Once positioned, the data transfer takes place.

Non-Volatile vs. Volatile Memory ICs: Within each group of memory ICs a further clasification can be distinguished:

a) *Volatile Memory.* The binary data stored in the memory array is lost when power is removed from the IC. Upon reapplying power to the IC, the data must be rewritten into the memory array. As a general rule, higher speeds of operation can be used for the read/write cycles.

b) *Non-Volatile Memory.* The binary data stored in the memory array is retained when power is removed from the IC. Upon reapplying power to the IC, the data is preserved and can be instantaneously utilized. As a general rule the speed of the write cycle is slow and that of the read cycle is fast.

D.11.1 Dynamic Random-Access Memory (RAM)

DESCRIPTION

The dynamic RAM IC stores binary data words. The design of dynamic RAM allows the user to access any memory location by setting the address inputs to the selected binary value. Once selected, data is written into or read from the memory cells. The dynamic RAM is volatile. Removal of power results in the loss of data in the memory array. Figure D.11.1 illustrates the operation of the dynamic RAM.

"Refresh" Mode: Higher bit density, lower power dissipation, and faster read/write operations are achieved with capacitive type of memory cells, but these capacitors must be recharged or "refreshed" periodically to retain the data in the memory array. Internal or external clock generators with suitable control logic must be provided for reliable "refresh" operation.

Write Operation: The address bus to the dynamic RAM IC is decoded by the column and row address decode logic. The outputs select a specific memory location. When the chip select and write command is activated, the data value at the data inputs is written into the selected memory location.

Read Operation: The read operation is similar to the write operation except that a read rather than a write command is generated. This results in routing the contents of the selected memory location through the data I/O circuits to the data output bus.

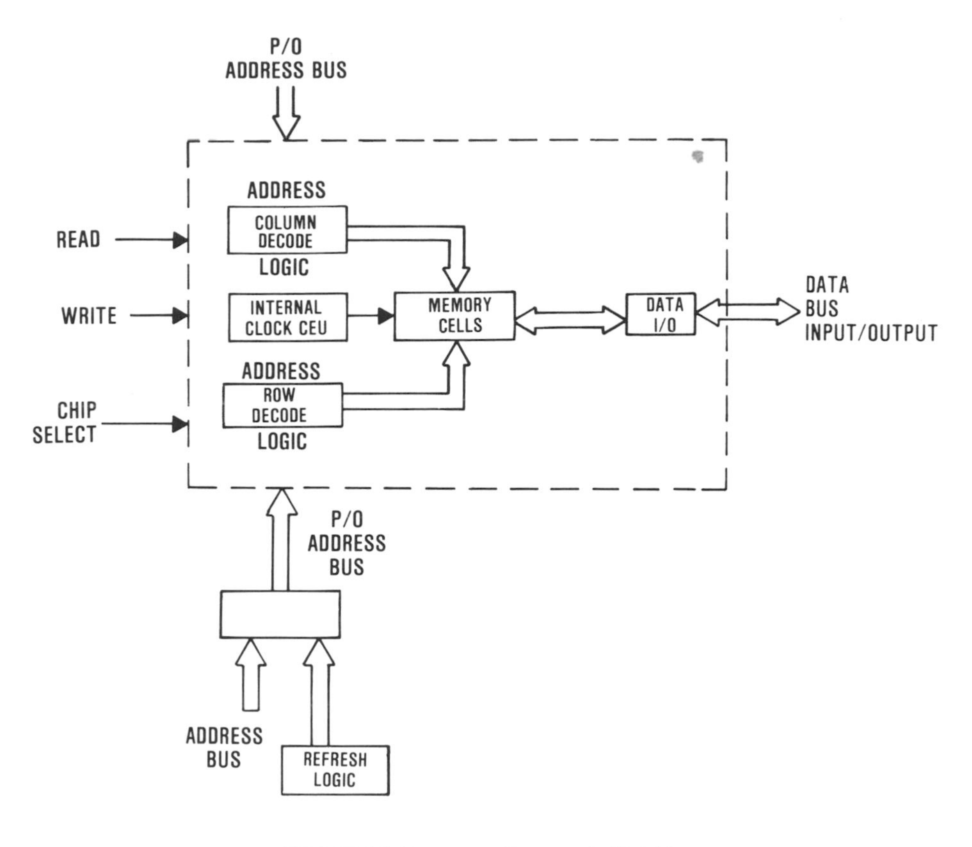

FIGURE D.11.1. Dynamic RAM

KEY PARAMETERS

a) *Read cycle time.* The time required from the leading edge of the read command until the contents of the selected memory are available at the output of the data I/O buffers. Typical values are 300 to 600 ns.
b) *Write cycle time.* The time required from the leading edge of the write command until the data bit(s) at the data I/O buffer are written into the selected memory location. Typical values are 300 to 600 ns.
c) *Access time.* The time between the application of the column and row address bit control signals and the start of the read or write signals. Typical values are 100 to 300 ns.
d) *Refresh time.* The maximum time allowed to activate all row address lines in order to retain the data stored in the memory cell(s). Typical maximum time is 2 ms.

e) *Power supply.* The voltage(s) required to operate the dynamic RAM as a memory element. Typical voltages for MOS dynamic RAMs are -5 V, +5 V, +12 V.

f) *Power dissipation.* This value, in milliwatts, is the product of the supply voltage and the current consumption by the IC.
   Operating: The power dissipation of the IC when the IC is being operated as a memory element. Typical values are 300 to 600 mW.
   Standby: The power dissipation of the IC when the IC is not active but the data stored in the memory cell(s) is retained. The ranges of values are 50 to 40 mW.

g) *Memory organization.* The structure of the memory IC depending on the number of words and the number of binary bits for each word. Typical sizes are 1024 words by 1, 2 and 4 bits to 16,384 words by 1 bit.

APPLICATIONS

The dynamic RAM IC is used where large amounts of changing binary data bits are to be stored, as in computers, data processors, word generators, etc.

COMMENTS

Because of their design, dynamic RAM memory systems require more logic elements than do static RAMs due to the "refresh" requirements for the memory cells. Care must be taken to insure that the refresh timing function does not interfere with the normal read/write function. The various input/output lines to the dynamic RAM may operate at different interface voltage levels.

D.11.2 Static Random-Access Memory (RAM)

DESCRIPTION

The static RAM IC allows random selection of any memory location in the memory array. Static RAMs are volatile and allow the storage of the data in the memory array without the need of clocks or "refresh" logic (see Dynamic RAM). Static RAM operation is described below and illustrated in Figure D.11.2.

Write Operation: Data to be stored in the memory array are connected to the data input/output lines of the RAM IC. The location at which data will be stored is determined by the encoded word at the address inputs to the RAM. This value is decoded by the address decoder logic. The result is the

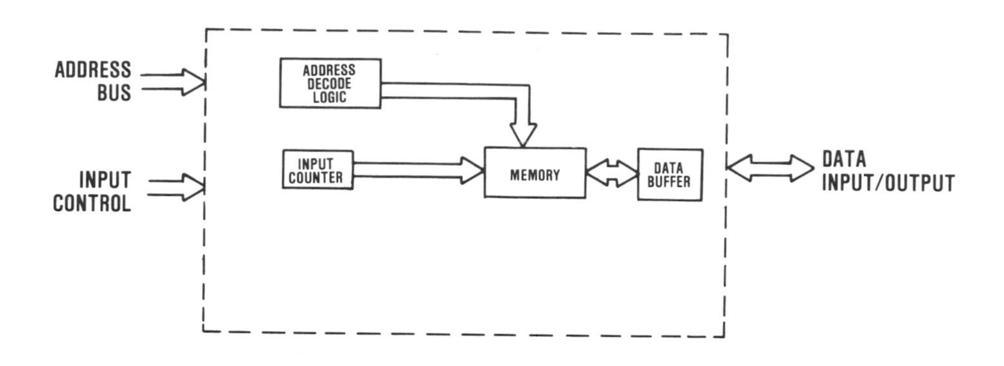

FIGURE D.11.2. Static RAM

activation of one of the memory select lines connected to the desired location in the memory array. The chip select line is activated, connecting the data input lines to the memory array, and the input control logic is enabled. The write command is then generated, causing the contents of the selected memory location to be updated to the new data word.

Read Operation: The selected memory location is determined by the decoding of the binary input by the address decode logic. This activiates the selected memory location. When the chip enable line is activated and the read command is generated, the contents of the selected memory location are connected to the data output lines.

KEY PARAMETERS

a) *Read cycle time.* The time required from the leading edge of the read command until the contents of the selected memory location are transferred to the output of the data I/O buffers. Typical values are 40 nano-seconds to 450 ns.
b) *Write cycle time.* The time required from the leading edge of the write command until the binary data at the data I/O buffer are written (stored) into the memory cells. Typical values range from 35 to 500 ns.
c) *Access time.* The time required from the beginning of the chip select and the address signals to the start of the read/write commands. Typical values are 10 to 100 ns.
d) *Power supply.* The voltages required to operate the static RAM as a memory element. Typically, a single voltage is used at +5 V.
e) *Power dissipation.* The product of the current consumed by the IC and the specified voltage. Due to the wide variety of memory sizes

available, this range of values is expressed on a per-bit basis. The power dissipation (PD) can vary from .04 mW/Bit to .15 mW/Bit. To determine the total power dissipation use the formula: total power dissipation = (number of words) × the (bits per word) × (PD per bit). Note: Those static RAMs designed with MOS technology have a standby mode which significantly reduces the power dissipation when the memory is not accessed.

f) *Memory organization.* The structure of the RAM IC is expressed as the number of words and the numbers of bits per word. Memory sizes range from 16 words by 4 bits to 16,384 words by 1 bit.

APPLICATIONS

Used to store changing binary data in equipment which requires high-speed operation such as cache memories, computer main frame memory, buffer storage, etc.

COMMENTS

The static RAM requires no clocks or "refresh" circuits.

D.11.3 First-In/First-Out Memory (FIFO)

DESCRIPTION

The FIFO consists of serial memory which stores binary data. The data can be written into the IC at one frequency, while simultaneously data is read from the memory at a different frequency. Control logic within the FIFO IC keeps track of the data once it is stored in the serial memory. This includes processing the data and setting status indicators when the FIFO is empty or full. Most FIFO memories are volatile.

Write Operation: Data on the inputs of the FIFO can be written into the serial memory when the input ready line is activated. Data is transferred into the serial memory when the input ready signal is enabled and the shift input pulse is generated. The internal control logic will shift the data towards the output of the serial memory. Shifting will continue until the data word is in the memory location adjacent to the previously stored data word. The data, once written into the FIFO, cannot be read until all previously stored data words are shifted out.

Read Operation: If data is stored in the FIFO, the output ready line will be activated. If the FIFO was empty, the output ready signal line will be

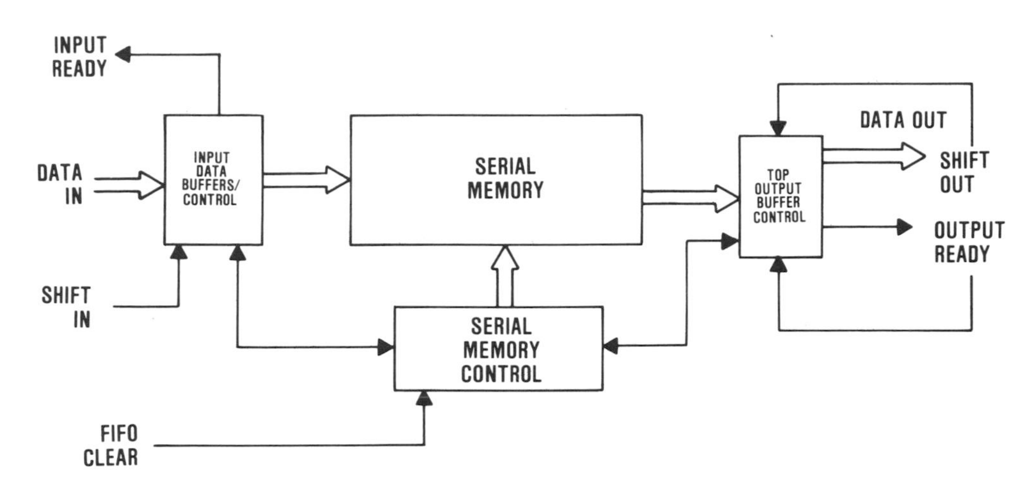

FIGURE D.11.3. FIFO

deactivated. Generation of a shift output pulse when the output ready is enabled will result in the transfer of the data from the FIFO through the data out buffer. It is a destructive read since the data is shifted by one location. The activation of the clear input to the FIFO results in erasure of all data in the FIFO.

KEY PARAMETERS

a) *Shift-in frequency.* The speed at which the data entering the FIFO can be transferred into the main memory for automatic processing. Typical ranges are DC to 15 MHz.b)
b) *Input ready.* The input ready signal indicates that data can be transferred to the serial memory. It is deactivated 50 ns after a shift input pulse is generated. When the input ready signal is disable, the serial memory is full.
c) *Shift-out pulse frequency.* The frequency at which data can be transferred from the main memory to the output buffers. DC to 15 MHz.
d) *Output ready.* The output ready signal is activated whenever data is available in the serial memory. When the serial memory is empty the output ready signal is disabled.
e) *Latency time.* The time it takes to shift the binary data through the serial memory to the output buffer. This can vary from 2 to 6 ns.
f) *Master FIFO clear.* The FIFO clear, when activated, clears all the FIFO memory cells out. Typical time required is 30 ns to 2 ms.

g) *FIFO organization.* The organization of the FIFO is based on the number of bits per word. Typical structures are 64 words of 9 bits each. Note: The FIFOs are designed for expansion to any number of words and any number of bits.

APPLICATIONS

FIFOs are used between two systems which operate at two different frequencies. For example, the input buffers are connected to a disc controller operating at 4 MHz while the output buffers are connected to main memory operating at 2 MHz.

COMMENTS

Expansion of the number of words in a FIFO system is achieved by connecting the data output buffers of one FIFO IC to the data input buffers of another FIFO. The output ready and shift-out of the first FIFO IC is connected to the shift-in and input ready lines of the next FIFO.

Expansion of the number of bits per word is achieved by connecting all the shift lines and input ready lines together. This is also done for the shift output ready lines. Access to the binary data stored in the serial memory is not available to the user until it ripples through the serial memory and all previous data is shifted out.

D.11.4 Serial Shift Registers

DESCRIPTION

In this type of memory the data is stored in serial form. Access to selected binary bits is accomplished by generating the appropriate number of shift pulses until the bit is at the output line. Shift register memories are volatile. Refer to D.15 for details on shift registers. The typical block diagram for shift registers (SR) is illustrated in Figure D.11.4.

Write Operation: The binary word to be stored in the shift register is connected to its data inputs. The recirculate line must be deactivated in order to allow new data to be stored in the SR. After the data word is stabilized, the shift pulse results in the transfer of the data word into the memory cells of the SR. This cycle is repeated for each data word. At the output of the SR the data is transferred to its data out lines.

Read Operation: Since the SR is used as a serial memory IC, data is shifted through the register until it reaches the last set of memory cells where it is connected to the data out lines. In order to prevent losing the data, the recirculate line connects the output of the SR last set of cells back to its inputs. Shift pulses cause the output data to be shifted to the input memory cells. New data cannot be transferred into the SR during the recirculate mode.

External logic keeps track of the data block stored in the SR. Bit tracking counters are used as indicators to where the start of the data block is inside the register. In addition, a word length counter is used, indicating the length of the data block.

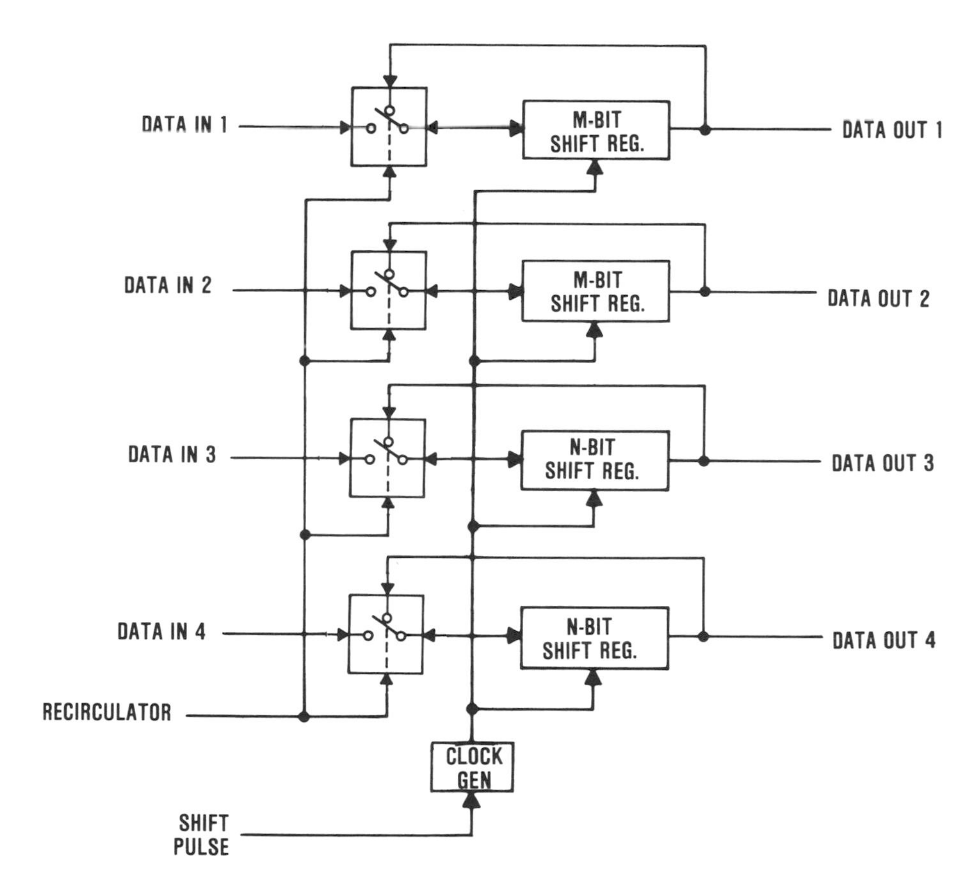

FIGURE D.11.4. Serial Shift Register Memory

KEY PARAMETERS

a) *Shift pulse frequency*. The range of pulse repetition frequencies at which the shift pulses shift the data through the register. The frequency range is up to 4 MHz for static SRs.

b) *Set-up time.* The time required for the data to be at the input to the SR prior to the generation of shift pulse. Minimum setup time is 60 ns.
c) *Data hold.* The time the data must hold at the input to the shift registers while the shift pulse is generated. Minimum time is 60 ns.
d) *Shift register (memory) organization.* As in RAMs, the SR IC is organized by the number of words and the number of bits per words. Typical organizational structures range from 64 words by 8 bits to 1024 words by 1 bit.
e) *Shift register technology.* Two types of serial shift registers are available—static and dynamic.

Dynamic SR: This type of shift register requires a shift pulse to be generated at least once every 2 ms ("refresh" rate) to prevent the data from being lost. The design of the dynamic shift register is more complex than the static registers. Due to the "refresh" logic required, the trade-off is higher bit density, lower power dissipation, and faster operation.

Static SR: There is no requirement to refresh the static shift register. This simplifies the design of the logic in terms of clock generators and refresh logic.

APPLICATIONS

Shift registers are widely used in video terminals, communication buffers, and other systems where random access to specified data words is not required.

D.11.5 Content Addressable Memory (CAM)

DESCRIPTION

The CAM IC is a memory device which compares the binary input word with all the words in the CAM. If any of the memory locations have the same binary bit pattern as the input word, an output signal is activated indicating that a comparison has occurred. In addition, the location which contains the same bit pattern is indicated. Figure D.11.5 illustrates the operation.

Write Operation: Information must first be written into the CAM IC prior to using the associated mode of operation. This information could represent status flags or addresses of where blocks of data are stored in the system. The writing of this information is accomplished by activating the appropriate address sensor/select line. When the write command appears, the binary bit pattern on the data bus is stored in the activated memory location.

Read Operation: If the user desires to read the contents of the various memory locations, he can activate the appropriate address line and then activate the read command. The contents of the selected memory location are read onto the data bus.

Associate Mode of Operation: In computer control systems with various memories there is a need to rapidly ascertain where the blocks of data are stored. In order to determine this at computer speeds, the associate mode is used by simultaneously comparing the binary bit pattern on the data bus with the contents of all the memory locations in the CAM. Those memory locations which have the same bit pattern will activate their associated address sensor/select line and activate the valid data command.

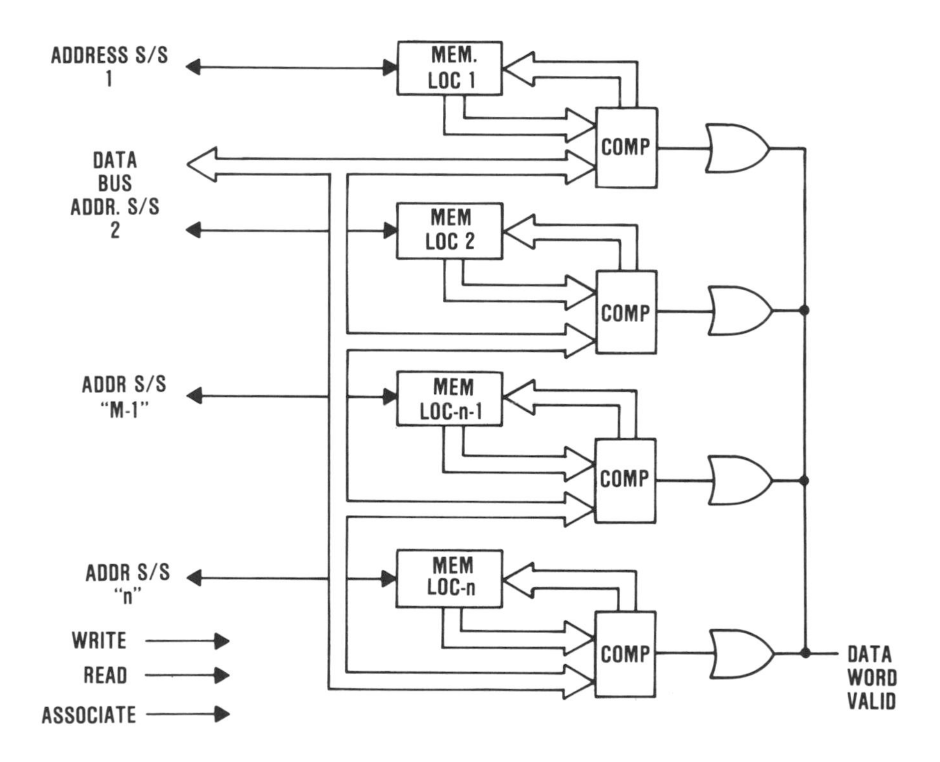

FIGURE D.11.5. Content Addressable Memory

The positional information pertaining to the activated address sensor/select line is used in the set for the block of data.

KEY PARAMETERS

a) *Read operation.* The time it takes to read data from the selected memory location. The minimum time is 15 ns.
b) *Write operation.* The time it takes to write the bit pattern on the data bus to the selected memory location. The minimum time is 15 ns.
c) *Associate mode.* The time it takes to compare the contents of the memory location to the activation of the address sensor/select, assuming a valid compare and the valid data word line. The minimum time is 15 ns.
d) *Memory organization.* Represented by the number of words and the number of bits per word. They vary from 8 words by 2 bits to 16 words by 8 bits.

APPLICATIONS

In large memory systems the CAM IC is used to identify the location of blocks of data/program. The address sensor/select activated (based on a valid compare) indicates the location of the block of data/program. This technique allows slow-speed memory systems to have high-speed characteristics.

COMMENTS

The input/output lines are not necessarily TTL compatible. The address select/sensor lines and valid data word as a rule are tri-state to allow word/bit expansion. The address select/sensor lines are usually linear coded rather than encoded. Some CAM ICs have other modes of operation in addition to the read/write/associate modes.

D.11.6 Read-Only Memory (ROM)

DESCRIPTION

ROM ICs are used to store binary information in a permanent pattern which is part of the manufacturing process. The binary bit pattern is retained in the ROM when power is removed from the IC, making it a non-volatile memory.

The general ROM operation is described below and illustrated in Figure D.11.6.

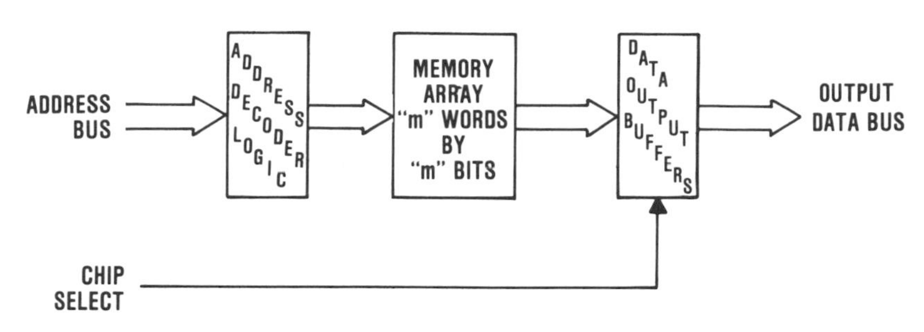

FIGURE D.11.6. Read-Only Memory (ROM)

Write Operation: The user of the ROM defines the information to be programmed on a set of computer cards. This set of cards is sent to the manufacturer where the information is used to create a "mask" for the ROM. The "mask" is bonded for assembly for each ROM purchased.

Read Operation: The binary address lines are decoded by the address decode logic to select a specific memory location (word). The data stored in the memory location is connected to the data output buffers. When the chip select is activated, the data word is connected to the output data bus.

KEY PARAMETERS

a) *Address set-up time.* The time the system must wait before the data is valid on the output of the ROM IC. This assumes the chip select is activated. Setup time ranges from 45 to 450 ns.
b) Chip select setup time. The time the system must wait before the data is valid on the outputs of the ROM IC after the chip select is activated. This assumes the address bus is at the selected value. Values vary from 45 to 200 ns.
c) *Chip disable.* The time the system must wait for the ROM IC to deactivate its data buffers after the chip select is disabled. The range is 30 to 200 ns.
d) *Memory organization.* The structure of the ROM IC in terms of the number of words and the number of bits per word. The size of ROMs varies from 256 words by 4 bits to 4096 words by 8 bits.

APPLICATIONS

ROMs are used for storing control programs for computer control systems, calculators, look-up tables, character generation, code converters, etc.

COMMENTS

ROMs should be considered as alternatives to PROMs when used in large quantities. ROMs are generally less expensive than PROMs. ROMs have a high masking charge, while PROMs are user programmable. The inputs and outputs of the ROM IC are usually TTL compatible.

D.11.7 Programmable Read-Only Memory (PROM)

DESCRIPTION

In the programmable read-only memory (PROM), binary bit patterns are permanently entered into the memory device. The user can program the PROM IC to his particular bit pattern, while ROMs must be programmed by the manufacturer. Once programmed, the PROM is used only to read selected memory locations for the binary bit patterns contained in each word. Removing power from the IC does not destroy the bit patterns programming procedure for entering each memory word varies from IC to the IC, the original binary informtion is again available to the user. This feature makes it non-volatile. As illustrated in Figure D.11.7 the PROM IC operates as follows:

Write Operation: The write operation can only be done once and the programming procedure for entering each memory word varies from IC to IC. In general, there are fusable links for each bit in the memory array. The address bus is set to the initial starting value of "0" and the address decode logic selects the first word in the memory array. Next, the chip select line is activated and each data output line is connected to a high-voltage, high-current device (HVHCD). The activation of each HVHCD is a function of whether or not 1's or 0's will be programmed into a particular bit in a memory location. As a general rule, the activation of the HVHCD will open the fusable link. Once the operation is completed, the next memory location is selected by changing the value on the address bus. The cycle is then repeated.

Read Operation: The binary value on the address bus to the PROM is decoded by the address decode logic. Each output of the decode logic is connected to a different memory location. Based on the way the given memory location was programmed, the selected bit pattern is connected to the data output buffers. If the chip select line to the PROM IC is activated, the binary value for the selected word is connected to the data out lines of the IC. If the chip select is not activated, the data-out lines are switched open, allowing other memory ICs to use the same data out lines.

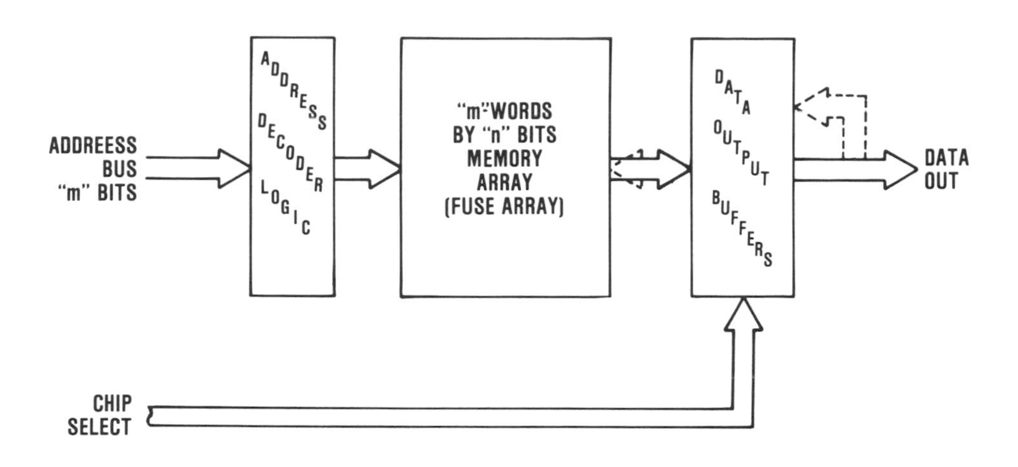

FIGURE D.11.7. Programmable Read-Only Memory (PROM)

KEY PARAMETERS

a) *Read access time.* The time from the beginning of the chip select and address lines to the time the data is to be read on the data output buffers. Minimum time requirement begins at 30 ns. If the address bus can be stabilized prior to activating the chip select line, the time to wait before the data is available at the outputs can be shortened.

b) *Chip disable time.* This parameter determines when other memory ICs can take control of the data bus. The system must wait a finite period of time after the chip select signal is deactivated. The minimum time is 20 ns.

c) *Power dissipation.* The amount of power consumed by the PROM IC depends on the technology used in the structure of the chip, the speed at which the IC can operate, and whether or not special circuits are used to remove power to the IC when the memory is not chip selected. The power dissipation can range from 30 microwatts to 150 microwatts per bit of storage. To determine the power dissipation (PD) of the PROM IC use the following formula: PD = (# of Bits for the PROM IC) times (PD for each bit.)

APPLICATIONS

PROMs are used in place of ROMs, wherever the number of devices is limited. The cost of mask-encoded ROMs is only warranted when thousands of identical ROMs are used. Most PROMs are used in mini- and microcomputers.

D.11.8 Electrically Alterable Read-Only Memory (EAROM)

DESCRIPTION

The EAROM IC illustrated in Figure D.11.8 is designed to write, store and read binary information. EAROMs are non-volatile. Power can be removed and reapplied without any loss of the data stored in the memory array. Unlike the ROM and PROM, repeated write operations are possible. Data can be written, erased, and read electrically. The EAROM allows in-system modification of the binary bit patterns stored in the memory array, but the number of read and write operations is limited. Each time the read/write/erase operation occurs, the non-volatile storage life is reduced. The erase operation can erase one or all memory locations. The write/erase operation is slow, while the read operation can be done at high speed. Access to the data stored in the memory array is random. The binary value on the address bus selects the memory location for reading/writing/erasing. Multiple power supplies are required.

Write Operation: The activation of the selected output of the address decode logic is determined by the binary value on the address bus. When the chip select line is enabled, the data word is routed through the data I/O logic to the specified location in the memory array. When the write pulse is generated for the specified period of time, the data word is stored in the memory location. Damage to the EAROM an occur if the signal is on too long.

Erase Operation: This operation deletes the binary bit pattern in a specified memory location or in the whole memory array. The erase operation must precede the write operation. When the erase word input is activated for the specified period of time, the data stored in the selected location in the memory array is deleted. If the erase memory array input is activated for the specified period of time, the contents of all memory locations are deleted. Exceeding the specified period of time for the erase operation can result in damage to the EAROM.

Read Operation: The read feature is similar in operation to other memory ICs. Which data word is read out of memory is determined by the

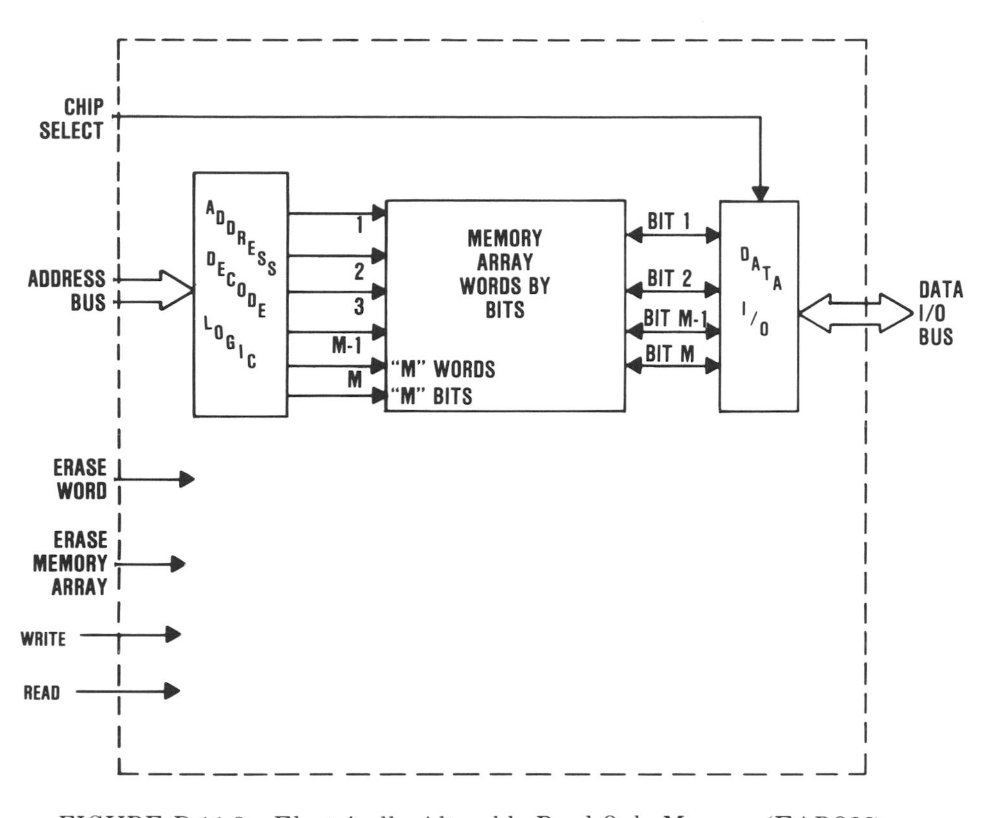

FIGURE D.11.8. Electrically Alterable Read-Only Memory (EAROM)

binary value on the address bus. If the chip select is enabled, the data stored in the specific word is connected to the data I/O bus.

KEY PARAMETERS

a) *Erase time.* The time requried to delete the binary bit pattern in the specified memory location or in the whole memory array. This time is approximately 14 ms. Exceeding the manufacturer's time limit can result in damage to the EAROM.
b) *Write time.* The time required to write a data word into a specified memory location. The approximate value is 2 ms. Exceeding the manufacturer's time limit can result in damage to the EAROM.
c) *Read access time.* The time required until data is available at the data I/O bus, once the address bus and chip select lines are

activated. This time varies among different manufacturers. The range is 0.5 to 12 μs.

d) *Life of EAROM.* The guaranteed life of the memory array is determined by the number of erase/write and read operations.
Typical values are: erase/write operations: 100 operations per word
read operation: several hundred billions operation

e) *Power supplies.* Most EAROMs require multiple power supplies. The ranges of power-supply voltages are ±5V, ±12V, ±26V, etc.

APPLICATIONS

Whenever non-volatility is required. The storage of data, parameters or tables, where the values must be maintained during power failures at turn-off, is a suitable application of the EAROM. Due to the limitation of the number of erase/write/read operations, it is not suitable for appplications where the read/write cycles are occurring frequently.

COMMENTS

The use of the EAROM in microprocessors is limited, except as a non-volatile storage device where its contents are transferred to a RAM. The system must be designed to execute the control program from the RAM rather than the EAROM. The timing requirements for the erase and write operations are critical. Generally, a sequency control timing generation is used to supply the erase/write/read signals to the EAROM. Activation of the sequence would be controlled by the microprocessor or computer system.

D.11.9 Electrically Programmable Read-Only Memory (EPROM)

DESCRIPTION

The EPROM is a non-volatile memory IC which is used to store binary information. Power can be removed without loss of data. Upon reapplying the power, the original binary data is still retained. Special circuits must be used to program the binary information into the memory array. This write function is done outside the system. Once programmed, the EPROM is placed back into the system and activated. One feature of the EPROM is that it can be reprogrammed to a new binary bit pattern. This is accomplished by exposing the EPROM to an ultra-violet (UV) source in order to erase the old binary information. A UV transparent lid on the chip allows this erasure to occur.

Figure D.11.9 defines the operation.

Erasure: In order to erase the EPROM, the UV transparent lid must be exposed to a UV source for approximately 10 to 20 minutes.

Write Operation: First the chip select line is deactivated in order to switch the data outputs to inputs. The address inputs are set to a starting value and the desired data word is connected to the data inputs of the EPROM. The program pin is then enabled. During the time of the program pulse, the data is written into the selected location of the memory array. This cycle is repeated for each location.

Read Operation: The read operation is similar to that used for RAMs, ROMs, PROMs, etc. The memory location, selected by the memory decode logic, is determined by the binary value on the address inputs. The data programmed in the memory array is connected to the data output buffers. If the chip select is activated, the data word is connected to the data bus. The data outputs are in the off-state when the chip select is disabled.

KEY PARAMETERS

a) *Program pulse.* The program pulse writes the data word into the selected memory location. It must be held active between 100 ms to 3 ms.
b) *Address set-up time.* The period of time the system must wait for the

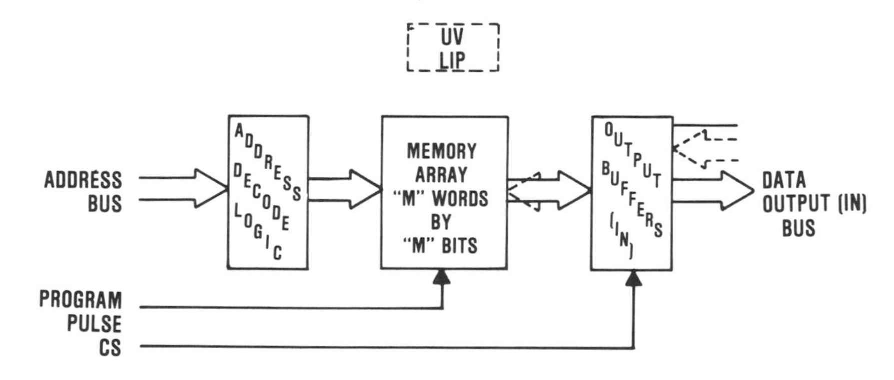

FIGURE D.11.9. Electrically Programmable Read-Only Memory (EPROM)

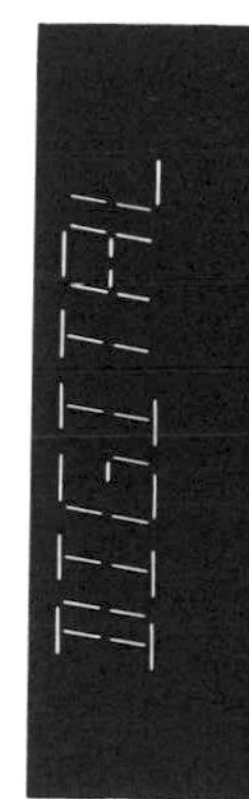

data to be valid on the data output lines. This assumes the chip select is activated. The range is 300 to 700 ns.

c) *Chip select time.* The period of time the system must wait for the data to be valid at the data output lines. This assumes the address lines are at the desired value. Typical delays are 100 to 175 ns.

d) *Output disable.* The time the system must wait for the data output lines of the EPROM to be deactivated once the chip select is disabled. This time ranges from 120 to 400 ns.

APPLICATIONS

The EPROM is used when the binary bit patterns are changed periodically. Look-up tables which are not the same from system to system are an example. Each table type must be programmed into a different EPROM. The EPROM is also used in microprocessor applications where the control program (set of instructions) is stored.

COMMENTS.

EPROMs are temporarily used for the debugging of the control programs in the microprocessors. Once the program is debugged, the EPROMs are replaced with the less expensive ROMs or PROMs.

D.11.10 Magnetic Bubble Memory (MBM)

DESCRIPTION

The IC shown in Figure D.11.10 is an extremely high-bit density and moderate speed mass-memory device. It has the capability of storing up to 2,000,000 binary bits of data. It is non-volatile. Power can be removed and reapplied without any loss of the stored data. Data stored in the MBM is accessed serially. If a specific block is to be moved out, it must first be moved to the bubble detectors before it can be read out. The data blocks are stored in the minor loops (serial memory block). The number of minor loops varies with different MBM Ics, from 100 to 1000 bits. At the present state of the technology all the minor loops in the IC cannot be guaranteed. A list of the defective minor loops is supplied to the user when the MBM IC is purchased. This information is used in the design of the memory system. Functionally, the operation of the MBM is similar to a recirculating shift register. The data blocks transferred in and out of the IC are stored in all of the minor loops. Each binary bit is stored as a bubble on the magnetically reactive material. A group of bubbles is stored in each minor loop. The presence of a bubble is a binary **1** and its absence is a binary **0.**

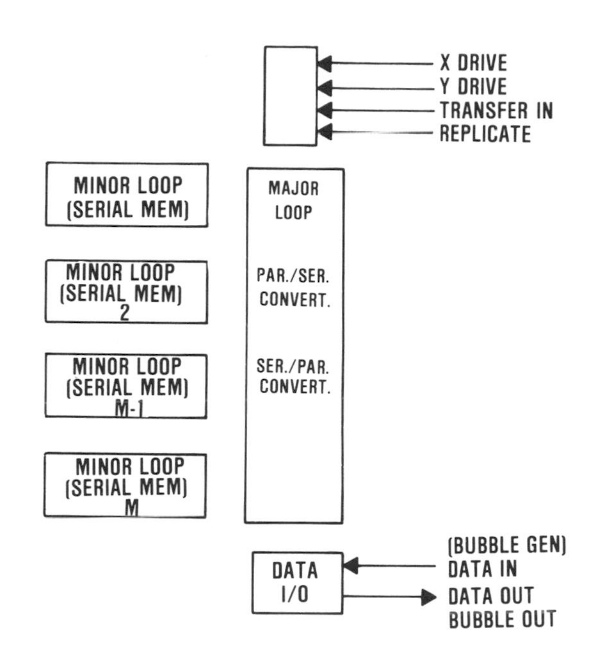

FIGURE D.11.10. Magnetic Bubble Memory (MBM)

Write Operation. The binary data block written into the MBM must be first shifted in the bubble generator input port. Each time the generator input port is pulsed, a bubble is developed. This stream of bubbles (or no bubbles) is moved along the major loop by the X and Y drive signals. The data block, consisting of the bubbles (1s, 0s), is shifted along the major loop. Once the data is aligned with all the minor loops, the transfer-in signal is generated. This signal causes the bubbles (data block) on the major loop to be transferred to the specific bit position for each minor loop.

Read Operation. The minor loops are rotated, using the X and Y drive signals. Bit positions in the minor loops are aligned with the parallel-to-serial converter (major loop). Once the alignment has occurred, the transfer-out signal is generated. The stream of bubbles is then shifted towards the data-out line (bubble detector). In order to read the data, a replicate pulse must be generated for each bubble position, and, as result, the bubble is duplicated. The major loop eventually transfers the bubble back to the same bit position in the minor loop. The other bubble is shifted to the data output line and converted into a binary level.

Control for the movement, writing and reading of the bubbles in and out of the IC is done by external hardware.

KEY PARAMETERS

a) *Latency time.* The time required by the MBM IC controller to transfer and read the selected data block. This time depends on the speed with which the control signals drive the IC and the number of bit positions in the minor loops. The access time is, on the average, 4 ms.
b) *Frequency of operation.* The clock frequency at which the bubbles can move from one bit position to another. This ranges from 100 to 150 kHz.
c) *Memory organization.* The number of bits in each loop and the number of minor loops determine the size of the memory. For example: An MBM IC is designed with 260 minor loops where each minor loop stores 1024 bits of information.

APPLICATIONS

The magnetic bubble memory IC is used wherever non-volatile, mass memory is required. Applications where disc or magnetic tape systems are used can use MBM ICs.

COMMENTS

Keeping track of where the data blocks are stored inside the MBM IC is accomplished by external logic. Timing and signal characteristics for the input/output lines are complex. Most manufacturers of MBM ICs have designed complete mass memory systems with provisions for interfacing the device with a computer. Synchronizing the data blocks must be accomplished whenever power is removed or applied to the device.

D.11.11 Charge Coupled Device Memory (CCDM)

DESCRIPTION

The CCDM IC is an extremely high-bit density and high-speed memory IC. Functionally it is similar to the operation of a recirculating Shift Register. It consists of multiple serial memory. Address decode logic is included in order to select which serial memory block of data will be processed. Parallel-to-serial converters are included in order to connect the selected serial memory to the data output interface lines. In addition, serial-

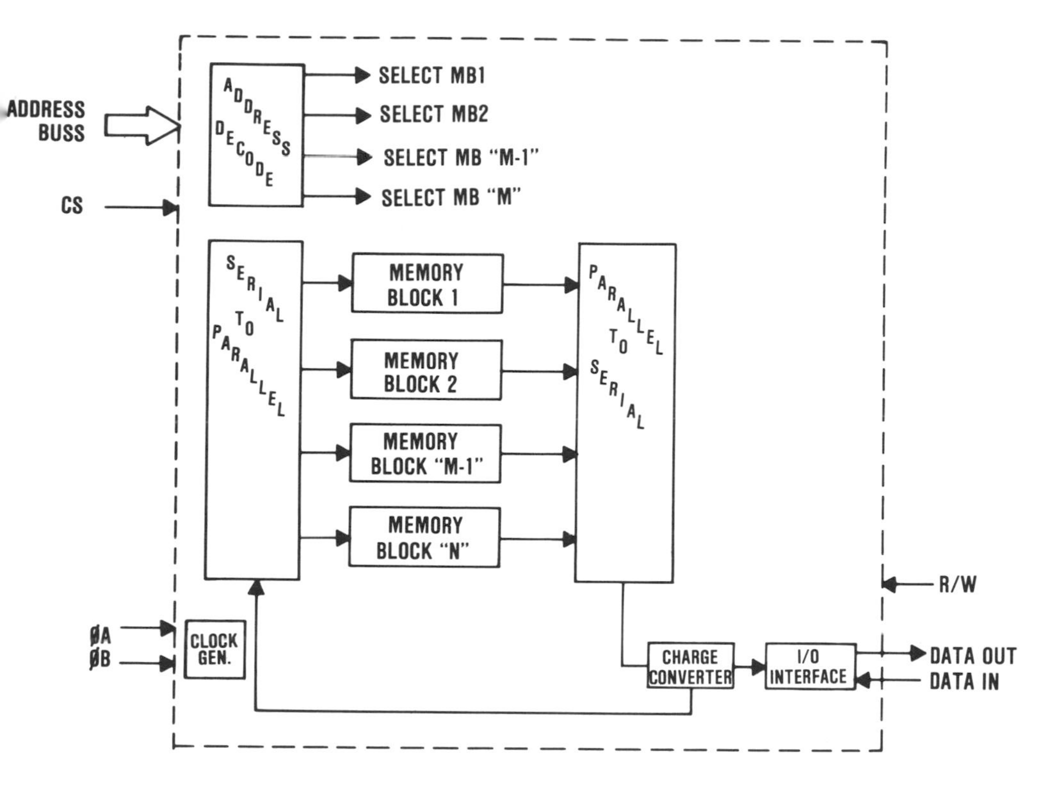

FIGURE D.11.11. Charge Coupled Device Memory (CCDM)

to-parallel converters are used to process the data, written into the CCD IC to the selected serial memory. Figure D.11.11 illustrates the operation of the CCD IC.

Read Operation: When the chip select line (cs) is activated and the read command is enabled, the binary value on the address bus is decoded by the address decode logic, activating one of the select memory lines. The select memory line steers the serial-to-parallel and parallel-to-serial converters to the selected serial memory array. This allows the selected serial memory to recirculate the stored data through the data input/output interface logic, thus allowing the read/write operation to occur. Pulsing clock $\Phi 1$, $\Phi 2$ lines shift the data from one memory cell to the next. One shift operation occurs for each $\Phi 1$, $\Phi 2$ cycle. As the data block is shifted through the charge converter, it is converted from a "charge" representing the binary bit, to voltage levels **1** or **0**. From the charge converter the data is recirculated back through the serial-to-parallel converter to its original serial memory and through the data output lines. External bit tracking counters are used to keep track of the start and length of the data block stored in the serial memory. When the data block is at the data output line it is read from the CCD IC.

Write Operation: If the data block is to be modified, a "read modify write" operation is required. A read operation is performed (see read operation discussions). When the selected data bit is at the data output line, the read command is disabled and the write command activated. This connects the data in line to the charge converter back to the selected serial memory. The modified data bit block is converted to a "charge" and steered into the original serial memory.

The CCD IC must be periodically "refreshed." This operation retains the data stored in the serial memory block: otherwise the data is lost. This is accomplished by generating Φ1, Φ2 for a finite period of time.

KEY PARAMETERS

a) *Clock frequency.* The speed at which the binary data stored in the memory block can be shifted through the memory cells. The range of ΦA and ΦB is from 1 MHz up to 5 MHz in operation.
b) *Latency time (LT).* The maximum time it takes to read or write a selected binary bit from the selected memory block. The formula is LT = (length of memory block in bits) × (time interval of the clock cycle). For example: Assume a memory block of 9182 bits long and the time period of the clocks is 0.2 μs. The latency time = (8192) ($.2 \times 10^{-6}$) = 1.683 μs. That is, if the user wants a particular data bit located at the very end of the data block and the first binary data bit of the data block is just beginning to be shifted out of the data out line, he would have to wait 1.638 μs before the desired data bit is at the data line. (Note: the tracking bit counter continuously indicates which bit of the memory block is at the data out line.)
c) *Read or write cycle.* The minimum time required to shift a binary bit through the data out line back to the data in line. The current technology is presently at a minimum of 200 ns.
d) *Read modify write.* The time required for selected data bits to be modified as they are recirculating through the memory block to the I/O interface ports. Present technology is 300 ns.
e) *"Refresh" time.* The period at which the memory block must be refreshed before data is lost. This varies from 0.5 μs to 2 μs. (Note: generally the data blocks are continuously recirculating.)
f) *Memory organization.* The number of memory cells per memory block. The number of blocks ranges from 16 blocks to 256 blocks. The number of bits per block varies from 256 bits to 4096 bits.

COMMENTS

All I/O lines to the CCD are not TTL compatible. A "clear" operation must be indicated whenever power is first applied or the data bits stored in the memory blocks may be lost due to deactivation of the clock. A clear operation consists of generating a minimum number of clock pulses to the CCD IC. Multiple power supplies are required.

D.12.0 Microcomputer and Microprocessor ICs

DESCRIPTION

The block diagram of Figure D.12.0.1 shows the key elements of any digital computer. Recent advances in large-scale integrated (LSI) ICs have made it possible to provide the central processing unit (CPU), the RAM and ROM memory and the input/output (I/O) port functions on a single IC. The capabilities of such a microcomputer (MC) are not as extensive as those of a computer consisting of a large number of ICs, but the essential functions are the same. When only the CPU is contained on a single IC, more capabilities become available and this microprocessor (MP) can be combined with memory and other digital ICs to approach the capabilities of a minicomputer. A microprocessor slice contains only some of the functions of a CPU, but performs those functions faster.

Like all digital computers, MCs and MPs perform certain binary functions according to the instructions contained in the control program. This program (software) is stored in some memory section, in binary form, and determines just what the CPU, or MP, does with the data, which is stored in another memory section. All of the functions of the MC and the MP are designed to move, deocde, and execute the instructions contained in the control program and manipulate the data accordingly. Both the program and the data must be entered, at some time, from an external device (peripheral).

MCs and MPs are often classified according to the number of parallel binary bits they can handle. Standard sizes are the 4-bit, 8-bit and 16-bit systems, and compatibility with bus (interconnection) systems having the respective number of parallel lines is essential. The greater the number of bits, the more powerful, and expensive, will be the entire computer system. Clearly a device able to handle four bits as address can address only 16 locations, while an 8-bit device can address 256 locations, at the same time, with the same clock frequency. Detailed descriptions concerning instructions, memory addressing, and other features will be found in the manufacturer's literature.

Hardware Configuration: ICs are available that perform all or part of the different computer functions. As illustrated in Figure D.12.0.2, we can identify three basic categories:

1. *Microcomputer ICs.* These ICs contain all the functions of a computer system, including the EPROM, ROM and RAM for storing the control program and data words to be processed. The controller and arithmetic logic unit (ALU) are the brains, while the input and output (I/O) sections are used to transfer information over the two data buses. The control program is limited by the size of the EPROM or ROM inside the IC, but some microcomputer ICs provide external memory expansion features, including an address bus.

An important characteristic of any microcomputer is the size of the data word that can be processed. There are 4, 8, and 16-bit microcomputer ICs. The number of bits determines the maximum value of the word that can be processed at any given time. For example, an "add" instruction executed by a 4-bit computer can add values up to 16, but with an 8-bit microcomputer values up to 256 can be added. (See Figure D.02(a).)

2. *Microprocessor ICs.* This IC is used where larger control programs and greater flexibility are needed. As illustrated in Figure D.12.0.2(b) the microprocessor contains no memory, but performs the key functions of the arithmetic logic unit (ALU) and the controller. It also contains a set of

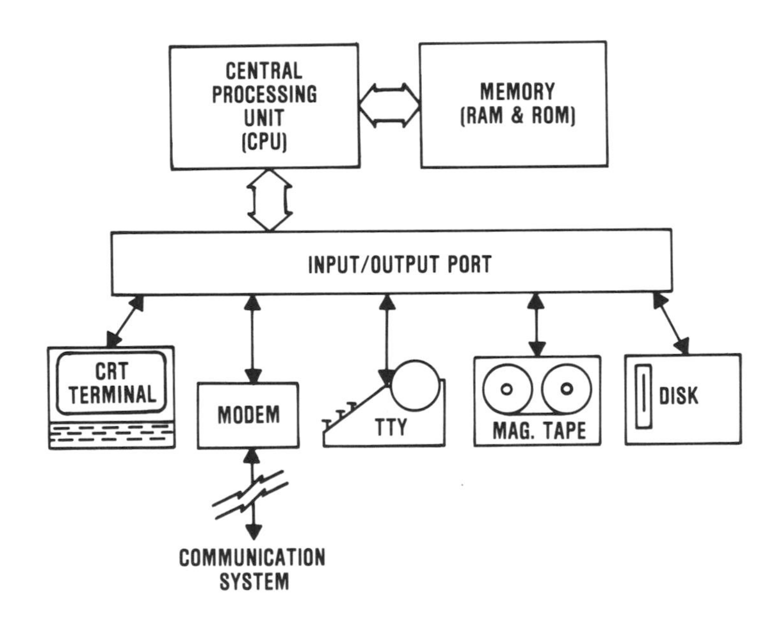

FIGURE D.12.01. Computer Block Diagram

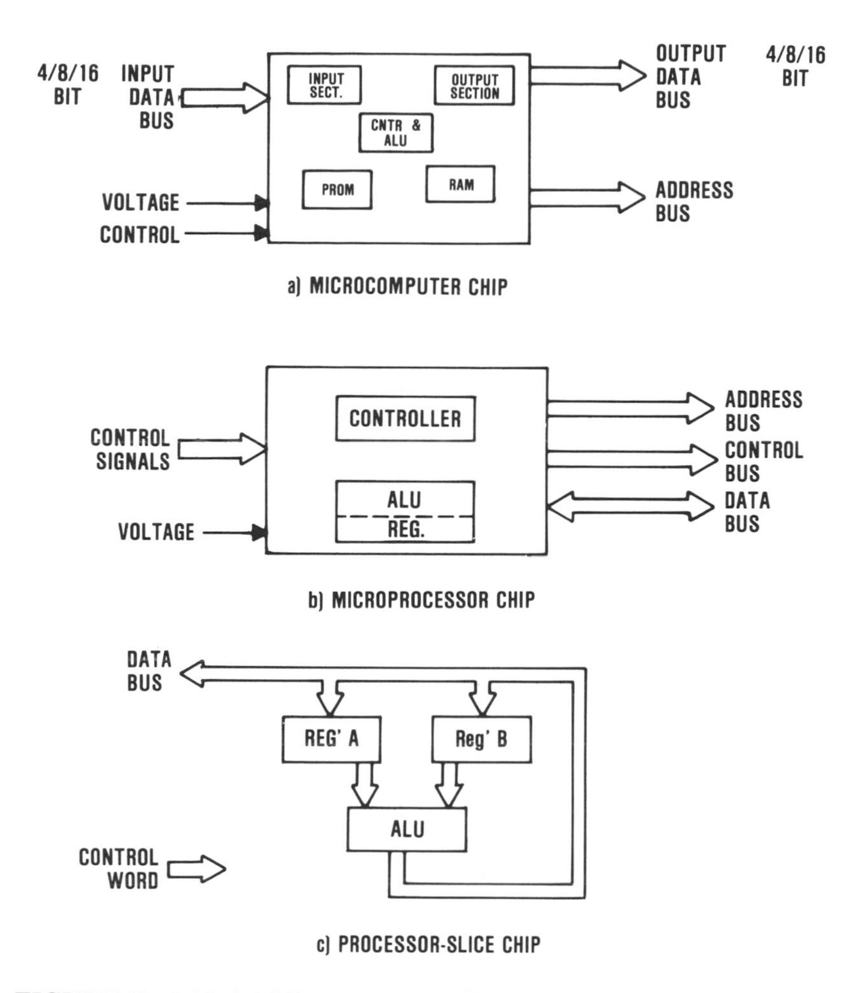

FIGURE D.12.02. (a) Microcomputer Chip (b) Microprocessor Chip (c) Processor-Slice Chip

internal registers and I/O ports to communicate with external computer elements. The address and data bus is used to access specific locations in the external memory during the "fetch" cycle. The ALU and register section is used to manipulate and provide temporary storage for the data words. Microprocessor ICs are available in 4, 8, and 16-bit versions, with 32-bit units expected in the early 1980s.

3. *Slice Processor ICs.* In many applications, flexibility and speed are the dominant considerations in the design of the system. Slice processor ICs are used when the number of ICs and cost are not important. The only function contained in this type of IC is the arithmetic logic unit (ALU)

and some registers to temporarily store the data words, as shown in Figure D.12.0.2(c). The controller, memory elements, and all other control functions must be provided external to the slice processor ICs. An external control word is used to select which register contains the data words to be processed and the type of function the ALU will perform on the data words. Slice processor ICs are available in 2, 4 and 8-bit slices. They can be cascaded for 8, 12, 16, 32-bit systems.

D.12.1 8-Bit Mirocomputer

DESCRIPTION

The 8-bit microcomputer illustrated in Figure D.12.1 is a complete computer system contained on a single IC. It contains a ROM/EPROM, RAM and a microprocessor, including the controller, program control, ALU and some registers. Using the 8-bit microcomputer rather than a 4-bit microcomputer allows the control program to be written with fewer instructions. Furthermore, larger numbers can be processed with the 8-bit microcomputer IC.

The control program, once written and debugged, is programmed into the ROM or EPROM. If a ROM microcomputer IC is used, the programming must be done by the manufacturer. If an EPROM is used, the programming is done at the user's facility. The decision as to which type to use is based on speed, cost, flexibility, etc. (See Section D.10 on EPROM or ROMs.)

When power is applied, the reset line is activated, causing the controller to clear the program counter. The first instruction is read from the **0** address of the ROM or EPROM to the instruction register where it is decoded and executed. The following types of instructions are available to the user. Arithmetic and logic instructions are used in conjunction with the internal RAM, accumulator, and arithmetic logic unit. Input or output instructions, written in the control program, transfer data words between the accumulator or RAM and the input/output port of the microcomputer. The balance of the instructions in the instruction set are used for manipulating the control counter. The interrupt steers the microcomputer to a different section of the control program. Provisions are included for external memory expansion.

KEY PARAMETERS

a) *EPROM/ROM size.* The maximum size of the control program that can be programmed into the EPROM/ROM. The size varies from 256 to 2048 locations. (Note: External memory expansion increases the size.)
b) *RAM size.* The maximum number of internal memory locations used to temporarily store information. The size varies from 16 to 128 locations.

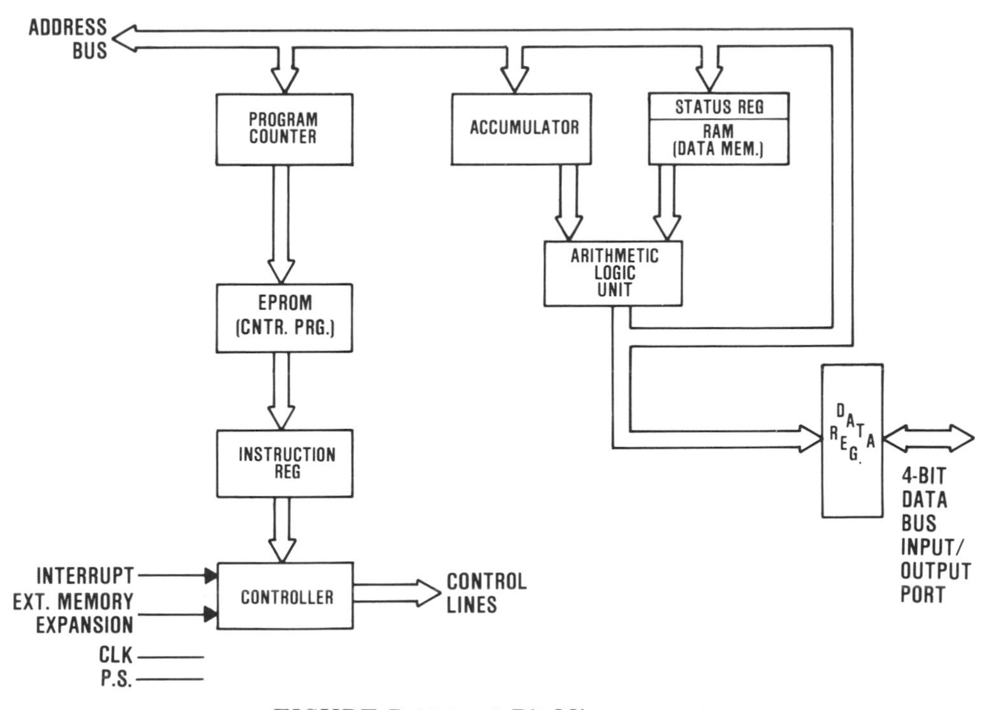

FIGURE D.12.1. 8-Bit Microcomputer

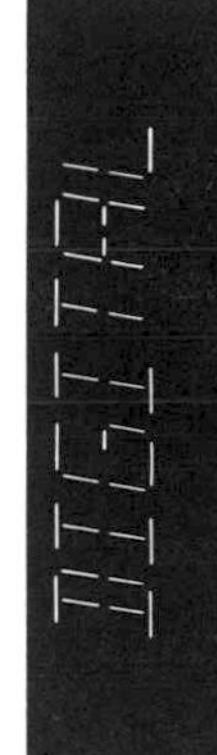

c) *Interrupt.* 1 to 6 interrupts are available with the present 8-bit microcomputers.
d) *Instruction set.* The number of instructions varies from 30 to 102 instructions.
e) *Clock.* The speed of the clock combined with the numbers of clock cycles per instruction determines the execute time of the microcomputer.

APPLICATIONS

As with the 4, 8, 16-bit microprocessors and 4-bit microcomputer, the 8-bit microcomputer can be used in microwave ovens, TV games, calculators, etc.

COMMENTS

Additional functions are included in some microcomputers. These functions include timers, UARTS, etc. The timer is provided to generate a real time clock and the UART permits serial data transfers.

D.12.2 4-Bit Microcomputer

DESCRIPTION

The 4-bit microcomputer illustrated in Figure D.12.2 is a complete computer system contained in a single IC. This computer system contains a ROM which stores the control program that is designed to perform a specific function. In addition, a RAM is included for temporary storage of data. The ALU and the associated logic elements execute instructions from the control program. The 4-bit microcomputer operates with the following two major cycles:

a) *Fetch Cycle.* When the microcomputer is this mode, the location of the ROM specified by the program control is read into the instruction register. Upon completion of this cycle the computer enters the execute cycle.

b) *Execute Cycle.* In this cycle the instruction stored in the instruction register is executed by the controller. If it was an add instruction, the binary word stored in the accumulator is added to the contents of the

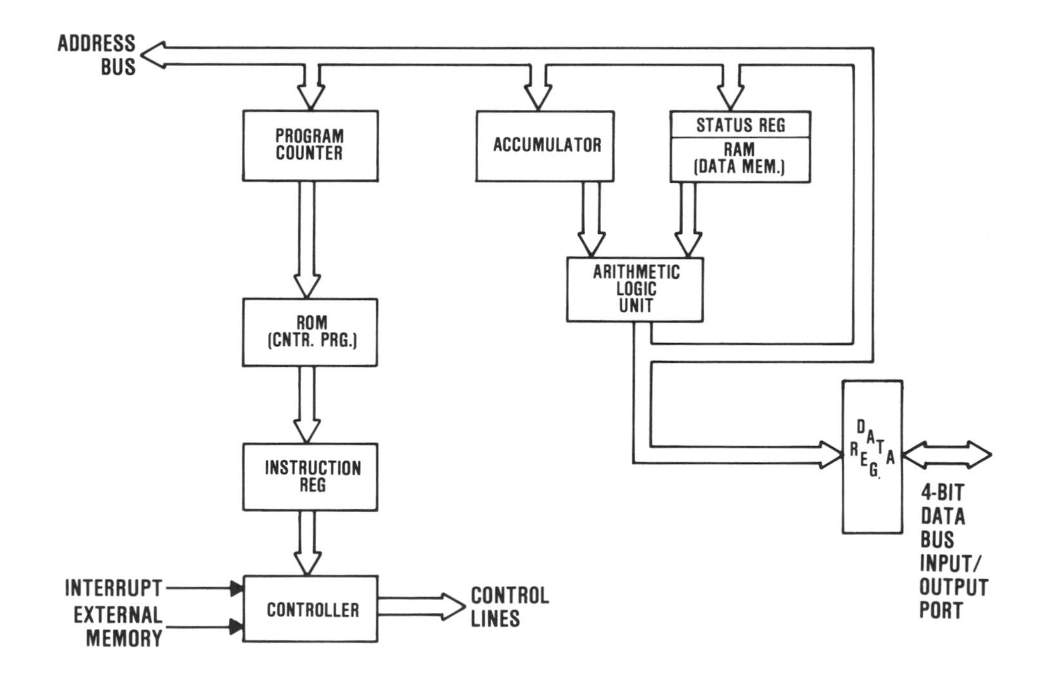

FIGURE D.12.2. 4-Bit Microcomputer

specified location. In addition, the status register is updated. If an output or input instruction is executed, a 4-bit data transfer occurs between the input/output data port and the accumulator or a specified location in the RAM.

KEY PARAMETERS

a) *Control program size.* The number of instructions stored in the internal ROM varies from 256 to 2048 instructions.
b) *Interrupt.* In most microcomputers either 0, 1, or 2 interrupts are allowed.
c) *Instruction set.* The number of basic instructions varies from 3 to 74 instructions.
d) *Clock.* Different microcomputers operate at a clock frequency from 300 kHz to 1.0 MHz.
e) *Internal storage locations.* The number of RAM locations varies from 16 to 160 words.

APPLICATIONS

The 4-bit microcomputer is used where simple dedicated functions will be performed. Cash registers, TV games, microwave ovens, are typical consumer products that use this IC.

COMMENTS

Consideration must be given to the size of the control program when using the 4-bit microcomputer. The size of the program is limited to the size of the internal ROM. Many of the 4-bit microcomputers do not have the capability to expand the ROM by adding external ICs. If the size of the control program exceeds the size of the internal ROM, the 4-bit microcomputer cannot be used.

D.12.3 16-Bit Microprocessor (MP)

DESCRIPTION:

The 16-bit microprocessor illustrated in Figure D.12.3 is similar in structure to the 4 and 8-bit microprocessor, but there are some differences:

a) Larger numbers can be manipulated within a single instruction cycle. Numeric values up to 65 thousands can be processed in one add cycle, while the 8-bit microprocessor is limited to numeric values up to 256 in one add cycle.

b) Fetch cycles are required in most instructions, but in the 8-bit MP two fetch cycles are required to read a 16-bit instruction.
c) The latest digital design techniques such as "memory to memory operations," "instruction queue," are used with 16-bit microprocessors, permitting faster execution times of the control program.

The operation of the microprocessor is characterized by the major cycles described below:

Fetch Cycle: During the fetch cycle the binary value stored in the program counter (PC) is routed through the 16-bit address register to the address bus. The PC contains the address of the instruction being executed at the present time. The address register stores the address long

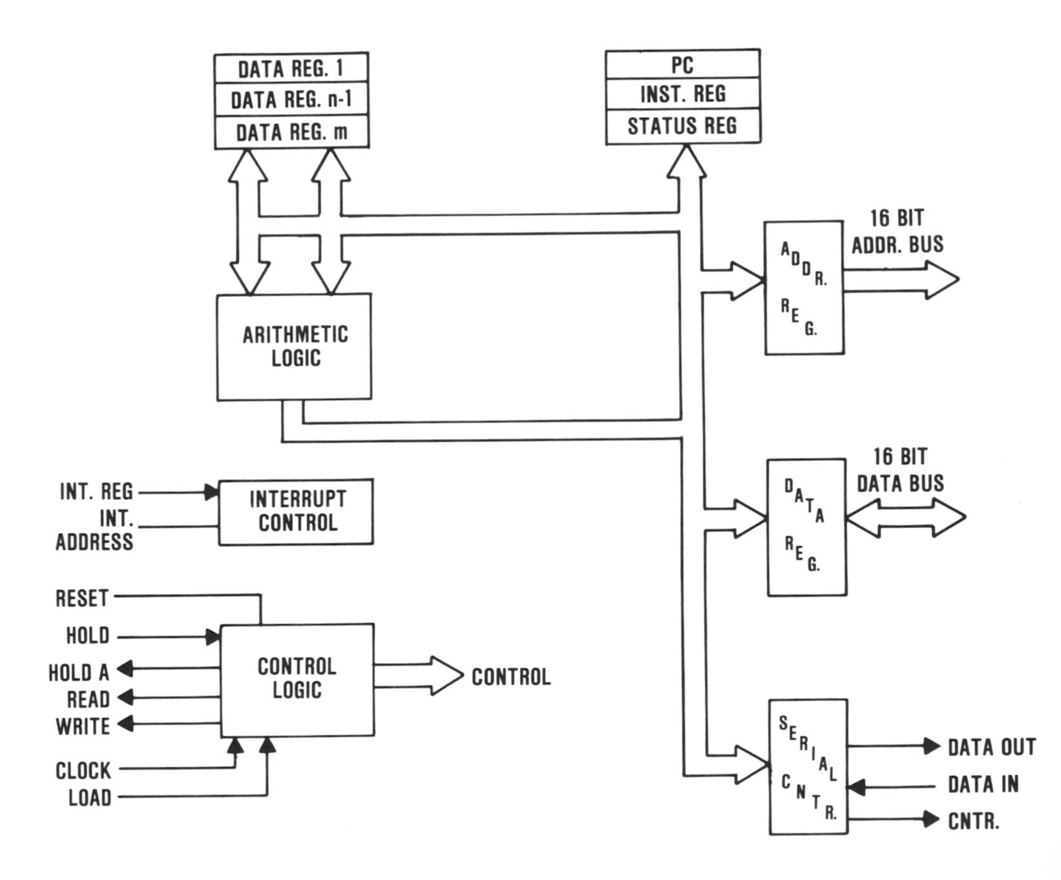

FIGURE D.12.3. 16-Bit Microprocessor

enough to access the external memory storing the control program. The binary value on the address bus is decoded by the memory IC in order to read the selected instruction onto the 16-bit data bus. This instruction is processed through the data register and written into the instruction register. The control logic generates all necessary timing signals for this cycle to occur.

Executive Cycle. When in the executive cycle, the control logic generates the necessary signals to perform the instruction. If an add instruction is being executed, the arithmetic logic unit (ALU) would be set to the ADD function. Other command signals from the control logic connect the selected registers to the ALU. Once the add function is completed, the result of the addition is written into one of the internal registers in the MP or a location in the external memory chip. The number of internal data registers varies between different microprocessor types, but some have no internal registers. In these ICs, memory address data registers are used. These MADRs are assigned locations in the external memory ICs. The memory location is determined by the instruction and the contents of an internal register which defines the starting address. Completion of certain types of instructions can result in flag bits being set in the status register. The status register consists of storage elements which are used by the control program. These storage elements indicate certain events that might have occurred during the execution of the instructions.

Interrupt Control. The interrupt control block is able to take control of the microprocessor and steer it to another part of the control program if certain events external to the MP have occurred. The decision to allow the interrupt to take control is selectable by the control program, and the number of different interrupts available varies between microprocessor models. If there is more than one interrupt, an interrupt request (Int Req) and interrupt address register are included in the MP.

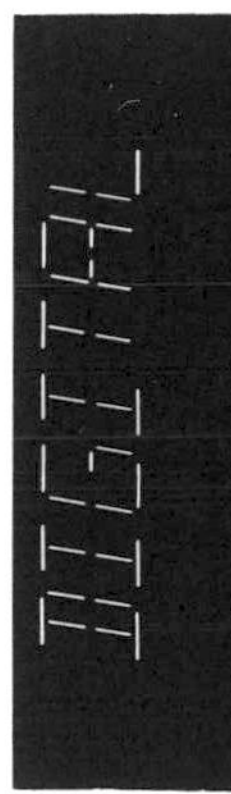

Once an interrupt occurs, the control logic stops what it is doing at the end of the present execute cycle. The control logic then generates the appropriate command signals which save the contents of the program counter, status register and the "data registers." This allows the microprocessor to return to its original program upon completion of the interrupt program. Once the information is saved in memory, the MP switches to a designated memory location, based on the interrupt value. The contents of the memory location define the starting location of the interrupt program.

Serial Control. Certain 16-bit microprocessors have, in addition to a parallel interface, a serial interface to the I/O. This control line is pulsed each time a data bit is sent. The decision to activate the serial control logic is determined by a program instruction.

In addition to these major cycles, most MPs also operate on a series of special control signals. These signals and the actions they control are described in D.12.3.4 for the 8-bit MP.

KEY PARAMETERS

a) *Direct addressing.* The number of locations that can be accessed directly by the microprocessor. From 32K words to 14 M bytes.
b) *Hardware interrupts.* The number of interrupts permitted in a particular microprocessor. The number ranges from 16 to 256 interrupts.
c) *Instruction set.* The number of instructions that the microprocessor executes. This varies from 33 to 153 instructions.
d) *Clock.* The highest frequency at which the microprocessor will operate. 2.0 to 14.0 MHz is typical.
e) *Internal registers.* Internal registers available to manipulate the data. They range from none up to 16.

APPLICATIONS

16-bit microprocessor MPs will provide faster throughput than 4 and 8-bit microprocessors. Present applications include TV games, airborne control systems, process control applications, personal computers and small business computers.

D.12.4 8-Bit Microprocessor (MP)

DESCRIPTION

The 8-bit microprocessor illustrated in Figure D.12.4 acts as controller in a microcomputer system. An 8-bit MP provides faster execution times and higher performances than a 4-bit MP. A 4-bit MP would require more than one add instruction when adding numbers greater than 16, but an 8-bit MP can handle numbers up to 256. The fetch and execute cycles of the 8-bit MP are essentially the same as described in D.12.3 for the 16-bit MP.

In addition to the fetch and execute cycle, the following control functions are included in most 8 and 16-bit microprocessors:

Reset Signal: All control programs have a starting point when power is applied. As a general guide, the starting address is location 0. When the input reset line is activated, the MP resets the program counter to "0" and enters a fetch cycle. Information from location 0 of the external memory is read into the microprocessor and executed. This information could represent a linking address to another section of the control program or an instruction.

Load Signal: The load signal operates in a similar fashion to the reset signal. The load signal is used to force the MP to another section of the

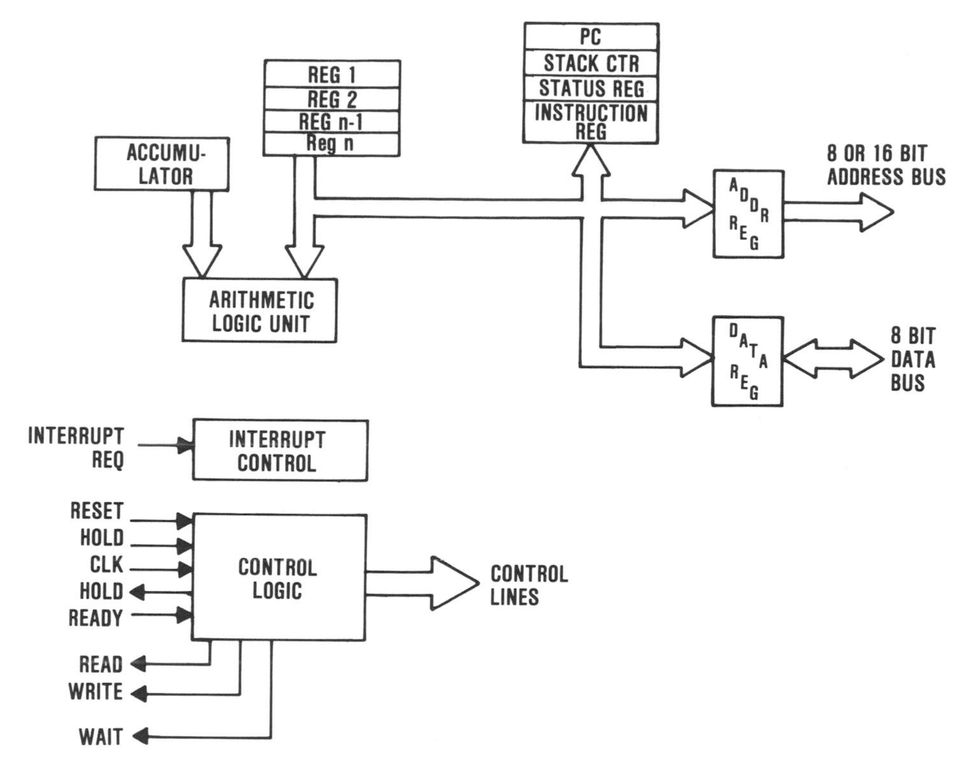

FIGURE D.12.4. 8-Bit Microprocessor

control program. When the load input is activated, the program counter (PC) is set to a specific value and connected to the address bus. The value on the address bus is used to specify the memory location to be read back into the PC.

Based on the design of the specific MP, this information is used as a link or starting address for the program.

Hold/Hold A: These signals are used to stop the MP from executing the control program in order to execute direct memory access (DMA) transfer. The DMA operation transfers binary information between the memory and other parts of the system. DMA is used because the MP cannot operate at the speeds required by the incoming data, such as information from a disc or a high-speed data link. Activation of a DMA transfer occurs when the hold signal is applied. The MP stops at the end of the next memory cycle and puts the address and data bus, write and read lines into a high impedance state. This allows the DMA controller to utilize these lines without interference from the MP. The DMA data and address word is then connected to the memory bus and a write signal is generated.

Stack: This section of the MP is used to store the address of routines which have been interrupted until the interrupt is completed.

KEY PARAMETERS

a) *Direct addressing.* The number of memory locations directly accessible by the MP. This varies from 2K bytes to 128K bytes.
b) *Interrupts.* 8-bit MPs have interrupt requests but no provisions to process multiple interrupts. (See Section D.12.3.) External hardware must be included in the system to provide a multiple interrupt function.
c) *Instruction set.* The number of instructions goes from 8 to 96.
d) *Clock.* The highest frequency at which the MP operates. For 8-bit microprocessors the maximum is between 2 and 10 MHz.
e) *Internal registers.* The number of general purpose registers available within the MP. They vary from 0 to 128.

APPLICATIONS

8-bit MPs are used in the majority of microcomputer systems on the market. They find application in electronic games, appliance controls, personal computers, automotive controls, and industrial controls.

COMMENTS

Some manufacturers of the 8-bit MPs are designing their 16-bit MPs to be instructional compatible with the 8-bit control programs.

D.12.5 4-Bit Microprocessor (MP)

DESCRIPTION

The 4-bit microprocessor illustrated in Figure D.12.5 is used to process binary and BCD information. This MP has two basic cycles, fetch and execute. While in the fetch cycle, the binary value in the program counter (PC) is connected to the 12-bit address bus and a read command is generated from the control logic. During the read command the instruction stored in the location specified by the information on the address bus is read via the data bus into the data register. This information is then transferred to the instruction register. The MP then enters the execute cycle.

In the execute cycle the instruction is decoded and the appropriate commands generated. If an add instruction is to be executed, a command

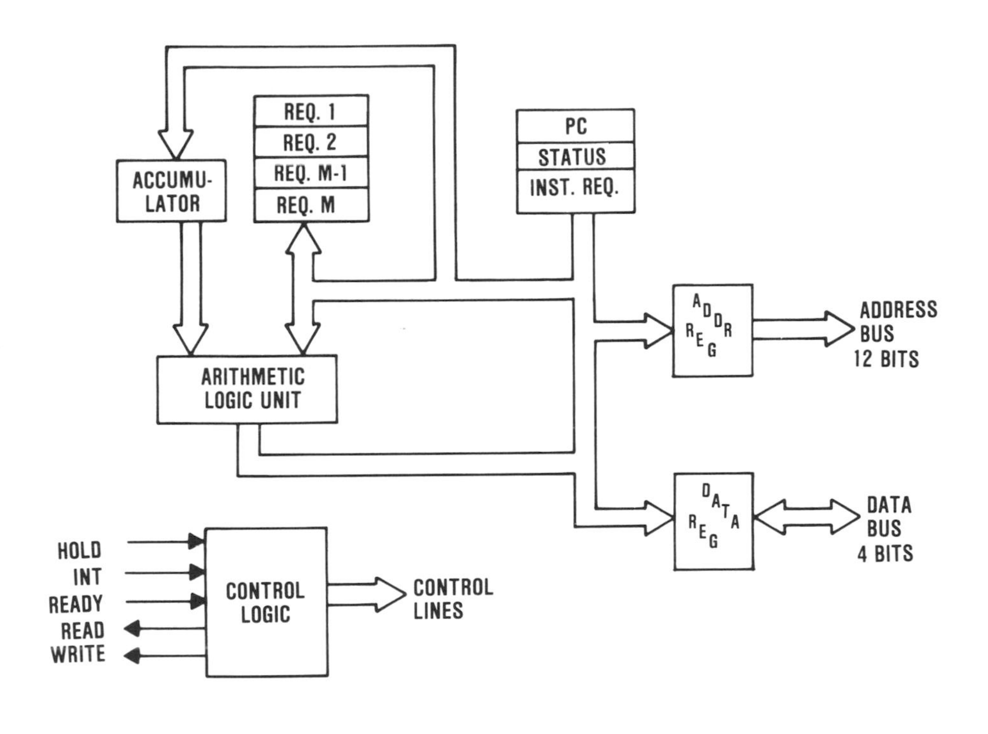

FIGURE D.12.5. 4-Bit Microprocessor

from the control logic connects the selected register to the arithmetic logic unit (ALU). The ALU is set up to perform an add function between the binary information in the accumulator and the selected register. The result of the addition is then transferred into the accumulator and the previous information stored in the accumulator is destroyed. If the resultant number is too large for the accumulator, the appropriate flag bits in the status register are set. The status register is used to store events such as overflow, equal, etc.—events that occurred as a result of executing the instruction. The flag bits set can be tested by succeeding instructions.

Another function provided in the MP is the interrupt. If the interrupt line is activated, the MP stops what it is doing and proceeds to a different section of the control program. The ready and wait signals are used when the MP is working with memory elements which are slower in speed. The memory section of the system is designed so that whenever the read or write signal is activated, the ready line to the MC is pulled down. This causes the MP to stop whatever it is doing and allows the memory elements to respond to the information on the address and data bus. During this time the wait signal from the MP is generated. Once the memory elements perform the read or write operation, the ready line is deactivated, allowing the MP to continue in the control program.

KEY PARAMETERS

a) *Direct addressing*. The number of memory locations that can be directly accessed by the microprocessor. These locations can be used to store instructions or data. The number of location varies from 4K to 8K words.
b) *Interrupts*. The number of interrupts that direct the MP to execute a different set of instructions. As a general rule there are from 1 to 8 interrupts.
c) *Instruction set*. The number of different instructions that can be executed by the MP. This number varies from 16 to 64 instructions.
d) *Clock*. The speed at which the microprocessor operates. Note: Multiple clock cycles are required to execute a single instruction. The clock frequency varies from 0.5 to 2 MHz.
e) *Internal register*. The number of registers internal to the microprocessor which can be used to store binary information. This number can vary from 1 to 24 registers.

APPLICATIONS

The 4-bit MP is used in slow speed, simple controlled consumer products, such as microwave ovens, TV games, etc.

D.12.6 4-Bit Microprocessor Slice (MPS)

The 4-bit microprocessor slice illustrated in Figure D.12.6 performs the basic function of arithmetic, logic and data manipulation operations. This is different in operation from regular MPs which not only perform the above operation but also supply all timing signals necessary to execute a control program. Regular MPs are designed for the additional feature and are slower for a given program than the MP slice which only operates on the data and does not supply the control signals for the entire system.

The MP slice is under the control of an external control word, which consists of two parts—the address and the operand. The address part of the word is connected to the address bus, while the operand word controls the arithmetic logic unit.

The address bus is activated in order to select the registers to read or

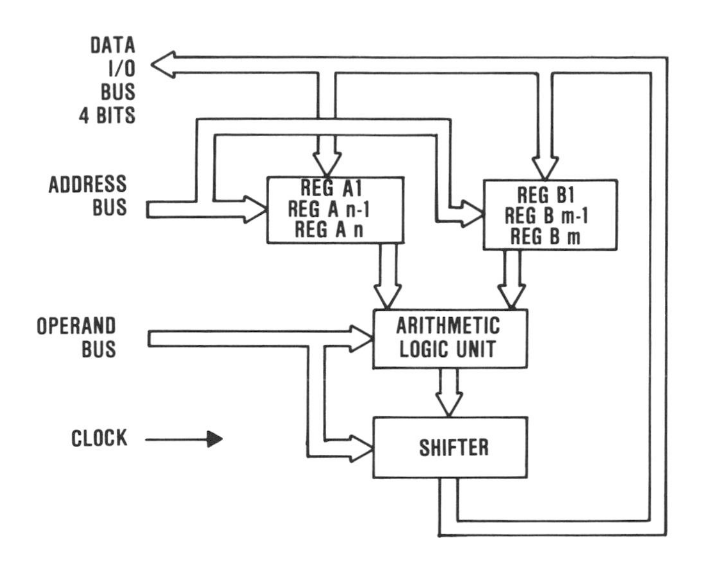

FIGURE D.12.6. 4-Bit Microprocessor Slice

write the data words. If data is in the register, the operand bus is used to select one of the functions designed into the ALU, such as add, subtract, complement, etc. If the data is to be shifted left or right, the appropriate code is sent to the shifter logic. The control word, consisting of the operand and address, is sent to the MP slice where it is then executed. If data is to be transferred between the MP slice and the external hardware, the data I/O bus is activated. The control words are stored in an external memory such as a PROM or ROM and executed sequentially.

KEY PARAMETERS

a) *Number of basic instructions (Operand Code).* Which instruction is executed by the MP slice is determined by the external control word. Their number varies from 8 to 42 instructions.
b) *Number of internal registers.* The number of registers used to store data varies from 0 to 20. In the case where no registers are included in the MP slice, external registers are designed into the system.
c) *Address bus.* The address bus for the MP slice relates specifically to the number of internal registers. The bus size varies from 0 to 4 bits.
d) *Clock.* The rate at which the MP slice will execute an instruction. The rate varies from 5 to 20 MHz.

APPLICATIONS

The MP slice is used in application where speed is the primary consideration in the design of the system. Cost and IC density would be a secondary consideration. MP slice systems are used in dedicated applications where through-put is important.

COMMENTS

MP slice ICs really execute micro-instructions. An add instruction might require five to ten micro-instructions and a specific sequence is required to generate the ten micro-instructions. The sequence generation is external to the MP slice. To perform multi-bit operations, several 4-bit MPs slice must be cascaded. To execute a 16-bit add instruction, four 4-bit MPs slice must be used.

D.12.7 2 and 8-Bit Microprocessor Slice (MPS)

The 2 and 8-bit MP slice illustrated in Figure D.12.7 is similar in operation to the 4-bit MP slice. A control word, external to the MP slice, activates the selected registers and routes them to the arithmetic logic unit (ALU). The operand field of the control word determines which specific function is executed. The modified data is routed back to a register or to the data I/O bus. The control word is divided into the following three fields.:

Operand Field: The number of encoded bits in this field defines the number of different functions in the ALU. A 4-bit operand field, for example, defines 16 functions in the ALU.

Source Register Field: Defines where the data words that will be routed to the ALU are stored.

Designation Register Field: This field performs a dual function. The register indicated in this field denotes not only which register will be selected to read data, but also which is used to store the results from the ALU. If an add instruction for S Reg. 3 and D Reg. 5 must be performed, the address field of the control word would be defined as follows: "1001001110." The first four bits on the left, "1001," would define the code for the ALU to perform an add operation. The next three bits, "011," indicate that source register 3 will be selected as the one of the data words. The next three bits, "101," indicate that designation register 5 will supply the second data word

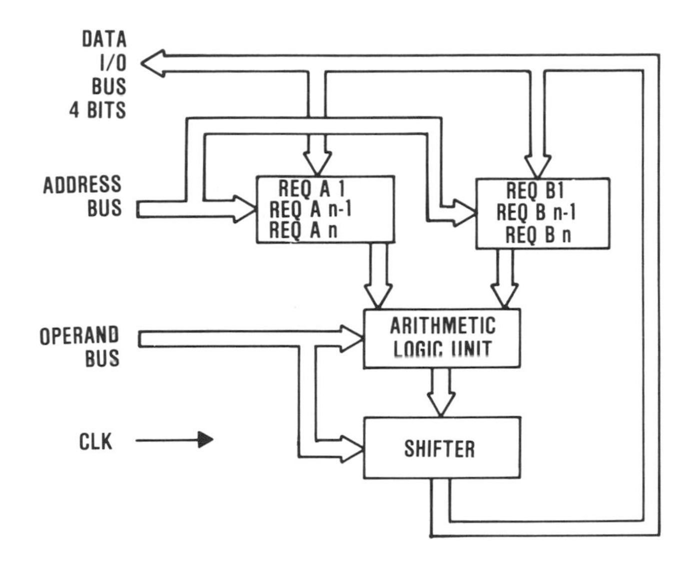

FIGURE D.12.7. 2- and 8-Bit Microprocessor Slice

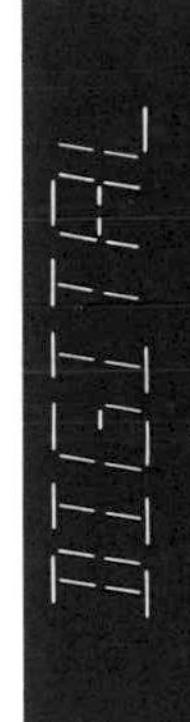

and where the results will be stored. The original contents of register 5 will be replaced with the new value.

KEY PARAMETERS

a) *Clock*. The rate varies from 10 to 20 MHz per micro-instruction cycle.
b) *Number of instructions*. The number of instructions varies from 27 to 40 different functions in the ALU.
c) *Address bits*. The range of addresses is up to 512 words.
d) *Internal registers*. The internal registers are used to store data temporarily. Their number varies from 9 to 16 registers.

APPLICATIONS

The 2 and 8-bit MP slice is used where speed and performance rather than cost and IC density are prime considerations. Many MP slices are used to emulate other computer systems, and where their higher speed permits rapid software development.

D.13.0 Microprocessor/Microcomputer Support Functions

DESCRIPTION

Neither the microprocessor (MP) nor the microcomputer (MC) ICs described in D.12 form a complete system. Both must be connected to peripheral devices, to additional memory, and to some very specialized support ICs. The interface ICs for external memory devices and for other peripherals are included in the section on Interface ICs.

The specialized MP and MC support ICs described in this section are representative examples of the most widely used types. In general, a particular MP or MC family includes a variety of support ICs, tailored to the MP and MC and configured for specific applications. In some instances two or more support functions are included in one IC. The clock generator and interrupt controller are functions that almost every MP and MC requires. The programmable interface and the I/O expansion are also frequently required, but the interval timer is not always used, since its function can be implemented by the program. Interval timers are very popular for personal computers where home appliance control is an important feature. The development IC is used primarily for designing small computer systems because the contents of read-only memories (ROM) or of programmable read-only memories (PROM) can be determined in a flexible manner.

KEY PARAMETERS

Support ICs are generally applicable only to the MP and MC families for which they are designed. Their key parameters will therefore be the same as for the particular MP or MC. Specific function parameters are listed in the following pages for each IC.

APPLICATION

Support ICs are used with the particular MP or MC for which they are designed as part of the total computer system.

COMMENTS

To avoid system problems be sure to consult the support IC data sheets, the respective MP or MC data sheets, and any special application notes available from the manufacturer.

D.13.1 Microprocessor Clock Generator

DESCRIPTION

Many manufacturers of microprocessors (MP) provide an associated clock generator IC capable of generating the proper timing signals for the MP.

Figure D.13.1 is the block diagram of an MP clock generator IC and shows a crystal connected to the oscillator input of the clock generator. A crystal is used to insure accuracy and stability. The frequency divider establishes the actual clock frequency of the MP and determines the duty cycle for the clock, such as 25% on, 75% off. This duty cycle allows the phasing of the clocks to be developed and minimizes noise to the MP. The output of the frequency divider is connected both to the phase generator and to the decoder. The phase generator sets up the timing as to when each clock output will be activated. Correct phasing of the clock is important since the MP performs certain basic functions at each phase. In a fetch cycle, for example, phase 1 sets the address bus to a memory location. Phase 2 might start the read or write signal. Phase 3 might be used to start the data transfer between the MP and the memory element.

The decoder logic is used to gate the clock from the frequency divider into the individual phases, and the number of decoder gates depends on the number of clock phases required. The drivers set the clock signals to the correct voltage levels.

KEY PARAMETERS

a) *Frequency*. Defines the frequency of the basic clock cycle. The frequency can be as high as 10 MHz.
b) *Phase*. The number of clock outputs from the clock generator IC. It varies between one and four clock outputs.
c) *Voltage level*. The voltage levels range from -12 to +18 Volts.

APPLICATIONS

The clock generator IC is usually designed to interface with a particular microprocessor family.

COMMENTS

Microprocessor ICs are usually TTL compatible for the address and data bus, but the clock inputs are not. Some clock generator ICs include additional functions such as the reset, load or interrupt signals.

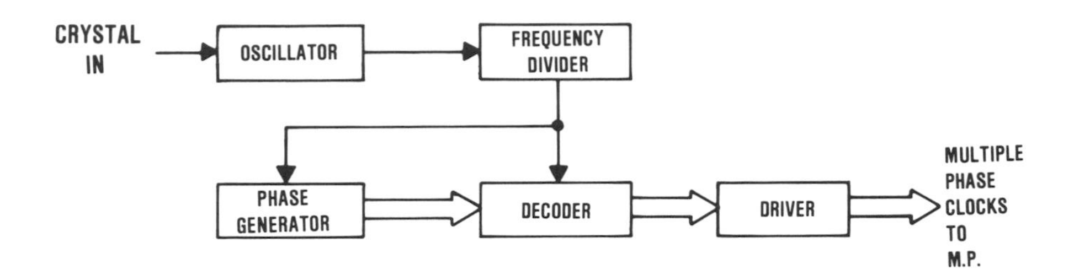

FIGURE D.13.1. Microprocessor Clock Generator

D.13.2 Microcomputer Input/Output Expander

DESCRIPTION

This IC is used with microcomputers (MC) to increase the number of data ports and allows the MC to interface with additional peripheral devices such as keyboards, printers, modems, etc. Figure D.13.2 represents the typical block diagram for this IC. The width of the data port may range from 4 to 16 bits. Specific instructions in the computer program control access to the IC expander, including an address field which determines the I/O port that will be used.

The address decoder generates the appropriate command signals to the selected port. When a "read I/O" instruction occurs, the data word on the selected input buffer is routed through the multiplexer to the external data bus to the MC. If the I/O instruction is a "write," the data word is transferred from the MC over the data bus through the demultiplexer to the latch of the selected port.

KEY PARAMETERS

a) *Port width*. The number of bits must be the same as in the MC.
b) *Number of ports*. The user assigns each port to a peripheral device. Most I/O expanders have four ports.

APPLICATIONS

I/O expanders are used in microcomputer systems when the number of peripherals is greater than the number of ports available in the MC itself.

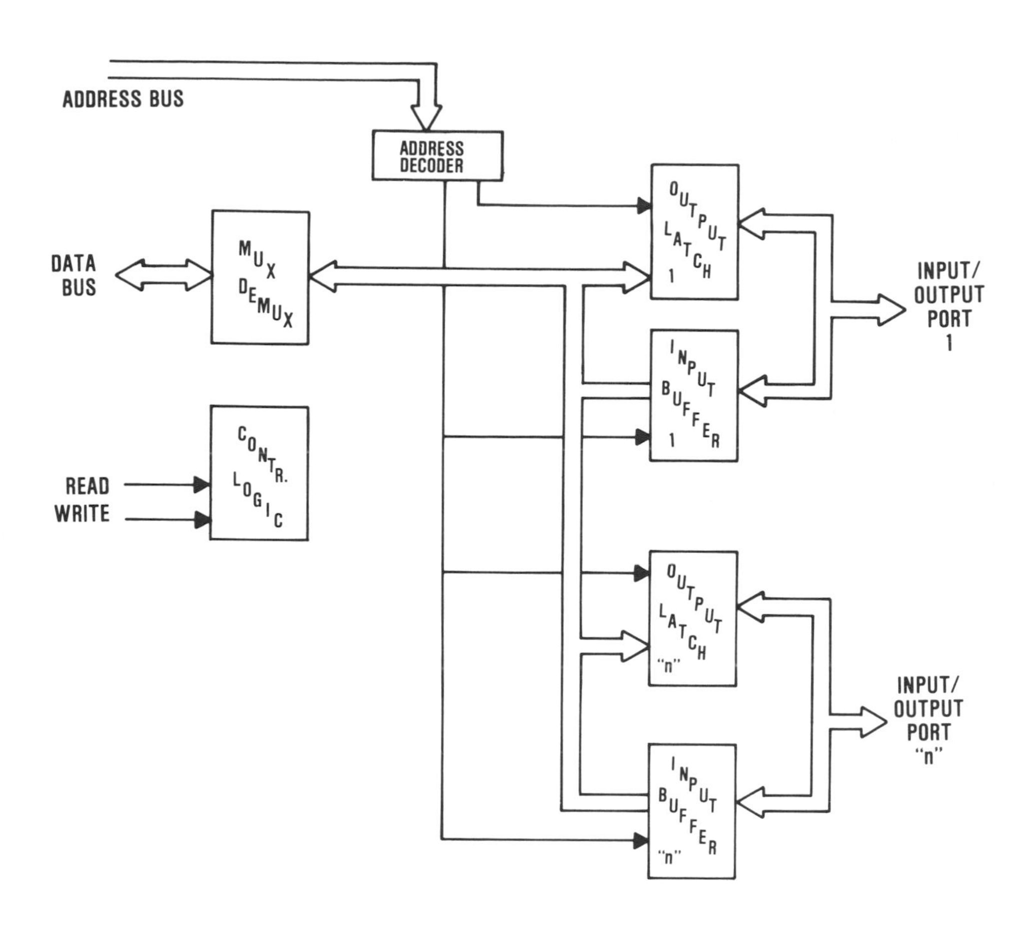

FIGURE D.13.2 Microcomputer Input-Output Expander

D.13.3 Interrupt Controller

DESCRIPTION

Interrupt controller ICs are designed to process and format the interrupt requests generated by the various peripheral devices such as keyboards, timers, printers, etc. Interrupt requests are then sent to the microprocessor (MP) or microcomputer (MC). These interrupt requests are used to steer the MP or MC to a different part of the control program, enabling the system to handle the various "real time" requests occurring periodically from peripheral devices. Figure D.13.3 represents a typical functional block diagram of an interrupt controller IC. Interrupt requests are received from the various peripheral devices in the total system.

When the operator depresses a key on the keyboard, for example, the assigned interrupt request is activated. This request is routed through the interrupt buffer to the masking gates. The masking gates are used to prevent selected interrupt requests from being processed. There are times when the MP or MC executes sections of the control program which cannot tolerate an interrupt request. In order to prevent these requests from being processed, the MP or MC will generate a control word which is sent to the interrupt controller over the control word bus. This control word is stored in the masking register. Each bit in the control word is assigned to an interrupt input. If the bit is set, the interrupt masking latch is enabled, causing the masking gate to inhibit any requests for that input. If the masking bit is disabled the interrupt request line is enabled, allowing any interrupt request to be processed by the MP or MC. The formatter encodes the interrupt request and generates the control signals sent to the MP or MC. This is done to minimize the number of interrupt control lines.

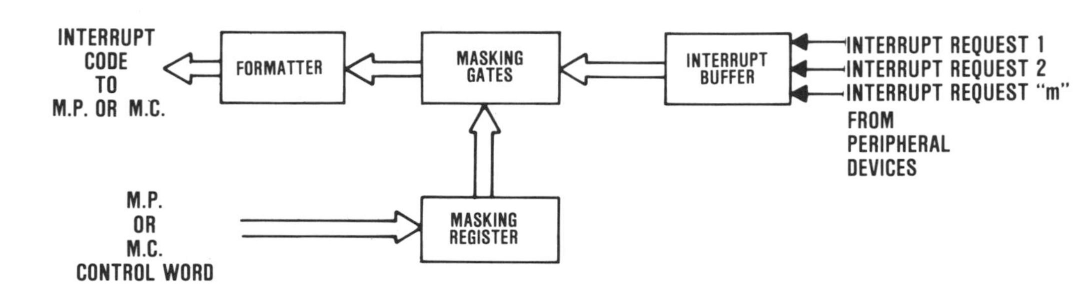

FIGURE D.13.3. Interrupt Controller

KEY PARAMETERS

a) *Number of interrupts*. The number of interrupts varies from one to as many as 16.

APPLICATIONS

Interrupt controllers are used wherever MCs or MPs are working with peripheral devices requiring real time service.

COMMENTS

Depending on MP or MCs used, the interrupt/address select logic is either internal or must be supplied by external logic. The interrupt address select logic activates a different starting address on the address bus for each interrupt. Polling is another method used to determine when a peripheral device requires servicing. It is less expensive to implement in hardware, but more time consuming.

D.13.4 Interval Timer

DESCRIPTION

The interval timer as illustrated in Figure D.13.4 consists of binary or BCD counters, connected to an external clock and the microprocessor (MP) or computer (MC). The external clock is usually an accurate and stable pulse train at a specific frequency. For each clock pulse the counter advances by one bit and generates one output pulse at a predetermined value. This predetermined value is established by the MP or MC. In a typical example an interrupt is required every two ms. In order to set up the two ms interrupt, a binary word from the MP or MC is transferred over the data bus and the data buffers to the selected counter. The particular counter is selected by the information on the address bus. The value of the binary information word is based on the frequency of the clock and the number of bits in the counter. The output pulse line is connected to an interrupt controller as described in D.13.3

The following calculation shows how the predetermined value of the data word sent to the counter is obtained:

> If a clock frequency of 1 MHz and an interval time of two ms is assumed, the least significant bit has a decimal value of one ms. bit #2 will be two ms, bit #3 will be four ms, etc.
>
> Using the decimal-to-binary conversion procedure, the binary word is 011111010000. This binary word is complemented and sent to the selected counter.

KEY PARAMETERS

a) *Number of counters*. Each counter in the interval timer can be used for a different time value. Interval timers are available with as many as four separate counters.
b) *Number of bits per counter*. The size of the counter and the frequency of the external clock determine the maximum time entered. Counter lengths vary from eight to 16 bits.

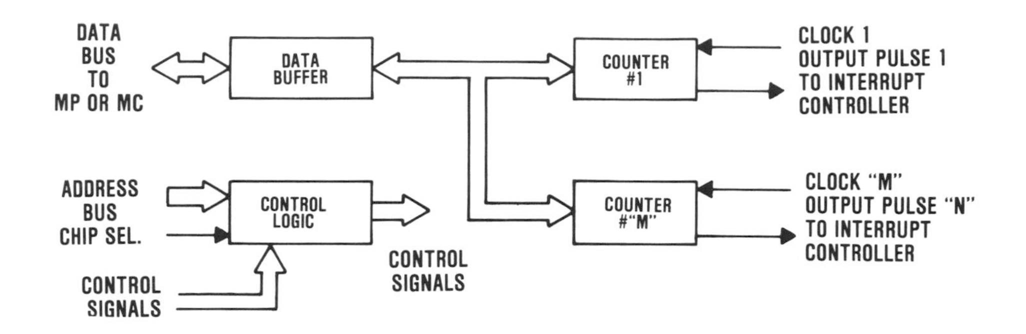

FIGURE D.13.4. Interval Timer

c) *Speed of operation.* The maximum allowable clock frequency determines the accuracy, resolution, and smallest time internal. Clock frequencies of up to 3.0 MHz are typical.

APPLICATIONS

Used in computer based systems where real time processing is required. Microwave ovens and TV games are typical examples.

COMMENTS

All characteristics of the interval timer are programmable and this function can be achieved by software. While most of the counters are straight binary, there are some interval timers that provide binary-coded decimal (BCD) operation.

D.13.5 Microprocessor Programmable Interface

DESCRIPTION

Most microprocessors (MP) provide only one data and one address bus, which operate at the speed of the microprocessor. There are many applications where the peripheral devices connected to the MP must operate at different speeds. In addition, the size of the data word may vary between particular peripherals. The programmable interface can be used to increase the number of interfaces and accommodate different word lengths and data

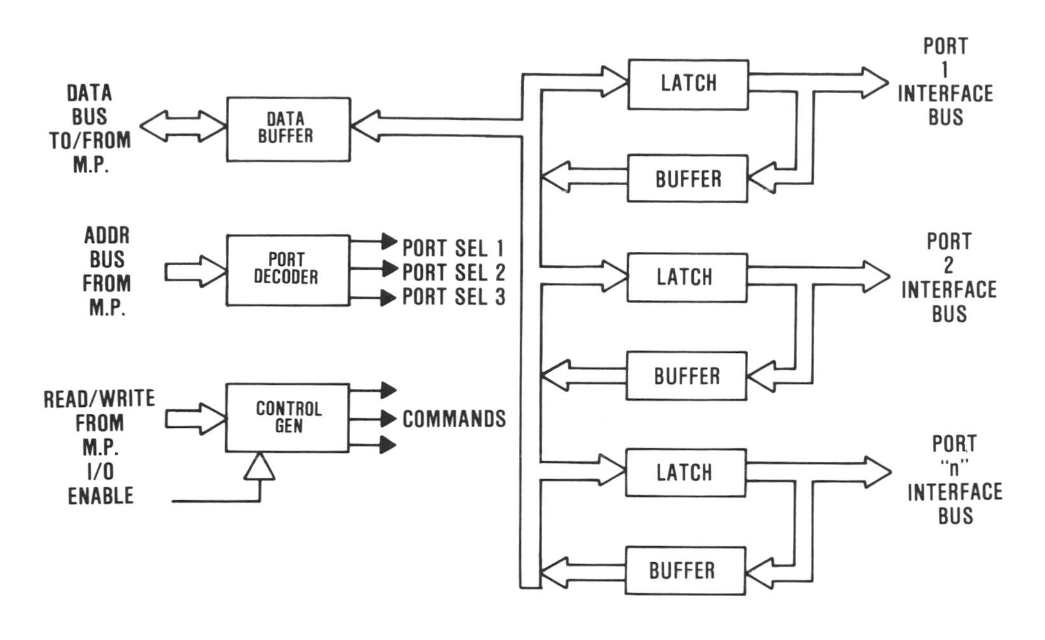

FIGURE D.13.5. Programmable Interface IC

rates. As illustrated in the block diagram of Figure D.13.5, instructions in the control program determine which port is selected for a data transfer. Data words can be transferred either from the MP to the selected port or from the selected port to the MP. In addition, ports can be assigned to a communications channel, video display, keyboard, etc.

When a "write I/O" instruction is executed, the data word from the MP is placed on the data bus to the programmable interface. The binary word on the address bus defines which activated port will receive the data. When the write signal is received, internal control signals are generated, causing the data word to be loaded into the selected latch. The interface bus for the selected port will be configured for the data word from the MP. When data is transferred to the MP it sends a "read I/O" instruction. The number of the selected port is sent out over the address bus to the port decoder. When the read and I/O enable signals are activated, the information on the selected port interface bus is routed through the input data buffers over the internal data bus through the main data buffers to the MP. Each port is bidirectional, i.e., it can be programmed as an input or output port.

KEY PARAMETERS

a) *Number of ports*. The number of ports varies from one to four.
b) *Number of bits per port*. The size of the port, in bits, is often the same as the MP. Port sizes range from 4 to 16 bits.

APPLICATIONS

This IC extends the I/O capability of MPs.

COMMENTS

Programmable interface ICs can also include other features such as interrupt controller, interval timers, etc. In some of these ICs an individual bit can be programmed for both sending and receiving data bits.

D.13.6 Microcomputer Development IC

DESCRIPTION

Used during the system development phase of such microcomputer controlled systems as electronic toys, appliance controls and industrial devices, this IC contains all of the essential portions of a typical microcomputer (MC), as described in D.12.0 and D.12.1. Most mass-produced MCs contain read-only memories (ROM) which store the program for a particular application. If this program has to be changed, the MC becomes useless. For this reason the MC development IC is used during the design and development stages.

MC development ICs are identical to the MC for which they act as intermediary, except that the ROM is replaced by a programmable read-only memory (PROM). This permits the system designer to evaluate and optimize the program before it is finalized for incorporation into the ROM of the MC. Most MC development ICs use ultraviolet erasable PROMs.

Figure D.13.6 shows the functional block diagram of the MC development IC. Except for the type of memory used (UV PROM rather than ROM), it is identical to the architecture of the MC. During the development stages, an application program is written which is then entered into the UV PROM. Once the program is written into the UV PROM, the MC development IC is then inserted into the system. The program is executed, with the program counter first set to zero. This causes the first instruction to be read out of location zero of the UV PROM. The instruction is loaded into the instruction register and executed by the control logic. The program counter is then updated to point at the next instructions. The logic control then generates the necessary commands to execute the instruction. This cycle is repeated for the execution of the entire program. The RAM and registers are used to temporarily store data, and the address and data buffers isolate the MC from the external logic. If an instruction is executed which addresses the external logic, the buffers are activated. When a bug is found in one of the instructions or revision must be made to the program, the

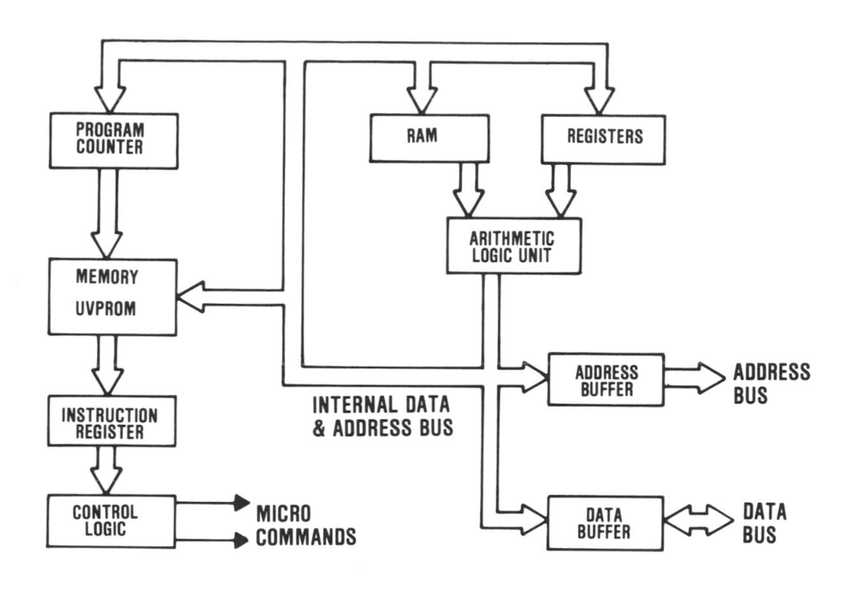

FIGURE D.13.6. Microcomputer Development IC

MC development IC is removed from the system and connected to the read/write electronic circuits used to change the contents of the UV PROM. The UV PROM is erased and the corrected program written into it. This process is repeated until the system operates correctly. The finished program is then sent to the manufacturer of the MC where the program is masked into the MC ROM.

KEY PARAMETERS

a) *Memory*. The size of the UV PROM is the same as the ROM used in the final product. Typically it can be 1 K by 4 or 8 bits.
b) *RAM*. The size is identical to the RAM used in the final IC. Typically it would be 64 words by 8 bits.
c) *I/O lines*. The number of outputs used to drive the external logic. Typically it can be 8 to 24 lines.

APPLICATIONS

The IC is used in applications where the software program is likely to change during the design stage of a microcomputer controlled system.

COMMENTS

In applications where only a small number of systems are built, the MC development IC can be used directly, instead of a specially programmed MC.

D.14.0 Multiplexer, Quad, 2-Input

DESCRIPTION

One of the most widely used digital multiplexer configurations is the quad, 2-input version illustrated in Figure D.14.0. This particular IC is noninverting and consists of eight AND and four OR circuits, arranged so that two sets of 4-bit inputs can be selected by the data select signal. As shown in Figure D.14.0, each of the Q outputs is, in effect, a tri-state OR circuit controlled by the output enable signal. This means that any of the Q terminals can be either **0**, **1**, or a high impedance.

Other digital multiplexer ICs are available that invert the input, with or without tri-state outputs.

KEY PARAMETERS

a) *Power dissipation.* Typical power dissipation during operation is 150 to 250 mW for standard TTL and 15 to 24 mW for low-power TTL ICs.

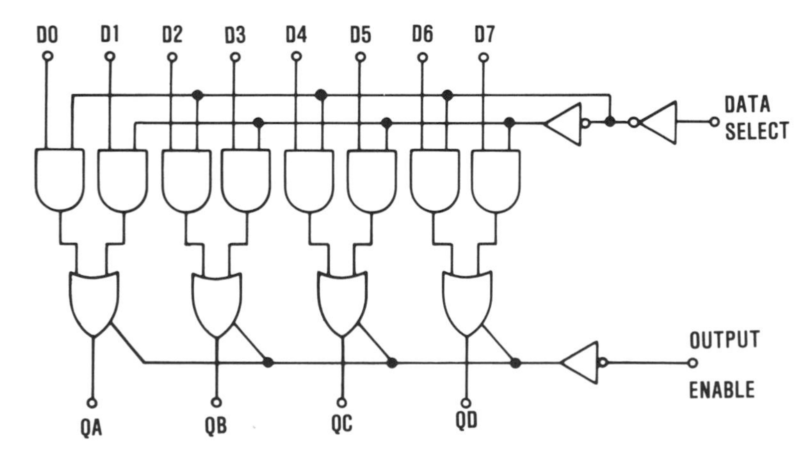

FIGURE D.14.0. Logic Diagram—Quad 2-Input Multiplexer

b) *Propagation delay time*. Time required from data input to output. 8 to 10 ns is typical for standard TTL ICs and 40 to 60 ns for low-power TTL ICs.

APPLICATIONS

The principles of time division multiplexing in any digital system can be implemented by using multiplexer ICs. A quad 2-input multiplexer can also be used to generate four functions of two variables and is useful for multiplexing dual data buses.

COMMENTS

Quad 2-input multiplexers are also available with FF storage. The combination of multiplexing with flip-flop (FF) latches is particularly useful in the input/output (I/O) ports of digital computer systems.

A variety of different multiplexer configurations is available in each of the digital IC families.

D.14.1 Multiplexer, Dual, 4-Input

DESCRIPTION

This IC consists of an array of logic gates capable of multiplexing eight data lines into two. As illustrated in the functional block diagram of Figure D.14.1, the data lines are arranged in groups of four, data *A* and data *B*. Each of the two multiplexers (MUX) receives the same four data select signals, derived from the two data select inputs. The binary equivalent of the two inputs and their four possible states selects one of the four input lines in each multiplexer and permits one of them to appear at the output. If data group *A* alternates with data group *B* outputs, *A* and *B* will alternate at the same rate.

The dual 4-input multiplexer is available with inverting or noninverting features, tri-state outputs, and storage latches.

KEY PARAMETERS

a) *Power dissipation*. Typical power dissipation is 170 mW for standard TTL ICs and 31 mW for low-power TTL ICs.
b) *Quiescent current*. The current used by the IC when no operations take place. 5.0 nA at 5 V is typical for low-power CMOS.

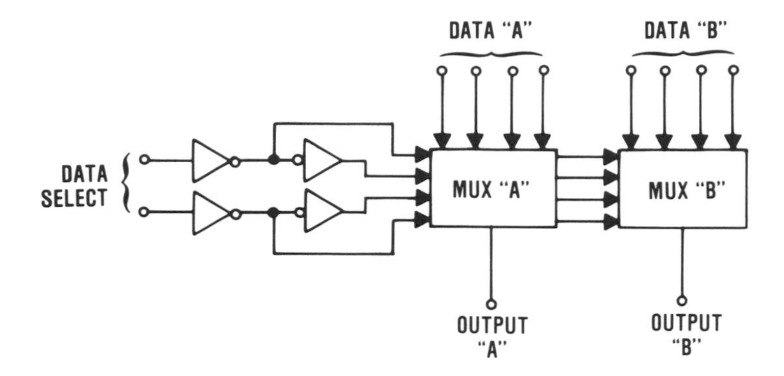

FIGURE D.14.1. Block Diagram—Dual 4-Input Multiflexer

c) *Propagation time delay, data input to output.* Time required from data input to output. 200 ns is typical at 5 V for low-power CMOS, 11 ns is typical for standard TTL, and 14 ns is typical for low-power TTL. ECL ICs have a 4.5 ns propagation delay time.
d) *Propagation delay time, data selection.* The time required from the data selection input until the output changes. 225 ns at 5 V is typical for low-power CMOS, 20 ns for standard TTL, and 22 ns for low-power TTL. ECL ICs require only 6.0 ns.

APPLICATIONS

Whenever time division multiplex is used to combine digital data. This IC can also perform parallel-to-serial conversion and can be cascaded in other digital multiplexing configurations.

COMMENTS

All types of multiplexers are available in every major digital IC family.

D.14.2 Multiplexer, 8-Input

DESCRIPTION

In this IC, one out of eight data inputs is selected for output at the Q and $\overline{Q}$ terminal. Three "data select" lines provide the binary input which can have up to eight different values. As illustrated in the function block of Figure D.14.2(a), there is also an enable signal which controls the tri-state output. The truth table of the same illustration shows that, when the enable signal is **1**, the Q output will be a high impedance. The data selection by inputs A, B, and C can take place only when the enable signal is **0**.

8-input multiplexers are available with temporary storage and without the tri-state output. In the TTL IC family there are also dual 8-input ICs, using a 24-pin DIP package to provide selection of one of the 16 data inputs.

KEY PARAMETERS

a) *Power dissipation*. Typical power dissipation for the entire IC is 150 to 200 mW for standard TTL ICs, 30 mW for low-power TTL, and 300 mW for low-power CMOS ICs.
b) *Quiescent current*. The total current drawn by the IC when no operations take place. 5.0 nA at 5 V is typical for low-power CMOS, 50 nA is typical for TTL, and 15 nA is typical for low-power TTL ICs.
c) *Propagation delay time, data input to output*. The time required when the data selection is completed until data appear at the output. 10 ns is typical for TTL ICs, 12 to 14 ns is typical for low-power TTL ICs, and 300 ns is typical for low-power CMOS at 5 V.

APPLICATIONS

8-input multiplexers are used whenever digital time division multiplexing methods are applied. These ICs can also be used in data routing, parallel-to-serial conversion, signal gating, number sequence generation, and as part of a Boolean function generator.

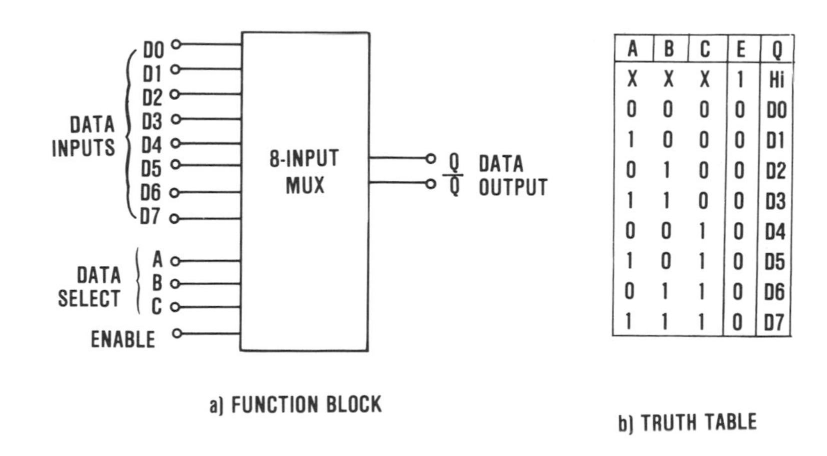

A	B	C	E	Q
X	X	X	1	Hi
0	0	0	0	D0
1	0	0	0	D1
0	1	0	0	D2
1	1	0	0	D3
0	0	1	0	D4
1	0	1	0	D5
0	1	1	0	D6
1	1	1	0	D7

b) TRUTH TABLE

FIGURE D.14.2. (a) Function Block
(b) Truth Table

D.15.0 Multivibrator, Monostable (One-Shot)

DESCRIPTION

Multivibrators are a form of flip-flop circuit in which an R-C time constant is used to determine the rate of change of state (toggling). In the monostable or one-shot multivibrator (MV), an external trigger signal starts the change of the state of this MV and the external R-C time constant determines the time required from the beginning to the end of this one-shot oscillation.

A basic monostable MV is shown in Figure D.15.0(a). The key elements are the two trigger inputs, the reset input, and the values of the external R-C time constant. Note that the OR circuit which triggers the MV has a small circle on one of its inputs, indicating that it can accept either a positive or a negative edge trigger. The edge-triggering ability is produced because this particular OR circuit is combined with a Schmitt-Trigger effect. In some ICs this type of circuit has a hysteresis characteristic and is referred to as "transmission gate."

The operation of the monostable MV requires that it be first reset so that the Q output is **0** and the $\overline{Q}$ is **1**. When either a positive or a negative trigger signal is entered, the Q immediately changes to **1** and the $\overline{Q}$ to **0**. After a period of time, determined by the R-C time constant, the MV returns to its original state, having generated one pulse. Whenever the reset signal occurs, the MV will return to the original state where Q is at **0**. In retriggerable monostable MVs, any trigger signal that occurs during the period when Q is at **1** prolongs the duration of the pulse beyond the time period determined by R-C.

The function block of a typical emitter-coupled logic (ECL) monostable MV is illustrated in Figure D.15.0(b) and shows some of the various features that are available in monostable MV ICs. Trigger signals are applied to the trigger input and the external +enable or –enable signals determine whether the MV will accept positive or negative going edges. Internal Schmitt-Trigger circuits make the trigger input insensitive to rise and fall times. Although there is an external R-C time constant, there is also an input for external pulse-width control. With an external resistor, a control voltage can be used to vary the pulse width. When a control current is used, the resistor is not required. In addition, this ECL IC has a special "high-speed trigger" input which bypasses the internal Schmitt-Trigger circuits and permits a particularly rapid response.

Monostable MVs are available that can be retriggered during the pulse generation, with the number of the triggers that occur within a given time period increasing the pulse width according to a fixed ratio. Other monostable MVs include a preset feature which can be combined with retriggering to generate specific pulse waveforms.

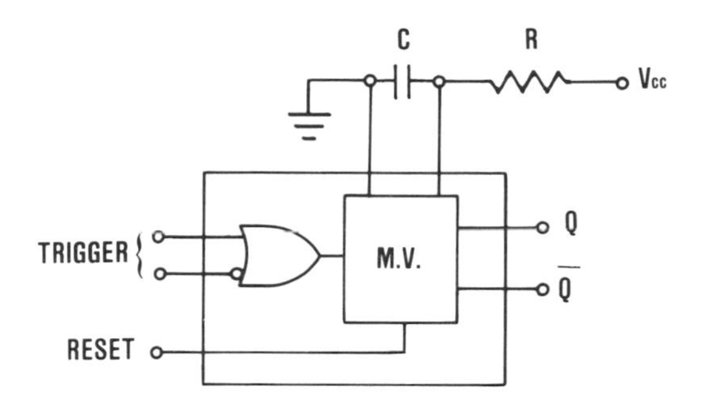

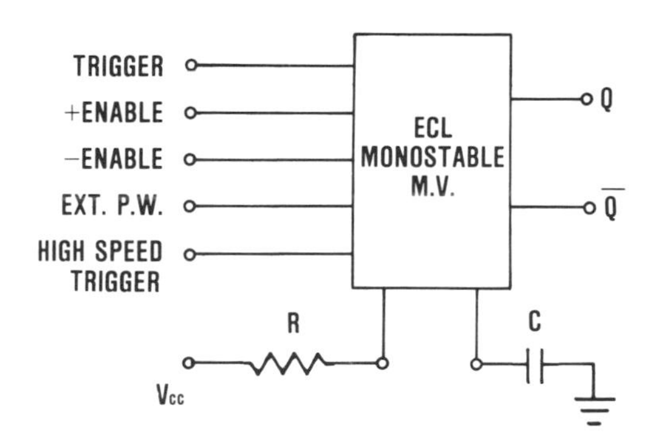

FIGURE D.15.0 (a) Monostable Multivibrator (One-Shot) (b) Function Block

KEY PARAMETERS

a) *Power dissipation.* 150 mW is typical of standard TTL ICs, 30 to 60 mW is typical of low-power TTL ICs, 250 mW is typical for low-power CMOS ICs, and 415 mW is typical for ECL ICs without any external load.
b) *Quiescent current.* The current drawn by the IC when no operations occur. 5.0 nA at 5 V is typical for low-power CMOS, 12 mA is typical for TTL, and 4.0 mA is typical for low-power TTL ICs.
c) *Propagation delay time, trigger input to change of output.* The time required from the trigger edge instant until an output change takes

place. A typical value for ECL ICs is 4.0 ns and 2.0 ns at the high-speed trigger input. TTL and low-power TTL ICs have a time delay of approximately 40 to 50 ns. 300 ns is the typical value for low-power CMOS ICs at 5 V.

d) *Minimum input pulse width.* 2.0 ns is typical for ECL ICs, 40 to 50 ns is typical for standard or low-power TTL ICs, and 35 to 80 ns is typical for CMOS ICs at 5 V.

e) *Minimum reset pulse width.* 40 to 50 ns is typical for standard and low-power TTL ICs; 100 ns is typical for low-power CMOS ICs.

f) *Minimum output pulse rise time.* 1.5 ns is typical for ECL ICs, 10 to 15 ns is typical for standard and low-power TTL ICs, and 180 ns is typical for low-power CMOS at 5 V.

g) *Minimum output pulse fall time.* 1.5 ns is typical for ECL ICs, 10 to 20 ns is typical for standard and low-power TTL ICs, and 100 ns is typical for CMOS ICs at 5 V.

APPLICATIONS

Monostable MVs are used in a variety of timing circuits and wherever pulses of specific duration are required. They are found in a large variety of digital systems and form one of the building blocks in timing and sequencing sections.

COMMENTS

A variety of features can be combined in the basic monostable MV as described above. Many of the standard ICs contain two identical monostable MVs on a single package.

D.16.0 Shift Register, Static

DESCRIPTION

Flip-flops (FF), the subject of Section D.7.0, can be connected as counters or shift registers. When used as shift register, the FF functions as temporary memory or storage element, rather than as the "divide-by-two" element used in counters. The basic shift register operation is described in Figure D.16.0 which shows the logic diagram of a 4-bit static shift register and the truth table of that shift register during clocked operation. The data stream, a series of square waves or pulses, is applied at the data input, where two inverters provide buffering and drive the D-input to the first FF. This FF will transfer the **0** or**1** condition present at the D-input only during that short period of time when the clock pulse edge rises. If a **1** is entered at D, this

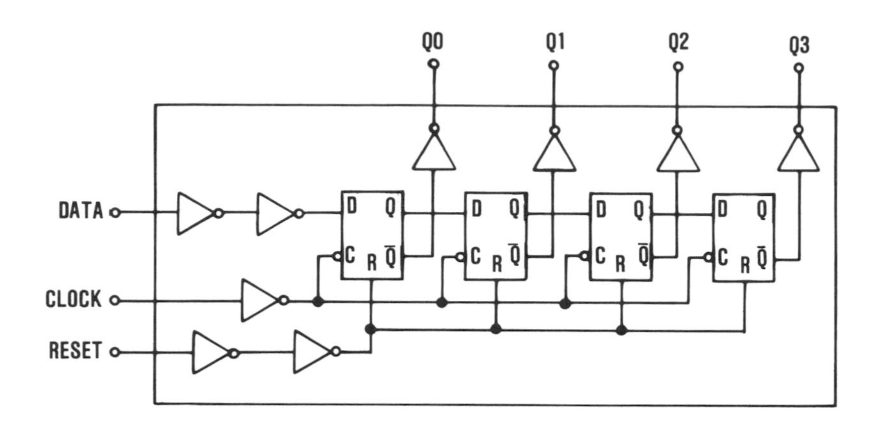

a) LOGIC DIAGRAM — 4-BIT STATIC SHIFT REGISTER

R	D	Q_n	Q_{n+1}
1	X	0	0
0	0	0	0
0	0	1	0
0	1	0	1
0	1	1	1

X—DON'T CARE

$Q_{n+1} = D_n$

b) TRUTH TABLE—CLOCKED OPERATION

FIGURE D.16.0 (a) Logic Diagram—4-Bit Static Shift Register
(b) Truth Table— Clocked Operation

will appear as a **1** at the Q output. An inverter is used here as output buffer and it simply uses the $\overline{Q}$ signal to provide the correct Q output. During the next clock pulse rising edge, the **1** is transferred to the Q output of the second FF. During subsequent clock pulses, this **1** logic information, or bit, moves from left to right through each of the FFs. It takes four clock pulses to move a bit from the first FF to the last. Of course, other information, either **0** or **1**, will appear at the D-terminal of the first FF and will move into the shift register.

If only four clock pulses are present, the four bits of information will be loaded into each of the four FFs and will appear as parallel data at the Q0, Q1, Q2 and Q3 output terminals. When a **1** signal appears at the reset terminal, all of the four FFs will change to **0** at their Q output. This is indicated in the first line of the truth table of Figure D.16.0. Note that in this truth table the change from one data bit or one clock pulse to the next is indicated by the Q_n and Q_{n+1} columns.

A typical static shift register IC will contain two identical 4-bit registers on a single IC. The following pages describe shift registers that can shift both left to right and right to left, and can provide parallel input as well as output in a variety of combinations.

KEY PARAMETERS

a) *Power dissipation.* Either the maximum or the typical power dissipation of an entire IC is given.
b) *Quiescent current.* The current drawn by the IC when no operations take place.
c) *Toggle rate.* The highest clock frequency that can be used for this shift register.
d) *Output loading (fan-out).* The number of digital inputs that can be driven from a particular output without impairing the operation of the driver.
e) *Output rise time.* The time elapsed between 10% and 90% of the amplitude of the leading edge of the output pulse.

APPLICATIONS

Shift registers are used throughout all digital systems to manipulate and rearrange digital data, to provide temporary storage of digital data, and to perform mathematical functions such as accumulation, comparison, etc.

D.16.1 Shift Register, Serial-In/Parallel-Out—Parallel-In/Serial-Out

DESCRIPTION

Two specific 8-bit shift register ICs are described and illustrated in Figure D.16.1. In Figure (a) the partial logic diagram of a serial-in/parallel-out register is shown. Note that there are two data inputs, combined in an AND circuit. This permits gating or control of the data at terminal A through appropriate signals at terminal B. Eight identical FFs are contained in this IC, controlled by a common clock and a common reset signal.

The logic diagram of an 8-bit parallel-in/serial-out shift register is shown in Figure D.16.1(b). Information is entered in parallel at the data

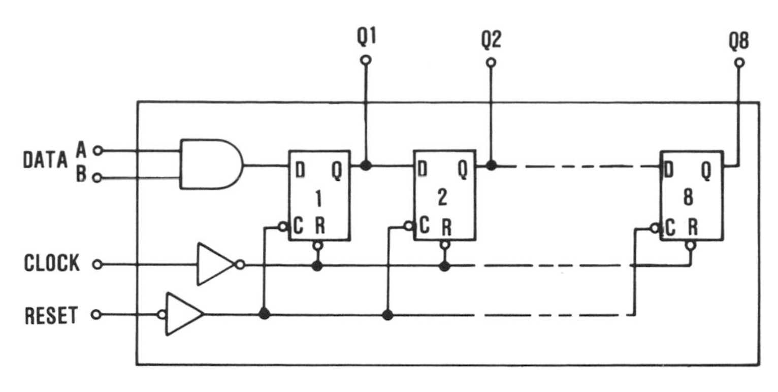

a) LOGIC DIAGRAM — 8-BIT, SERIAL IN/PARALLEL OUT SHIFT REGISTER

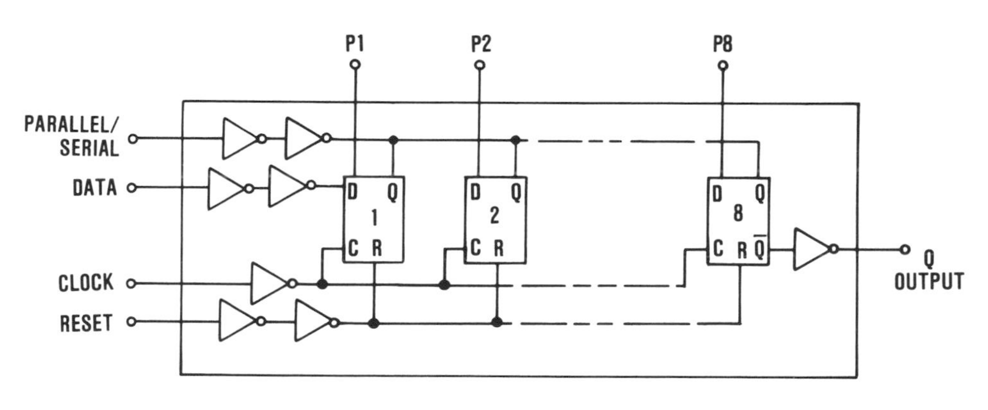

b) LOGIC DIAGRAM — 8-BIT, PARALLEL IN/SERIAL OUT SHIFT REGISTER

FIGURE D.16.1. (a) Logic Diagram—8-Bit, Serial In/Parallel Out—Shift Register
(b) Logic Diagram—8-Bit, Parallel In/Serial Out—Shift Register

inputs P1 through P8 which preset FFs 1 through 8. As clock pulses move the information from left to right, the output of the eight preset FFs will appear in serial form at the Q output terminal. It takes eight clock pulses, after each of the eight P-terminals has been set, to shift out this 8-bit byte of information.

In another mode of operation, this same shift register can enter data serially and output them serially. The parallel/serial terminal acts as a control and provides the gating for each FF. When the signal at that terminals is **0**, the IC will operate with parallel input and serial output, but when that control terminal is at **1**, the IC will enter data serially and shift them out in the same manner. The simpler serial-in/parallel-out shift register shown in Figure D.16.1(a) does not have this control, but if only the Q8 output is used, it will automatically act as serial output.

KEY PARAMETERS

a) *Power dissipation.* Typical power dissipation is 120 mW for low-power TTL ICs, 300 mW for standard TTL ICs, and 350 mW for low-power CMOS ICs.
b) *Toggle rate.* The highest clock frequency for standard and low-power TTL ICs is 45 MHz. For CMOS ICs 2.0 MHz is typical at 5 V power supply.
c) *Propagation delay, clock-to-output.* 30 ns is typical for low-power TTL ICs.
d) *Propagation delay time, reset-to-output.* For low-power TTL ICs, 25 ns is typical.
e) *Clock pulse width.* 20 ns is typical for low-power TTL ICs.
f) *Reset pulse width.* 20 ns is typical for low-power TTL ICs.

APPLICATIONS

Used in all aspects of digital systems, these shift registers are particularly useful as parallel-to-serial, or serial-to-parallel converters.

COMMENTS

In some versions of these shift registers, special input or output terminals are provided for the first three or other selected FFs. Consult the manufacturer's data to meet specific requirements.

D.16.2 Shift Register, 4-Bit Universal

DESCRIPTION

This shift register can perform both serial and parallel input and output as well as left and right shift. Universal shift registers are avilable in all digital IC families and use the same basic R-S flip-flop (FF) as the simpler shift registers. Their ability to accept data in serial as well as parallel form, to provide both serial and parallel outputs, and to allow the data to be shifted either to the left or to the right, is due to a specific array of logic gates, as illustrated in the partial logic diagram of Figure D.16.2. The input of each FF is controlled by an arrangement of four 3-input AND gates and one 4-input OR gate. These logic gates are controlled by four separate control lines, based on the 2-bit input selection lines, S0 and S1. These two binary signals correspond to four control lines, representing the four modes of operation. In one mode, the information in the FFs remains fixed and parallel inputs and outputs are available. In the second operating mode, the data is shifted from left to right. In the third mode, it is shifted from right to left, and in the fourth mode parallel operation of the inputs is enabled.

As illustrated in the logic diagram, the parallel outputs are available at all times. The clock and reset signal is common to all FFs. In all other respects the universal shift register operates in exactly the same way as the standard shift registers described in Sections D.16.0 and D.16.1

KEY PARAMETERS

a) *Power dissipation.* Typical power dissipation is 47 mW for low-power TTL and 425 mW for ECL under no-load conditions.
b) *Quiescent current.* The current drawn by the IC when no operations take place. 5.0 nA is typical at 5 V for low-power CMOS.
c) *Toggle rate.* The maximum clock frequency. 3.6 MHz at 5 V is typical for low-power CMOS, 200 MHz is typical for ECL, and 40 MHz is typical for standard and low-power TTL ICs.
d) *Output rise and fall time.* The time required for the 10% to 90% amplitude of the leading and trailing edge of the output signal. 100 ns is typical for low-power CMOS. 12 ns is typical for low-power TTL, and 2 ns is typical for ECL ICs.
e) *Propagation delay time, clock-to-output.* The time delay from the clock pulse edge until a change in output occurs. 235 ns at 5 V is typical for low-power CMOS, 2.0 ns is typical for ECL, and 20 ns is typical for low-power TTL ICs.
f) *Clock pulse width.* 140 ns is typical for low-power CMOS, 20 ns is typical for low-power TTL, and 2 ns is typical for ECL ICs.

APPLICATIONS

Universal shift registers are used in digital data communications and in digital computers. They are used for accumulators in the arithmetic units of central processors of digital computers and other applications where digital data must be shifted in the four possible modes.

COMMENTS

The application of universal shift registers requires timing and other control signals external to the shift register, and for this reason manufacturer's application notes should be carefully consulted.

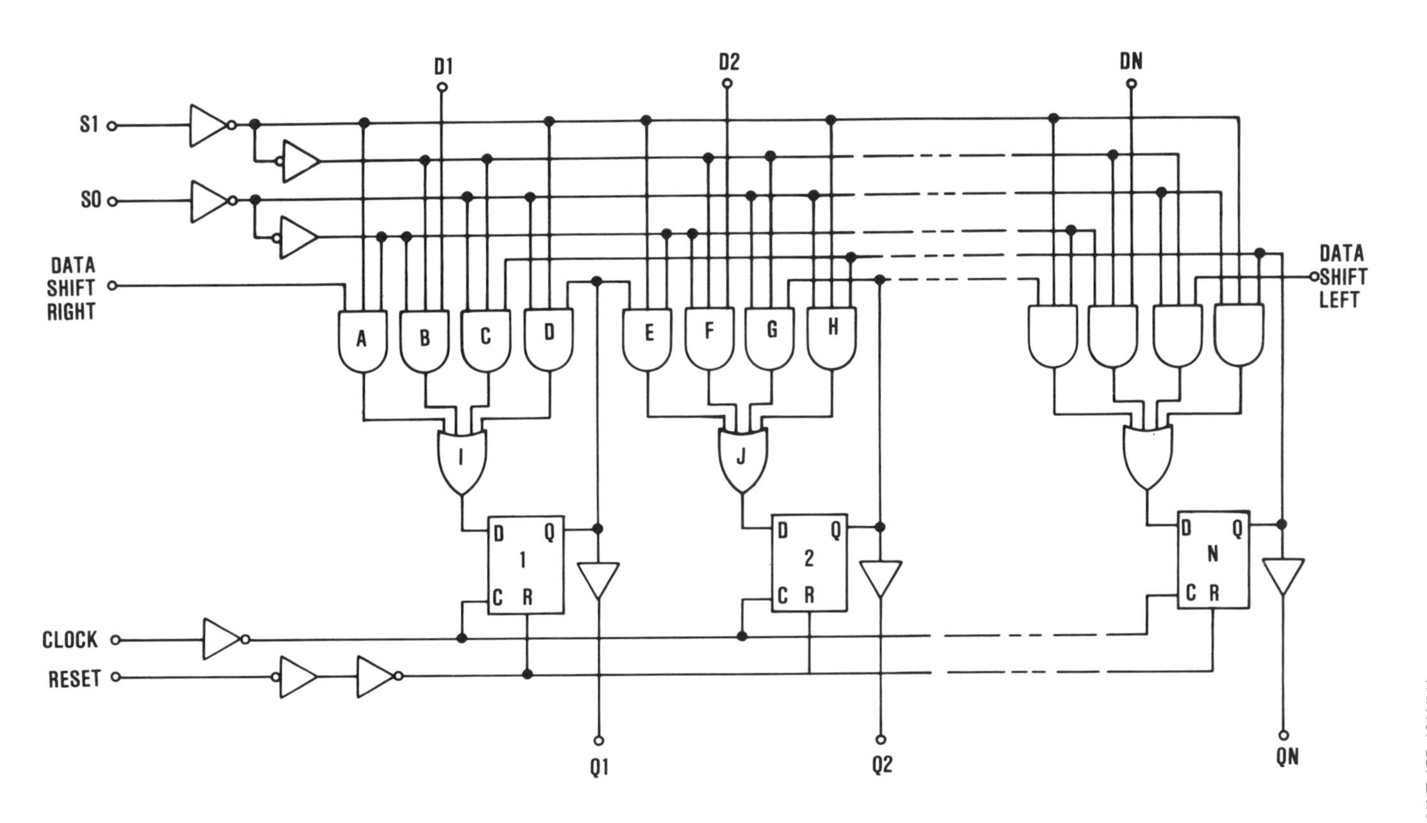

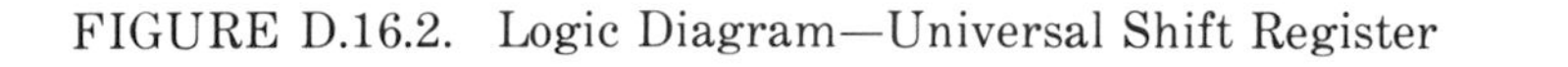
FIGURE D.16.2. Logic Diagram—Universal Shift Register

D.16.3 Shift Register, 8-Bit Universal

DESCRIPTION

8-bit universal shift registers operate in essentially the same way as the 4-bit universal shift registers described in Section D.16.2. The main difference between the 4 and the 8-bit universal shift register ICs is the use of a single terminal for parallel input and parallel output, made necessary by the limitations on the number of terminals available in the standard 16-pin DIP package. Figure D.16.3 illustrates the method used to permit a single pin to be used as both output and input for parallel data. Q1 through Q8 are used as outputs when the output control signal is in the **1** state. When the tri-state output buffers are disabled, Q1 through Q8 is in the high impedance mode. This means that signals can be applied to these Q terminals to act as parallel inputs for the respective stages.

In all other respects the logic control section operates in the same manner as the multiple AND/OR gate array shown in Figure D.16.2. Two mode selection signals, S0 and S1, provide the four available operating modes.

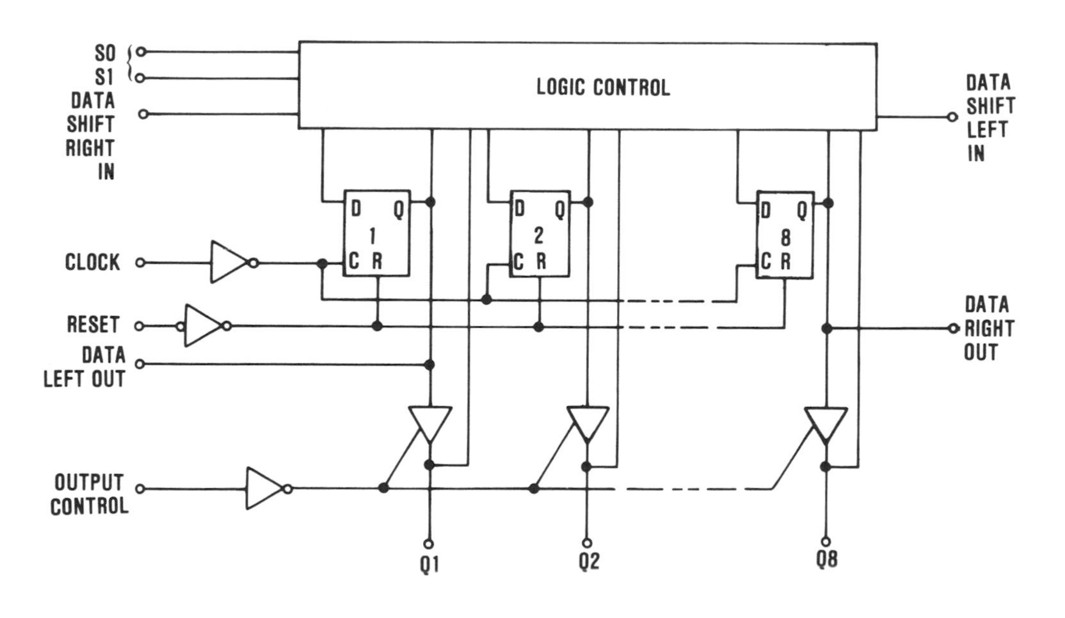

FIGURE D.16.3. 8-Bit Universal Shift/Storage Register

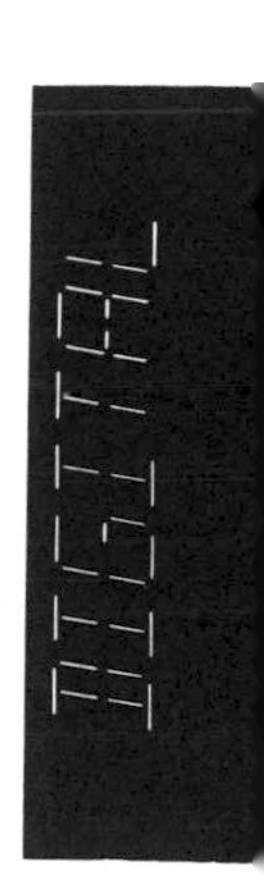

KEY PARAMETERS

All of the key parameters listed for the 4-bit universal shift register in Section D.16.2 are applicable to the 8-bit universal storage register.

APPLICATIONS

This IC is used in the central processing unit of digital computers where it can perform arithmetic functions and also act as accumulator. Because it has an 8-bit capacity, it is suitable for single byte operations. In many applications, 8-bit universal shift registers are cascaded to provide full word length functions. Digital communications systems and control systems that require operations involving left and right shift are also among the applications for this IC.

D.16.4 Shift Register, Successive Approximation

DESCRIPTION

This type of shift register is specifically designed for the successive approximation required in analog-to-digital (A/D) converters. The function block of Figure D.16.4(a) illustrates a serial input (D), a serial output (QO), and eight parallel outputs including the complement of Q7. Although a single clock input is shown, internal circuitry changes it to a 2-phase clock, which controls the internal operations. The ability to perform the successive approximations and store the results as indicated in the truth table, is provided by two flip-flops per shift register stage. A master and a slave FF are arranged so that the two non-overlapping, complementary clock signals control the shifting of information serially as well as in the parallel storage of the slave FF. There is a start input and a special output indication that the conversion is complete (CC).

The truth table of Figure D.16.4(b) demonstrates the principle of successive approximation. At t_0 the start input is **0**, and nothing happens. When the start input changes to **1**, Q7 changes to **0** and all other inputs, including conversion complete (CC), become **1**. In subsequent clock pulse periods, the individual inputs D7 through D0 get successively shifted into the respective places. Note that once D7 has reached the serial output (QO), it also is stored in Q7. It remains in Q7 until all of the conversions have been done. D6 is shifted through next and once it reaches Q6, it remains there. This is the successive approximation principle which is necessary to use a D/A converter, in connection with an analog input and a comparator, to perform A/D conversion. A more detailed discussion of A/D conversion is contained in the section on Interface Circuits. When all of the outputs are

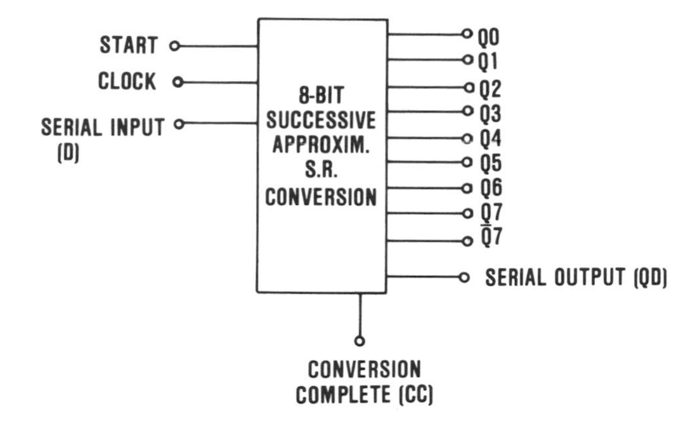

a) FUNCTION BLOCK

T	INPUTS		OUTPUTS									
tn	D	S	QD	Q7	Q6	Q5	Q4	Q3	Q2	Q1	Q0	CC
0	X	0	X	X	X	X	X	X	X	X	X	X
1	D7	1	X	0	1	1	1	1	1	1	1	1
2	D6	1	D7	D7	0	1	1	1	1	1	1	1
3	D5	1	D6		D6	0	1	1	1	1	1	1
4	D4	1	D5			D5	0	1	1	1	1	1
5	D3	1	D4				D4	0	1	1	1	1
6	D2	1	D3					D3	0	1	1	1
7	D1	1	D2						D2	0	1	1
8	D0	1	D1							D1	0	1
9	X	1	D0	↓	↓	↓	↓	↓	↓	↓	D0	0
10	X	1	X	D7	D6	D5	D4	D3	D2	D1	D0	0

b) TRUTH TABLE

FIGURE D.16.4 (a) Function Block
(b) Truth Table

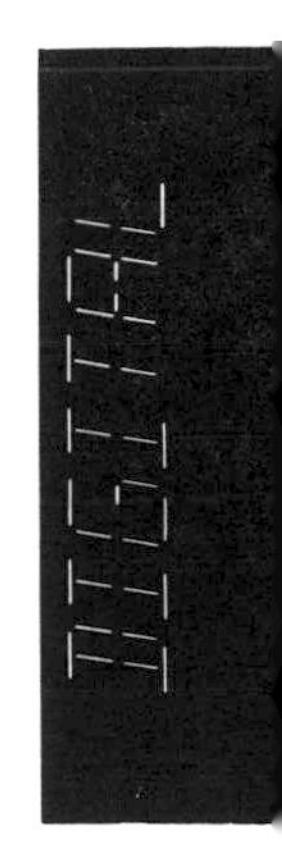

filled with their respective inputs, the conversion complete (CC) signal becomes **0**, indicating to external circuitry that no further information inputs are required. Although an 8-bit successive approximation shift register is illustrated here, these ICs are also available in 12-bit or in larger size, for specific A/D converter applications.

KEY PARAMETERS

a) *Power dissipation.* Typical power dissipation for standard TTL ICs is 300 mW and 75 mW for low-power TTL ICs.
b) *Toggle rate.* The maximum clock frequency. 21 MHz is typical for standard TTL ICs, and 15 MHz is typical for low-power TTL ICs.
c) *Propagation delay time, clock-pulse-to-output.* The time required from the clock pulse triggering edge until the output changes. 26 ns is typical for standard TTL ICs and 30 ns is typical for low-power TTL ICs.
d) *Clock pulse width.* 30 ns is typical for standard and low-power TTL ICs.

APPLICATIONS

Successive approximation registers are used primarily for analog to digital conversion systems. These ICs can also be used as serial-to-parallel converters, as ring counters, and as the storage and control element in recursive digital routines.

D.16.5 Shift Register, 18-Bit Static

DESCRIPTION

This IC is a simple static shift register without master reset, that consists merely of five separate, shorter registers as illustrated in the block diagram of Figure D.16.5. Note that separate inputs and outputs are provided for the shift register containing the first four bits and bits 10 through 14. Bits 5 through 9 and 14 through 18 are obtained by a 4-bit shift register added to a 1-bit, allowing an input or output at the common junction, Q8 or Q17 respectively. In spite of the relatively large number of bits, this shift register operates in the same basic mode as the simple static shift register described in D.16.0. The main difference is the absence of a reset signal, and this means that all 18 bits must be shifted through to clear the register. To set this register to 0, a set of 18 zeroes must follow the data.

KEY PARAMETERS

a) *Quiescent current.* The current drawn by the entire IC when no operations are performed. 5 nA is typical at 5 V for low-voltage CMOS.
b) *Output pulse rise and fall time.* The time period from the 10% to the 90% amplitude of the output pulse. 180 ns is typical for low-power CMOS.
c) *Toggle rate.* The highest clock frequency. 5.0 MHz is typical at 5 V for low-voltage CMOS.

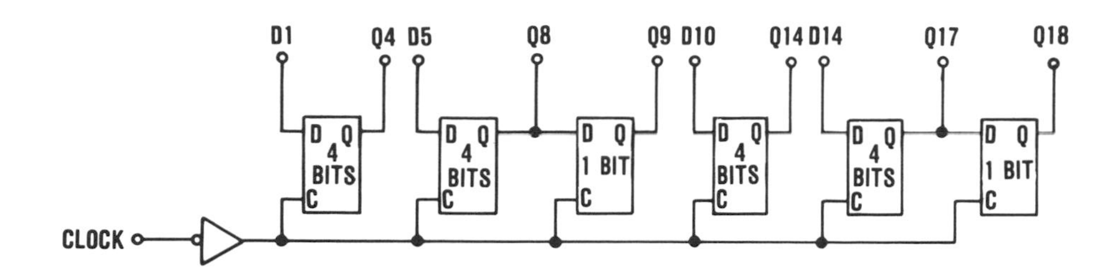

FIGURE D.16.5. Block Diagram—18-Bit Static Shift Register

d) *Clock pulse width.* 100 ns is typical for low-power CMOS.
e) *Propagation delay time, clock pulse-to-output.* The time delay from the edge of the clock pulse until change takes place at a particular output. 300 ns is typical at 5 V for low-power CMOS.

APPLICATIONS

This type of IC is particularly useful when fixed time delays, in terms of the number of shift pulses, are required. It can also serve as small recirculating memory and, because of the flexiblity of the terminals available, a number of other special digital computer functions can be implemented with this IC.

D.16.6 Shift Register, Dual 64-Bit Static

DESCRIPTION

This IC consists of two 64-bit static shift registers which have a parallel output every 16 bits, as illustrated in Figure D.16.6. Each register has separate clock and write enable inputs as well as separate data inputs and Q outputs. The serial data input and the clock output operate in the conventional, static shift register method. The purpose of the write enable signal is to control the tri-state buffers in each of the Q outputs. When the write enable input is **0**, all of the output stages will be at high impedance. This permits data to be entered at these terminals, just as in the eight-bit universal shift register described in D.16.3. Because of the tri-state output, this IC can be used directly with the bus system.

KEY PARAMETERS

a) *Quiescent current.* The current drawn by the IC when no operations are performed. 10 nA is typical at 5 V for low-power CMOS.

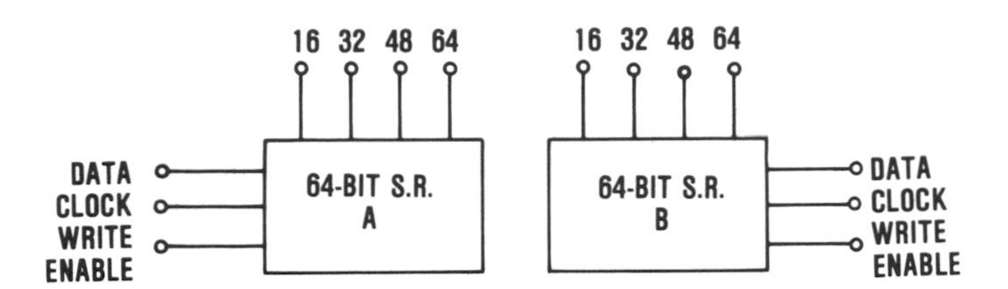

FIGURE D.16.6. Function Block—Dual 64-Bit Shift Register

b) *Toggle rate.* The maximum clock frequency. 3.0 MHz is typical for low-power CMOS at 5 V.
c) *Output rise and fall time.* The time between the 10% and 90% amplitude of the leading and trailing edge of the output signal. 100 to 180 ns is typical for low-power CMOS.
d) *Propagation delay time, clock pulse-to-output.* The time period required between the clock pulse triggering edge and the change in the output. 475 ns is typical for low-power CMOS.
e) *Clock pulse width.* 170 ns is typical for low-power CMOS.

APPLICATIONS

This IC is used in digital systems for temporary data storage circuits, time delay circuits, special purpose bus interfaces, and similar digital uses.

D.16.7 Shift Register, 128-Bit Static

DESCRIPTION

This shift register IC provides 128-bits worth of temporary storage in a single shift register. Parallel data outputs are available every 16 bits from 16 through bit 128. The function block illustrated in Figure D.16.7 shows only a single data and a clock input, without any reset, write enable, or other control features. This means that this 128-bit shift register IC can be used only in a continuous shifting mode.

KEY PARAMETERS

a) *Quiescent current.* The current required by the IC when no operations take place. 10 nA is typical for low-power CMOS.
b) *Toggle rate.* The maximum clock frequency 1.9 MHz is typical at 5 V for low-power CMOS.
c) *Output rise and fall time.* The time required from the 10% to the 90% amplitude of the leading and trailing edge of the output. 100 to 180 ns is typical for low-power CMOS ICs.

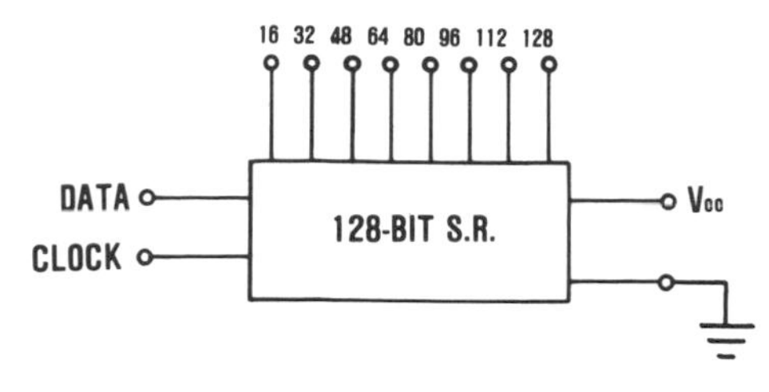

FIGURE D.16.7

d) *Propagation delay time, clock pulse-to-output.* The time required from the triggering edge of the clock pulse to the change in state at the output. 600 ns is typical for low-power CMOS ICs.
e) *Clock pulse width.* 300 ns is typical for low-power CMOS ICs.

APPLICATIONS

The 128-bit static shift register is used as temporary data storage in a variety of digital communications and computing equipment.

D.16.8 Shift Register, 32-Bit Left/Right

DESCRIPTION

This IC provides 32-bits of serial shifting, without any parallel input or output capability. The function block of Figure D.16.8 illustrates two data inputs, one for shift right and one for shift left, and two data outputs, for shift right and shift left, respectively. In addition to the clock input, there is also a clock inhibit terminal which permits control of the clock pulses. When the clock inhibit terminal is at **1**, no shifting can occur. The left/right control must be in the **1** state to have all shifting done to the left and in the **0** state to have the shift register shift to the right. The recirculating control enables the connection from the first to the last shift register bit, allowing recirculation of the shift register's contents when this terminal is **0**.

These control inputs make it possible to use this shift register IC as recirculating memory, as a "last-in/first-out" (LIFO) or "first-in/first-out" (FIFO) memory device.

KEY PARAMETERS

a) *Maximum power dissipation.* The maximum power that can be dissipated at 25°C. 200 mW is typical for high-voltage CMOS ICs.
b) *Quiescent current.* Current drawn by the entire IC when no operations take place. 40 nA is typical for high-voltage CMOS ICs.

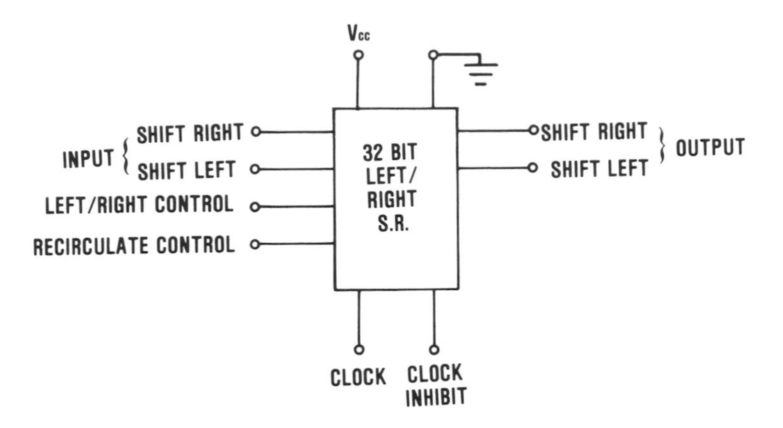

FIGURE D-16.8

c) *Toggle rate.* The maximum clock frequency. 1.0 MHz is typical at 5 V for high-voltage CMOS
d) *Propagation delay time, clock-to-output.* The time required from the triggering edge of the clock pulse until a change takes place at the output. 60 ns is typical at 5 V for high-voltage CMOS ICs.
e) *Clock pulse width.* 450 ns is typical for high-voltage CMOS ICs.

APPLICATIONS

This IC can be used for a variety of data storage functions as well as the normal serial shift register applications. It is particularly useful in digital computers to provide a so-called "stack" storage in the form of the last-in/first-out (LIFO) operation. It can also be used as digital delay line, time delay circuit, and recirculating shift register memory.

D.16.9 Shift Register, 1 to 64-Bit Variable Length

DESCRIPTION

This IC is a static, clocked serial shift register that can be varied in length from one bit up to 64 bits, by entering a 6-bit digital number. As illustrated in the function block of Figure D.16.9, there are six inputs, labeled 1, 2, 4, 8, 16 and 32, at which the length selection word is entered. The effective length of the shift register is determined by the binary number entered at these six terminals.

Data is entered at terminals A and B, with the A/B select input

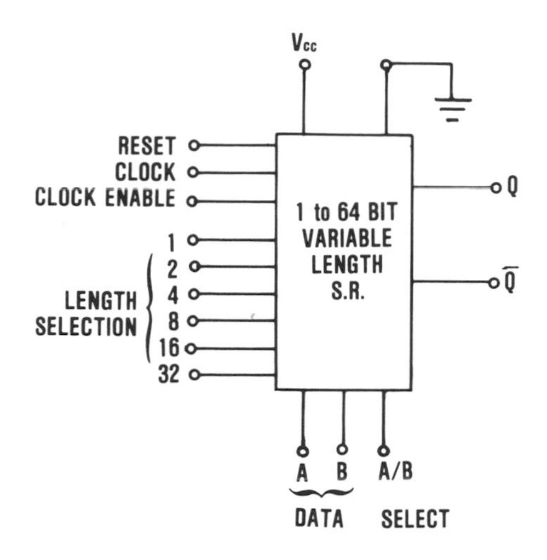

FIGURE D-16.9

determining which one will be shifted through to the serial outputs Q and its complement. The other controls include a master reset, a clock and a clock enable terminal. Because serial data may be selected from either the A or B data input by the **0** or **1** state of the A/B select signal, it is possible to use this shift register as recirculating shift register.

KEY PARAMETERS

a) *Quiescent current.* Current drawn by the IC when no operations take place. 10 nA is typical at 5 V for low-power CMOS.
b) *Toggle rate.* The maximum clock frequency. 2.5 MHz is typical at 5 V for low-power CMOS.
c) *Output rise and fall time.* The period between the 10% and 90% amplitude points on the leading and trailing edges of the output pulse. 180 ns is typical at 5 V for low-power CMOS.
d) *Propagation delay time, clock pulse-to-output.* The time from the clock pulse triggering until the output changes. 500 ns is typical at 5 V for low-power CMOS.
e) *Clock pulse width.* 220 ns is typical for low-power CMOS.
f) *Reset pulse width.* 300 ns is typical for low-power CMOS.

APPLICATIONS

This IC can be used for variable length digital delay lines or simply for shift registers of a specific, odd length.

D.16.10 Shift Register, First-In/First-Out (FIFO)

DESCRIPTION

This IC contains sufficient storage and control circuits to implement some simple computer memory functons. The block diagram shown in Figure D.16.10 illustrates the 4-bit parallel input and 4-bit parallel output, together with the important control functions. The memory itself consists of four 16-bit serial shift registers. This means that four bits can be entered simultaneously and pass through 16 stages until they are available at the output. In other words, the 4-bit digital data group entered first would also appear at the output first, hence the name First-In, First-Out (FIFO). Note that there is no continuous clock but the shifting action is provided by the shift-in and the shift-out signal. When four bits worth of data are available at the input, the shift-in signal will provide one shift pulse to move them in. When the next 4-bit group appears, another shift-in pulse is used. In the same way, the shift-out signals control the output, one step at a time.

The tri-state control signal makes it possible to interface the output directly with a data bus since the output can be set to the high impedance state. The control section provides two status outputs that are important for interfacing micro- or minicomputer functions. One is the "data in ready" signal and the other is the "data out ready" signal. The computer or other controller responds with a shift-in or shift-out signal, as may be needed.

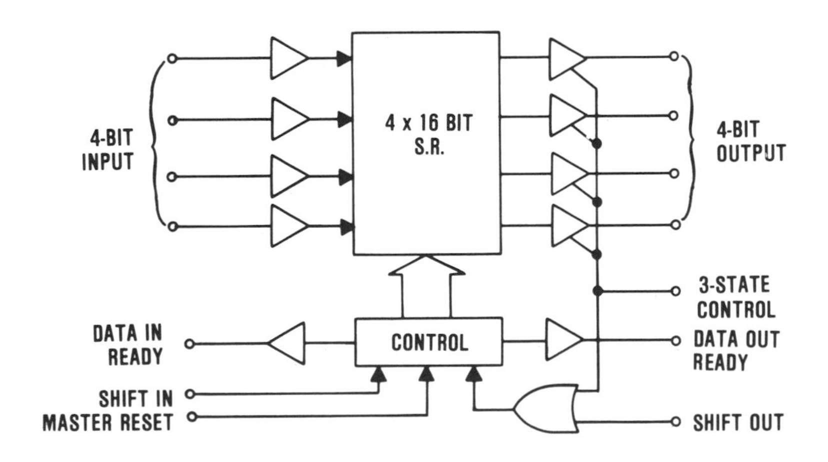

FIGURE D.16.10. Block Diagram—4×16 Bit FIFO S.R.

KEY PARAMETERS

a) *Maximum power dissipation.* Maximum power that this IC can dissipate at 25°C. 200 mW is typical for high-voltage CMOS ICs.
b) *Toggle rate.* The highest rate at which shift-in and shift-out pulses can be presented. 1.5 MHz is typical at 5 V for high-voltage CMOS ICs.
c) *Output rise and fall time.* Time between the 10% and 90% amplitude of the leading and trailing edge of the output signal. 15 ns is typical for high-voltage CMOS ICs.
d) *Propagation delay time.* The time from the trigger edge of the shift-in or shift-out pulse until the data output ready signal changes. 185 ns is typical for high-voltage CMOS ICs at 5 V.
e) *Shift-in pulse width.* 200 ns is typical for high-voltage CMOS ICs at 5 V.
f) *Shift-out pulse.* 360 ns is typical at 5 V for high-voltage CMOS ICs.
g) *Master reset pulse width.* 220 ns is typical at 5 V for high-voltage CMOS ICs.

APPLICATIONS

A FIFO register can be used for bit rate smoothing, buffering between the CPU and a terminal, and a host of other computer buffering applications. This IC is also used for automatic dialing circuits, in data communications, and in radar acquisition systems.

D.17.0 Schmitt-Trigger

DESCRIPTION

Schmitt-Trigger circuits are incorporated in many different digital IC configurations, but the basic function of amplifying at specific threshold levels is also available in the form of the basic inverter amplifier. Six of these amplifiers are usually available in a single IC.

The key characteristic of the Schmitt-Trigger is the hysteresis curve, as illustrated in Figure D.17.0(b). As the input voltage drops from two volts down to approximately 0.8 volts, the output voltage suddenly increases to about 3.4 volts and remains there. If the input voltage goes from zero toward two volts, the output voltage will drop down to near zero, once the input voltage goes past 1.6 volts. The arrows of the curve indicate the rapid rise and fall times available with the Schmitt-Trigger circuit.

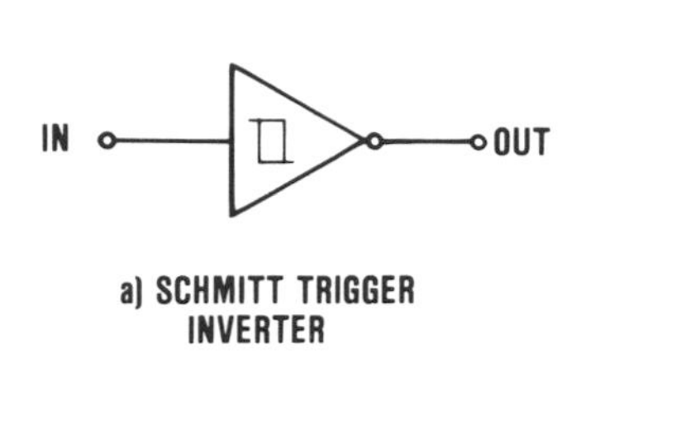

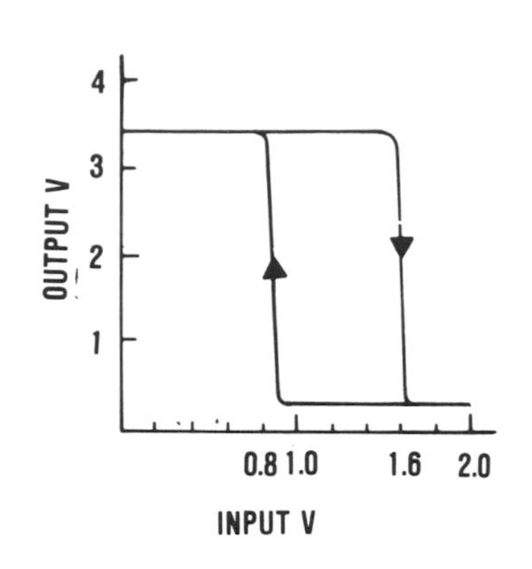

FIGURE D.17.0 (a) Schmitt Trigger Inverter
(b) Hysteresis Curve

KEY PARAMETERS

a) *Hysteresis voltage.* The difference in input voltage between a positive and negative going signal. In the curve of Figure D.17.0(b) the hysteresis voltage is approx. 0.8 volts. For low-power CMOS ICs, this value ranges from 0.85 to 3.4 V. For some low-power TTL ICs the hysteresis voltage may be 0.2 of the power-supply voltage. 0.8 V is a typical value for most standard and low-power TTL ICs. In some ICs the hysteresis voltage is temperature compensated for accuracy.
b) *Positive threshold voltage.* The positive going input voltage at which the output voltage drops to near zero. 3.3 V is typical at 5 V power supply for low-power CMOS. 1.7 V is typical for low-power and standard TTL ICs.
c) *Negative threshold voltage.* The negative going input voltage at which the output voltage increases to its maximum. 1.7 V is typical for low-power CMOS, and 0.9 V is typical for low-power and standard TTL ICs.
d) *Output rise time.* The time required for the output voltage amplitude to go from 10% to 90%. 180 ns is typical for low-power CMOS, and 20 to 30 ns is typical for standard and low-power TTL ICs.
e) *Output fall time.* The time required for the output amplitude to drop from 90% to 10%. 100 ns is typical for low-power CMOS, and 20 ns is typical for standard and low-power TTL ICs.
f) *Propagation delay time.* The time delay from input to output. 650 ns is typical for low-power CMOS, 200 ns is typical for low-power TTL ICs, and 20 ns is typical for standard TTL ICs.
g) *Quiescent current.* The current drawn by the circuit when no operations are performed. 0.5 nA is typical for low-power CMOS ICs, 80 μA for low-power TTL, and 0.5 mA for standard TTL ICs.

h) *Power dissipation.* 200 mW is typical for low-power CMOS ICs at maximum operating frequency, at 5 V, 40 mW is typical at low-power TTL, and 200 mW is typical for standard TTL ICs.

APPLICATIONS

Schmitt-Trigger circuits are used whenever slow transition input circuit signals must be changed into specific trigger levels. This speed-up of a slow waveform edge in interface receivers, level detectors and other logic input circuits, also helps to reduce the effects of noise. Schmitt-Trigger circuits are generally incorporated in digital computer bus interfaces, in transmission line receivers, and in telemetry where digital data is transmitted over a noisy medium.

COMMENTS

Many ICs specify a hysteresis action, without clearly stating that it is a Schmitt-Trigger circuit. To determine the fine difference between a level detector and Schmitt-Trigger, the reader is referred to the manufacturer's specific literature for each type of IC.

D.18.0 Timer, Programmable (Digital)

DESCRIPTION

The programmable timer described in the Analog Section, A.16.3, differs from the digital timer described here in that the former concentrates on the basic analog timing circuit while the digital version depends, primarily, on the divider. While there are more complicated programmable timers available, the block diagram of Figure D.18.0 illustrates the essential functions. The 24-stage divider is split into two portions. Stages 1 through 17 are fixed and have no parallel output. Only the last seven stages, from 18 through 24, have accessible parallel outputs. These last seven stages provide a change, in steps of two times, from 2^{18} (262,144) to 2^{24} (16,777,2216).

In this particular timer, two amplifiers, A and B, are available with separate inputs and outputs so that the user can connect them either as crystal oscillator, as R-C oscillator, or as the input buffer for some external clock signal. A single master reset returns all stages to **0** and disables the two amplifiers. If they are used as oscillators, the application of the reset pulse will prevent oscillation and thereby save standby power.

A more complex version of this type of timer contains a preliminary divider of nine stages, which can be bypassed on external command and has as many as 15 outputs, selectable by a 4-bit programming input. The basic operation of the timer, however, is the same.

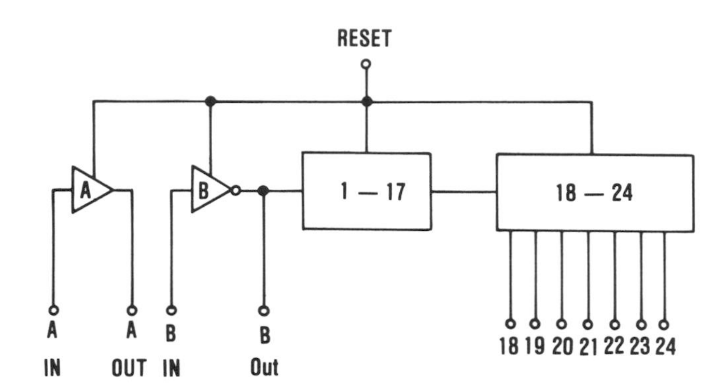

FIGURE D.18.0. Block Diagram—24-State Timer

KEY PARAMETERS

a) *Quiescent current.* The current drawn by the IC when no operations take place. 5.0 nA is typical at 5 V for low-power CMOS and 40 nA is typical for high-voltage CMOS ICs.
b) *Maximum clock pulse frequency.* 2.0 to 7.0 MHz for low-power and high-voltage CMOS types, respectively.
c) *Propagation delay time.* The time required from each clock input to a corresponding output. 2.2 to 4.5 ns is typical for high-voltage and low-power CMOS respectively.
d) *Minimum clock pulse width.* 100 to 150 ns for all types of CMOS ICs.

APPLICATIONS

This type of IC is used to provide accurate timing from a crystal oscillator, for timing applications such as clock synchronizing and digital timing references. These ICs may also be used to drive miniature synchronous motors, stepping motors, or power transistors which can then control larger power devices.

I. INTERFACE ICs

The integrated circuits described in this section vary greatly in their application, function, and key parameters because all types of interface ICs are included. Analog-to-digital converters (ADC) and digital-to-analog converters (DAC) are sometimes classified as a separate category, with many variations and combinations of features, accuracies, and operating speeds. Functionally, however, there is less variation among ADCs and DACs than is suggested by the large number of ICs on the market. There is little difference, for example, between an MOS DAC that has 10 bits and one that has 12, or between a 8-bit ADC that has a straight binary output and one with a binary-coded-decimal (BCD) output. The ADCs and DACs described in Section I are generally representative of the most widely used ICs and contain all of the features found—in various combinations—in all ADC and DAC ICs. Because this book is devoted entirely to monolithic ICs, we have not included hybrid and special purpose modules.

In some texts, line drivers and receivers, ICs used in data bus applications, are considered interface devices. These ICs are generally used as part of either a digital controller or computer or as part of a peripheral device, and in this book they are included in Section D, considered a digital IC rather than an interface function. Level shifters are included as interface ICs because they facilitate the interfacing of different digital IC families.

We have also included those memory interface ICs that are usually used to interface external or additional memory capacity with the microprocessor (MP) or microcomputer (MC). In general, these memory interface ICs work with IC memories, not with peripheral devices. Another classification, system interface ICs, deals with the peripheral memories such as a floppy disc or tape drive. Because there are hosts of special system interface ICs for each microprocessor or microcomputer family, we have included only representative examples of each widely used system interface function. As in the case of the ADC and the DAC, the differences between a particular system interface function, as implemented for different MPs, are minor and do not warrant separate discussion. All floppy disc controllers, for example, operate basically in the same manner, regardless of MP family, because the floppy disc control requirements are the same.

The reader looking for a particular interface function not found in this section is advised to refer to the index. Most likely the particular function is included in either Section A (Analog) or Section D (Digital).

I.1.0 Analog-to-Digital Converter (ADC)

DESCRIPTION

A large variety of analog-to-digital converters (ADC) is on the market for a host of different applications. Almost all ADCs operate on either one of the two principles described below, and while many of them are available in monolithic ICs, hybrid modules are frequently used for high-precision, special purpose applications. The examples described in this book have been selected as typical of the most widely used ADCs.

The comparison method of analog-to-digital conversion is illustrated in Figure I.1.0.1. The block diagram shows a counter driving a resistance ladder. Note that the values of the resistors increase in the binary fashion. The most significant bit (MSB), resistor R, drives the input of the op amp, with feedback resistor RF. Each less significant bit has an output resistor twice as high, making the least significant bit (LSB) output resistor 64 times the value of R. The analog input voltage is applied to both comparator 1 and comparator 2. Note that comparator 1 has a Schmitt-Trigger function and receives the summed output of the counter. The staircase waveform of Figure I.1.0.1(b) illustrates this comparison between the analog voltage input and the output of the counter through the resistance ladder network. Until the analog input voltage reaches the level of the staircase, output pulses, corresponding to the clock input, will pass through comparators 1 and 2, and the two NAND circuits to the serial digital output terminal. When comparator 1 senses that the voltage at the analog input is the same as the staircase, the pulses stop. The number of output pulses then corresponds to the analog input voltage. Most comparison ADCs contain more sophisticated circuitry than the one illustrated in Figure I.1.0.1. Parallel output as well as serial output and a method of speeding up the comparison by successive approximation is described in I.1.1.

The second method of analog-to-digital conversion uses a linear ramp which relates the analog input voltage to a time period. An integrator is used to generate this ramp, as illustrated in the block diagram of a dual ramp ADC illustrated in Figure I.1.0.2. The analog input voltage passes through an input amplifier, resulting in a current I_X. The integrator is a simple operational amplifier with a capacitive feedback network. The key to the accuracy of this system lies in the fact that the analog current I_X is allowed to generate the ramp for a fixed period of time (T1). At that point, the ramp control signal switches the input of the integrator to the reference current I_R. The time required for the ramp to reach the zero level is the time that is used for the digital conversion. At maximum analog input voltage, this constant slope I_R will take a maximum time T2. When the analog input voltage is lower, that descending ramp will be shorter. The time period T2 is recognized by the comparator, a Schmitt-Trigger circuit, which generates a

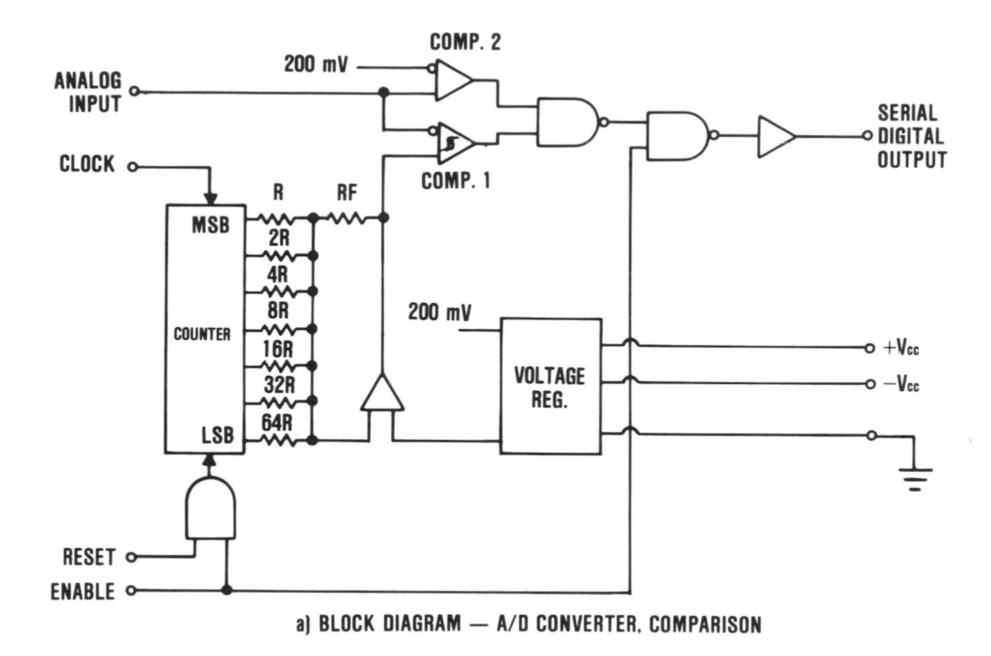

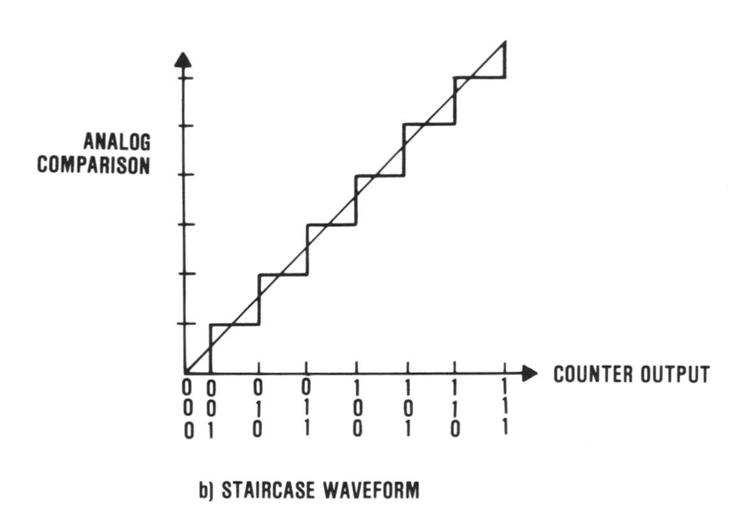

FIGURE I.1.0.1 (a) Block Diagram—A/D Converter, Comparison
(b) Staircase Waveform

gate corresponding to T1 plus T2. A fixed frequency clock signal is generated internally. The number of pulses that will be allowed to reach the output depends on the T2 gate. In other words, the analog voltage has been converted into a time period, and the time period has been converted into a digital number.

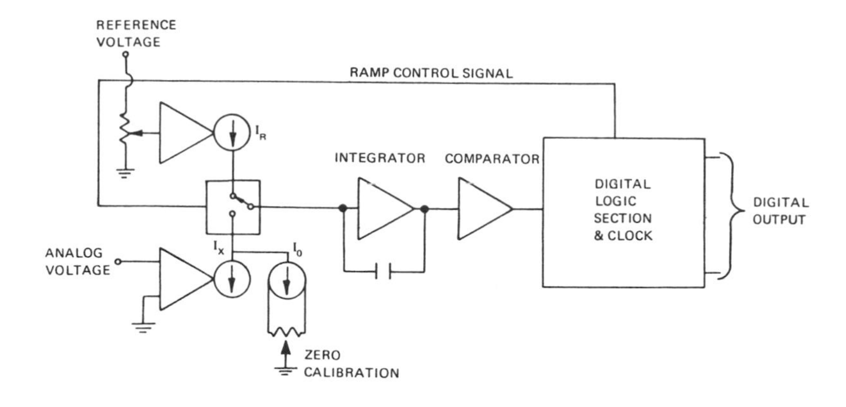

a) BLOCK DIAGRAM — DUAL RAMP A/D CONVERTER

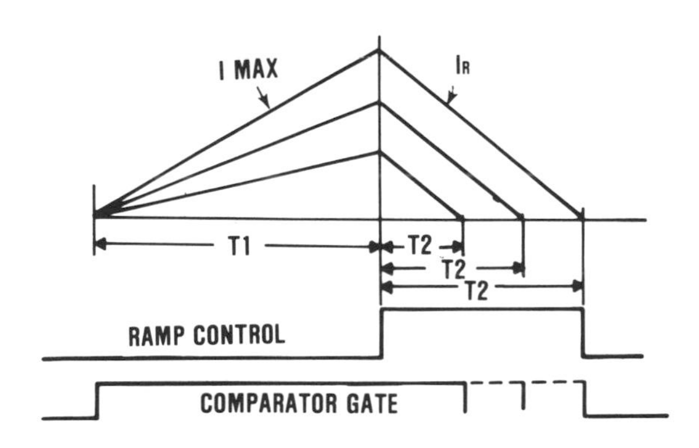

b) DUAL RAMP WAVEFORMS

FIGURE I.1.0.2 (a) Block Diagram of Dual Ramp A/D Converter
(b) Dual Ramp Waveforms

(*Complete Guide to Digital Test Equipment*, W.H. Buchsbaum, Prentice-Hall, Inc., 1977.)

In the comparison method, the accuracy of the system is limited by the number of bits in the counter and by the accuracy of the reference voltages. In the integrating ADC, the accuracy is limited by the reference voltage and by the internal clock frequency. Some comparison and some integrating type ADCs will be described in the following pages, including one featuring high precision and using a special quadruple ramp scheme. Most ADCs are "ratiometric" which means that some ratio can be introduced between the analog input and the digital output, other than the 1:1 relationship.

KEY PARAMETERS

a) *Relative accuracy*. Usually expressed in %, PPM, or fractions of the least significant bit (LSB). It describes the deviation of the analog value at any digital code, from its theoretical value at the full analog range of the device transfer characteristics. Relative accuracy can also be interpreted as a measure of nonlinearity.
b) *Conversion time*. The time required for a complete measurement of a specific analog voltage increment. This varies according to the number of bits, the method of conversion, and the type of IC technology used.
c) *Linearity*. Usually expressed in % or PPM of the full scale range or in fractions of the LSB, this describes the deviation of analog values from a straight line in a plot of measured conversion relationships. Sometimes expressed as nonlinearity.
d) *Power supply sensitivity*. Indicates the % change of analog input, relative to a % change of the DC power-supply output.
e) *Temperature coefficients*. Temperature instabilities are usually expressed in %/°C, PPM/°C, or as fraction of one LSB/°C. Temperature coefficients may be given for the converter gain, linearity, and reference voltage.

APPLICATIONS

ADCs are used in instrumentation, telemetry, computer controlled machinery, and other systems where an analog input signal must be used by a digital device. Most physical quantities, temperature, pressure, light, radiation, etc., can be measured by conversion into analog electrical signals and these must then be converted to digital signals for use by a digital process.

I.1.1 Analog-to-Digital Converter (ADC), 10-Bit, Successive Approximation

DESCRIPTION

This comparison type ADC uses a successive approximation shift register for high-speed A/D conversion. As illustrated in the block diagram of Figure I.1.1, there are three major functional blocks. The analog input voltage and the reference input are applied to a 10-bit digital-to-analog converter (DAC).

In the successive approximation method, the analog voltage is compared against a group of weighted reference voltages, the output of the DAC. Repeated rapid comparisons between the analog input and the output of the DAC are recorded and evaluated in the successive approximation logic. During the first approximation, half of the total DAC output voltage is used. If the analog input is larger than that, the next increment will be a quarter. If the analog voltage is now smaller, the quarter will be removed and one eighth will be added to the one half. In this successive method, the sum of the DAC output voltages that has been added up will be within one LSB of the actual value of the analog input. An external comparator, reference and clocking controls are required.

The particular IC described by the block diagram of Figure I.1.1 has special features which are suitable for connecting this ADC to a microprocessor. The start signal input directs the successive approximation logic to start the comparison process. There is a sync output which can be used in conjunction with the serial output, permitting a microprocessor to accept the serial data under clock control. The busy line, together with the busy enable line, controls the tri-state logic so that this IC can interface directly with a digital bus. The "hi" and "lo" enable lines control the two most significant bits (MSB) and the remaining eight LSBs. This permits the microprocessor to direct the interface to accept only an 8-bit byte. The SC8 input into the successive approximation logic permits the microprocessor to stop the clock after only eight bits, when this IC is used for only 8-bit conversion.

KEY PARAMETERS

The parameters presented below apply to a typical CMOS ADC. (For definition of parameters see I.1.0.)

a) *Relative accuracy.* ±½ LSB.
b) *Conversion time.* 20 ms is typical; 40 ms is the maximum.

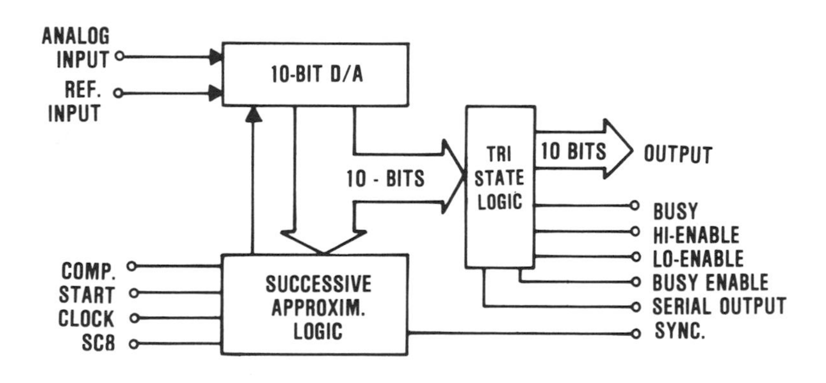

FIGURE I.1.1. Block Diagram—A/D Converter, Successive Approximation Type

c) *Linearity*. Stated as differential nonlinearity for this IC. 1 LSB is maximum.
d) *Maximum power dissipation*. 1 W up to 50°C.
e) *Internal clock frequency*. 100 KHz is typical, and up to 600 KHz is possible.

APPLICATIONS

This is a general purpose ADC, compatible with microprocessors and capable of eight or ten bits resolution.

COMMENTS

Similar ADCs are available in current injection logic and hybrid modules.

I.1.2 Analog-to-Digital Converter (ADC), Integrating, 8 to 12 Bits

DESCRIPTION

This IC is available with up to 12 bits of resolution and uses the integrating type of converter described in more detail in Section I.1.0. The block diagram of Figure I.1.2 illustrates the connection of an external integrating capacitor between the analog input terminal and the output of the integrator. Note the use of a current switch and external reference voltage. The control logic receives the start input and then provides all of the necessary internal functions. A conversion complete and a data valid signal are available for use by a microprocessor or some other digital interface. The data counter and output latches also provide the overrange indication. The choice of whether to use 8, 10, or 12 bits of resolution is made simply by the connection to the data output. All 12 bits are available, but only eight may be used.

KEY PARAMETERS

This IC is a low-power CMOS device.

a) *Relative accuracy*. ±1 LSB.
b) *Conversion time*. 1.8 ms for the 8-bit resolution up to 24 ms for the 12-bit resolution.
c) *Linearity*. ½ LSB.
d) *Power supply sensitivity*. 0.05% per 1.0% of the power-supply voltage is typical.

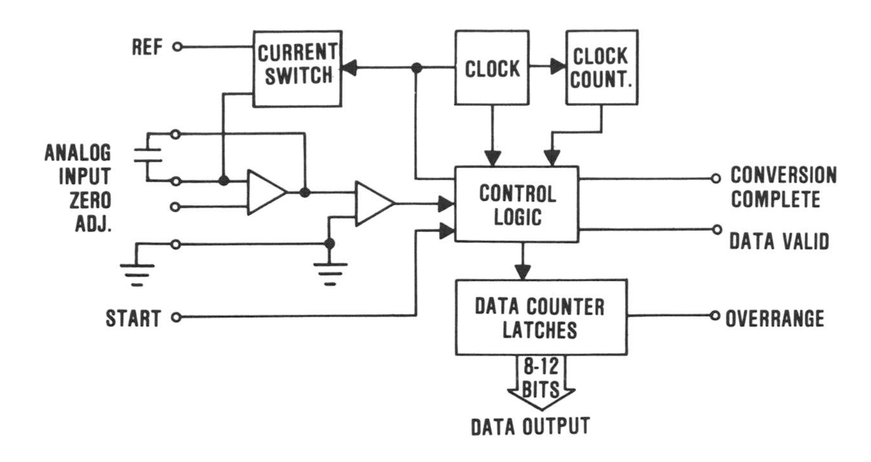

FIGURE I.1.2. Block Diagram—A/D Converter, Integrating

e) *Gain temperature coefficient.* 40 PPM/°C.
f) *Power consumption.* Typical power consumption is 20 mW for this low-power CMOS IC.

APPLICATIONS

Because of the relatively long conversion time, this IC is used in A/D applications where relatively slowly changing physical characteristics such as temperature, pressure, flow, etc., are used.

I.1.3 Analog-to-Digital Converter (ADC), Integrating, 3½ Digits

DESCRIPTION

A number of manufacturers produce ICs specifically designed for digital volt-ohm meters which perform both A/D and binary-to-decimal conversion. The block diagram of Figure I.1.3 illustrates the various functions of this type of IC. Both analog and digital circuits are constructed by the same low-power CMOS technology on the same chip. The analog subsystem receives the reference and the analog input voltage. The integrating R-C network and offset capacitors are the only external components required. The principle of the dual-ramp converter is explained,

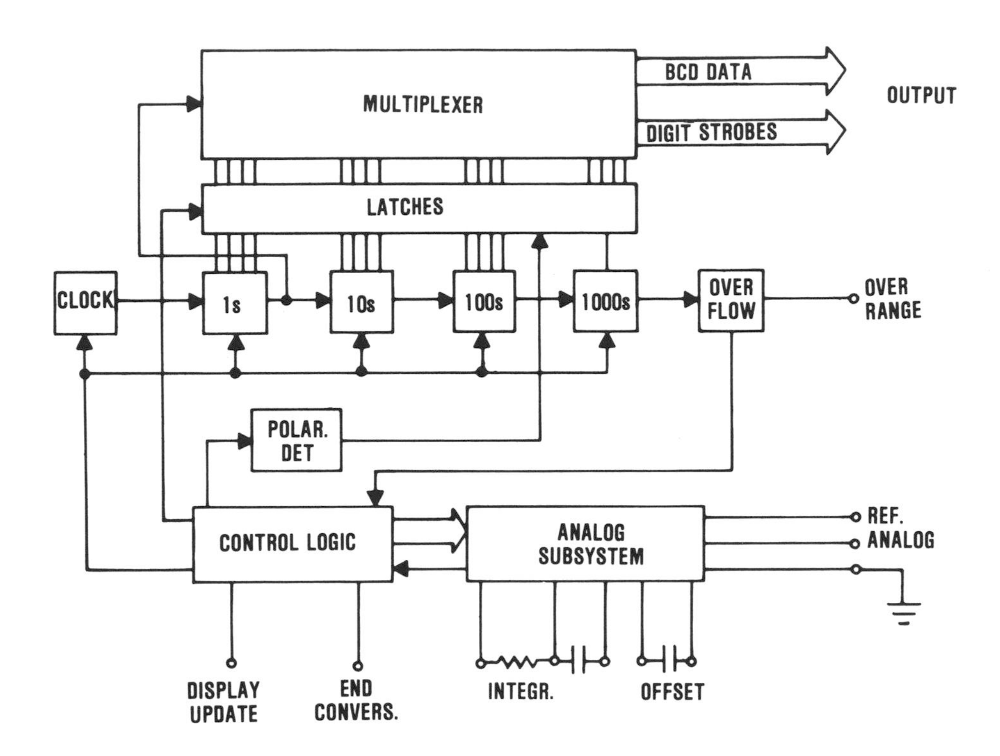

FIGURE I.1.3. Block Diagram—3½ Digit A/D Converter

in some detail, in Sections I.1.0 and I.1.2. During the gating period, when the clock is allowed to drive the four decade counters, the number of clock pulses will correspond to the analog voltage. Their output goes to the logic circuits and the multiplexer which, in turn, provides the final data output and digital strobes to the display. In some 3½ digit ADCs, the BCD-to-7-segment conversion is included.

A basic 3½ digit ADC, as illustrated in Figure I.1.3, provides outputs to indicate the overflow (overrange), the end of the conversion time to indicate completion, and a signal for the display to indicate that an update is required.

KEY PARAMETERS

a) *Relative accuracy*. ±0.05% of full scale reading, or ±1 LSB is typical for low-power CMOS and TTL devices.
b) *Conversion time*. Depending on internal clock frequency, ranges from 16 to 200 ms.

c) *Linearity.* ±0.05% of full scale reading or ±½ LSB.
d) *Power consumption.* Ranges from as little as 8.0 mW at 5 V for low-power CMOS to 25 mW for low-power TTL ICs.
e) *Clock frequency.* 66 KHz is typical for the low-power CMOS version. Up to 370 KHz is possible with low-power TTL ICs.

APPLICATIONS

This type of ADC is widely used in digital volt-ohm meters, thermometers, scales, and a host of other measuring devices where the number 1999 is sufficient in resolution.

COMMENTS

A 3-digit, 4-digit, 4½ digit, and even larger versions of this ADC are available. Their operation is essentially the same as described here, but their conversion times will be different.

I.1.4 Analog-to-Digital Converter (ADC), Integrating, High-Precision

DESCRIPTION

The key feature of this integrating ADC is a 4-phase ramp generation which is intended to improve the accuracy available with a dual ramp scheme, as described in I.1.0 and I.1.2. The block diagram of the 13-bit ADC shown in Figure I.1.4(a) shows the basic integrator and comparator inherent in all integrating type ADCs. The timing and control logic and the output counter block contain some unusual features. The key, however, to the accuracy of this ADC lies in the use of the four electronic switches, S1 through S4, which are controlled, in sequence, by the timing and control logic.

To understand the operation of this IC, we can refer to the quad ramp timing diagram of Figure I.1.4(b) and relate phase 1 with S1, phase 2 with S2, and so on. The sequential operation of these switches applies the inputs of the analog ground, reference voltage 1, the analog input voltage, and finally, reference voltage 2, in sequence to the integrator, creating the four ramps, phases 1 through 4 of the timing diagram. The dual ramps of phases 3 and 4 are the same as those in this standard dual ramp ADC, as described in Sections I.1.0 and I.1.2. The higher precision of this ADC is due to the count totalizing which includes phases 1 and 2, and therefore provides a more repeatable reference for the total pulse count of the A/D conversion.

One of the special features of this IC is the availability of the 2's

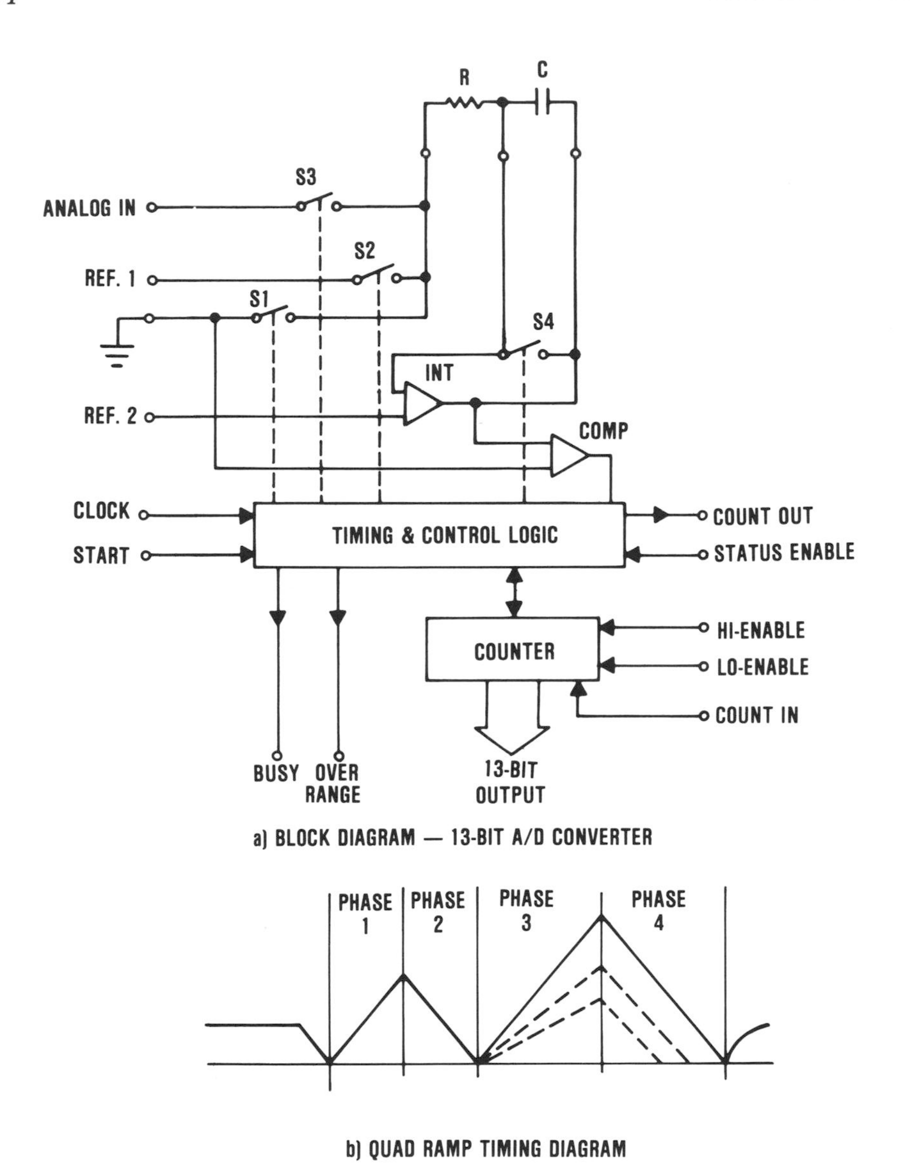

FIGURE I.1.4. (a) Block Diagram—13-Bit A/D Converter
(b) Quad Ramp Timing Diagram

complement at the counter output whenever the count-out and count-in terminals are connected together. The status enable signal provides tri-state control over the busy and overrange outputs. The hi-enable and lo-enable inputs provide tri-state control over the first eight output bits and the last five output bits respectively.

KEY PARAMETERS

a) *Relative accuracy.* ±½ LSB.
b) *Conversion time.* 40 ms with a 1 MHz clock.
c) *Maximum power dissipation.* 1 W up to 50°C for this high-density CMOS IC.
d) *Gain temperature coefficient.* 1 PPM/°C.
e) *Maximum clock frequency.* 1.2 MHz.

APPLICATIONS

This ADC is used where high resolution, up to 13 bits, and great relative accuracy and temperature stability are required. Precision measurements of temperature, pressure, flow, and similar physical parameters require ICs with these characteristics.

I.1.5 Analog-to-Digital Converter (ADC), Delta Modulator

DESCRIPTION

This unusual ADC is used for converting audio signals to digital data and digital data back to the analog audio signal. There is a variety of methods for performing the digitization of voice, but the all-digital, continuously variable slope, delta modulator provided in this CMOS IC is widely used in voice-to-digital communications systems.

The block diagram of Figure I.1.5(a) contains all of the functions to modulate and demodulate, transmit or receive, voice signals. The principle of this ADC is illustrated in Figure I.1.5(b), which shows the input voice signal, how it is encoded digitally and how that digital signal is decoded into the analog voice signal. The basic principle is that a pulse is transmitted whenever there is a change in the direction of the slope of the analog signal. Suitable filters must be connected at the input and output for the audio to limit it to its proper band.

The block diagram contains two unique audio functions. One is the signal estimate digital filter and the other is the syllabic digital filter. These two functions replace some of the analog loop filters normally required in delta modulation, use very low power, and do not need external timing components.

This IC is available in two versions: one is optimized for a 16 K Baud data rate and the other is optimized for a 32 K Baud data rate. The remainder of the block diagram is similar to that of most ADCs. As we can see, a 10-bit digital-to-analog converter (DAC) is used, indicating that the comparison method is the basis of delta modulation analog conversion. Note that an external signal sets the internal electronic switch which selects either the

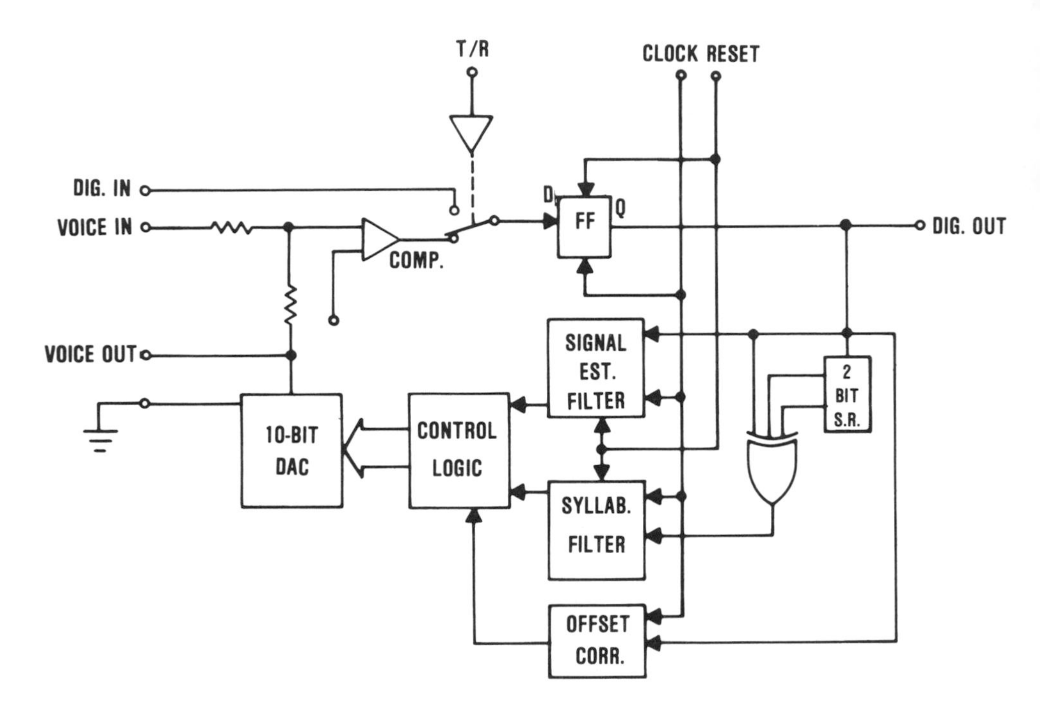

a) BLOCK DIAGRAM — A/D CONVERTER, DELTA MODULATOR

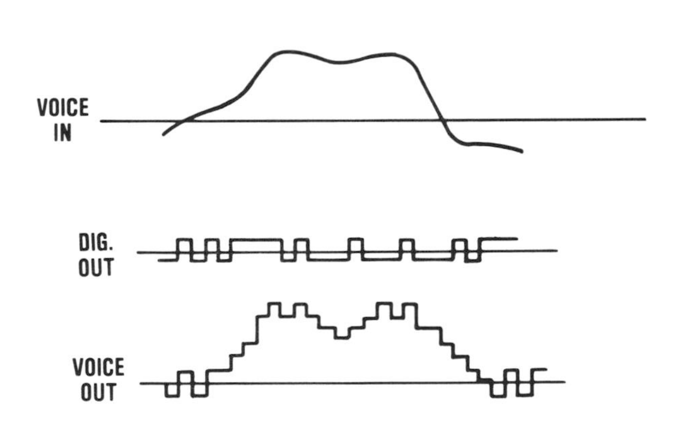

b) DELTA MODULATION WAVEFORMS

FIGURE I.1.5 (a) Block Diagram—A/D Converter, Delta Modulator
(b) Delta Modulation Waveforms

digital input or the output of the comparator, permitting either reception or transmission. The details of delta modulation and the continuously variable slope method of companding are beyond the scope of this book and can be found in the manufacturer's data and some digital communication reference handbooks.

KEY PARAMETERS

a) *Power consumption*. Typical power consumption is 6 mW at 5 V.
b) *Maximum clock frequency*. 64 KHz.

APPLICATIONS

This IC is used in digital voice transmission over data channels, voice encryption and scrambling systems, voice recognition systems controlled by computers, various audio signal manipulations, such as delay lines, time compression, echo generation and special effects. This type of IC is also used in telephone-type digital switching equipment.

COMMENTS

Most of the delta modulator ADCs are designed for specific applications, usually related to half-duplex voice telephone communications. In each application, the analog interface is carefully specified by the IC manufacturer. Consult manufacturer's application data for details.

I.2.0 Digital-to-Analog Converter (DAC)

DESCRIPTION

This function is available in monolithic integrated circuits as well as in hybrids and special, high-precision plug-in modules. Digital-to-analog converters (DAC) are all based on combining binary voltages or currents to produce the corresponding analog output. A host of special features are available for a wide range of applications, and the DACs described in this book are typical examples of the most frequently used DAC ICs.

Figure I.2.0(a) illustrates the fundamental DAC. A simple op amp with feedback resistor RF receives either a reference voltage or ground, depending on a series of switches. Each of these four switches is connected to the input of the op amp through a resistor. Note that the resistor values change according to the binary system. Clearly, the switch connected to 8 R will have 1/8 the effect of the switch connected to R, which corresponds to

the most significant bit (MSB). This binary weighted ladder DAC is used when only a limited number of bits is required and when the need for very large resistances does not present a problem. Most IC DACs, however, use the ladder network shown in Figure I.2.0(b). This is generally called an R-2R network because it consists of only two values, the basic resistor value and twice that much. In this scheme, current rather than voltage is used as the conversion medium. The switching circuit is essentially the same and the most significant bit (MSB) is the one providing the largest reference voltage to the input of the op amp.

In Figure I.2.0(a) and (b) the analog output voltage is obtained through an op amp, but many DAC ICs do not include this amplifier. As indicated in Figure I.2.0(c), the typical DAC symbol shows a positive and negative reference input and a positive and negative analog output, most frequently in the form of a varying current. If the reference input is a steady state voltage, the analog output will change only due to the changes in the digital input. When the reference voltage varies, as AC signal, the analog output will be the product of the reference signal and the digital input, and this type of DAC is called a "multiplying DAC." Almost all DAC ICs have this feature, but the multiplying frequency range is usually limited. The resistance networks are always constructed by thin film technology, but bipolar DACs operate as current switches while CMOS ICs operate as voltage switches, in a current steering mode. The characteristics of the switches themselves, basically analog switches as described in Section A.15.1, determine the speed and precision of the particular DAC.

KEY PARAMETERS

a) *Relative accuracy.* The relative accuracy error is usually expressed in %, PPM, or fractions of one least significant bit (LSB).
b) *Linearity.* Either given as linearity error in % or PPM or fractions of one LSB. Differential linearity indicates the relative linearity between adjacent digital codes. It is usually expressed in fractions of one LSB.
c) *Settling time.* The time required for the output of the DAC to reach and remain within a given fraction of the final value after a described data change has taken place.
d) *Slewing rate.* The maximum rate of change of the output voltage, usually limited by the output amplifier.
e) *Temperature coefficient.* Generally stated as %/°C or PPM/°C. Temperature coefficients can be assigned to the coverter gain, the linearity, and other parameters.
f) *Power consumption or maximum dissipation.* Generally stated in watts, this indicates the power requirements for a particular IC.

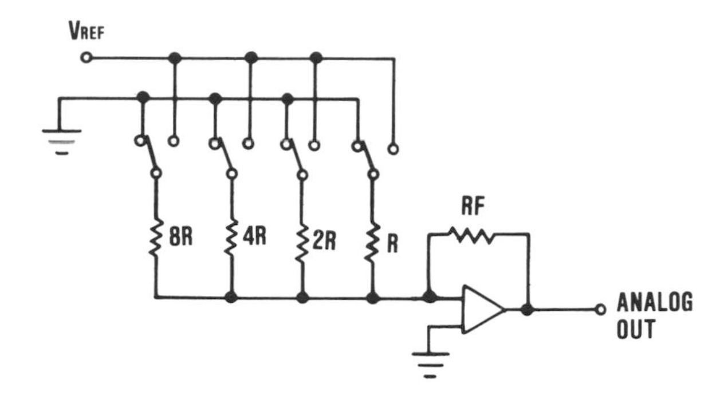

a) BINARY WEIGHTED LADDER D/A CONVERTER

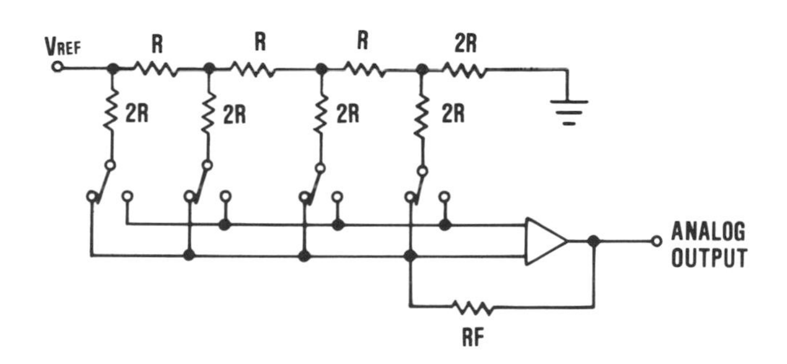

b) R-2R LADDER NETWORK, D/A CONVERTER

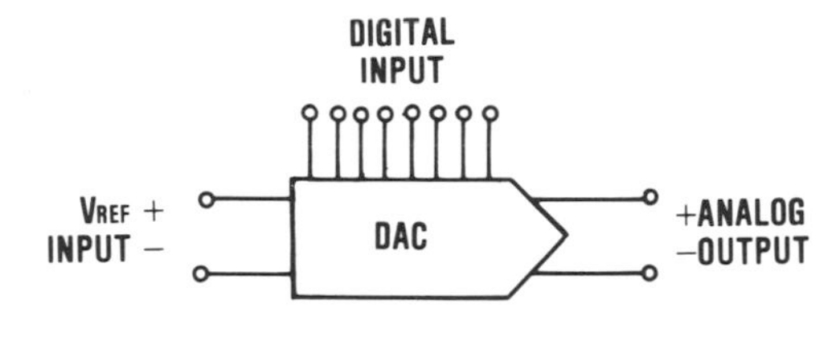

c) D/A CONVERTER SYMBOL

FIGURE I.2.0 (a) Binary Weighted Ladder D/A Converter
(b) R-2R Ladder Network, D/A Converter
(c) D/A Converter Symbol

APPLICATIONS

DACs are used in comparison-type A/D converters. They are also used in digital control systems where digital signals must be converted to analog signals for control of an analog device. Specific applications for certain types of DACs are listed in the following pages.

I.2.1 Digital-to-Analog Converter (DAC), 8-Bit, Basic

DESCRIPTION

The block diagram of the 8-bit multiplying DAC shown in Figure I.2.1 uses the R-2R ladder network, with a current switching source consisting of a transistor for each bit. The digital information is entered into the switch drivers which then actuate, as indicated by the dotted lines, the respective analog, solid state switch. These switches are connected to either the output current or the output of the complement current. The reference voltage is applied to an operational amplifier with external compensation which drives the bases of all of the individual transistor current sources. Note that the most significant bit (MSB) receives almost the entire negative supply voltage through its emitter resistor, while the least significant bit (LSB) receives its current through a series of eight series resistors. A separate bias

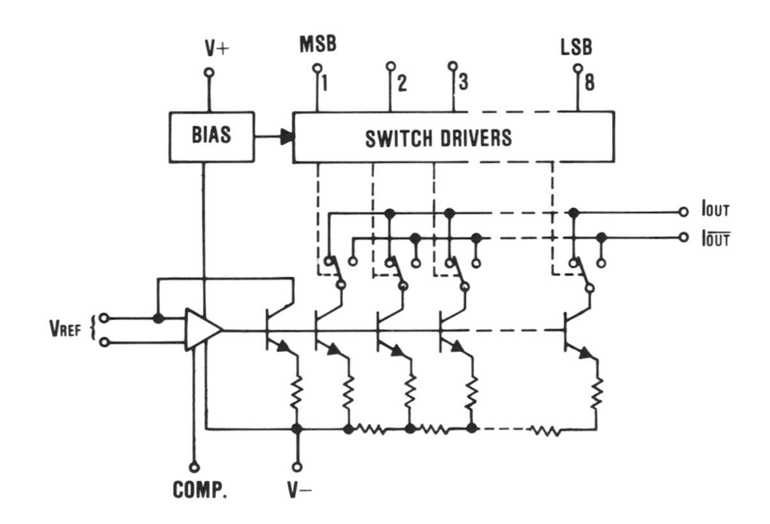

FIGURE I.2.1. Block Diagram of 8-Bit D/A Converter

and voltage regulator is contained on the IC to assure the proper voltages and stability for the switch drivers and the reference amplifier.

KEY PARAMETERS

a) *Relative accuracy.* ±0.1% is the maximum specified for this particular bipolar 8-bit high-speed, multiplying DAC.
b) *Differential linearity.* ±0.19% maximum is specified for this IC.
c) *Settling time.* With a maximum of 135 ns and a typical value of 60 ns, this IC is a high-speed device.
d) *Slewing rate.* A typical value is 8 mA/μs.
e) *Linearity temperature coefficient.* ±10 PPM/°C is typical for this IC.
f) *Maximum power dissipation.* 500 mW is the maximum for this bipolar IC.
g) *Power consumption.* The typical power required for normal operation. 100 mW is a typical power consumption requirement.

APPLICATIONS

This 8-bit DAC is used in fast A/D converters, variable gain amplifiers, digital waveform generators, programmable power supplies, and digital control systems.

I.2.2 Digital-to-Analog Converter (DAC), 8-Bit, Microprocessor Compatible

DESCRIPTION

Certain specific features make this DAC particularly suitable in conjunction with microprocessors. As indicated in Figure I.2.2, the major difference between this 8-bit DAC and the one described in I.2.1 is the addition of latches to the switch drivers and the inclusion of an output op amp. An 8-bit byte is entered into the eight digital inputs and stored there temporarily, when the latch enable terminal is activated. As soon as that occurs, each of the eight latches activates a switch driver which, in turn, activates the switches and controls the current sources in the R-2R ladder network. The analog output signal is then available through the output op amp. Note that external compensation is provided for this amplifier and that a separate analog ground terminal is used while the digital ground is part of the input circuit. The summing node is also brought out to a terminal. In all other respects this block diagram is the same as that of the 8-bit DAC of I.2.1.

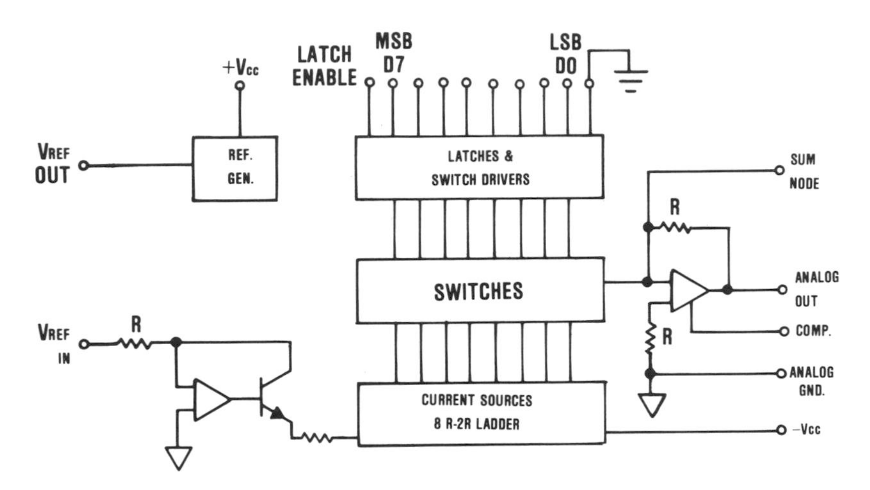

FIGURE I.2.2. Block Diagram—8-Bit, MP Compatible D/A Converter

KEY PARAMETERS

a) *Relative accuracy.* ±1/2 least significant bit (LSB).
b) *Differential linearity.* ±0.2% of full scale readings.
c) *Settling time.* 2 μs is typical. This IC is not a high-speed DAC.
d) *Latch enable pulse width.* 400 ns is typical. This corresponds to the pulse width available in many 8-bit microprocessors.
e) *Maximum power dissipation.* 800 mW.

APPLICATIONS

This DAC is used in microprocessor controlled digital-to-analog converters to generate analog signals for various appliances and devices and for analog displays. It is also used in programmable power supplies, in test and measuring equipment, and for analog-digital multiplication.

I.2.3 Digital-to-Analog Converter (DAC), High-Precision, High-Speed

DESCRIPTION

This DAC combines the precision inherent in 12-bit digital-to-analog conversion with extremely high speed. The thin film resistor network has been laser-trimmed for maximum precision, and a special dieletric isolation process has been used to reduce so-called "glitch" problems which occur during the rise and fall times of the switching process.

The basic circuit of this IC is the same as that illustrated for the 8-bit standard DAC of I.2.1. A few special features have been provided. As indicated in the function block of Figure I.2.3, there is one input which permits selection of CMOS or TTL compatibility. This controls a set of 12 level shifters, one for each input bit. Another feature is the availability of internal resistors for bipolar or monopolar operation. The 10 and 20 volt span outputs refer to internal resistors connected to the analog current output which permit use of an external amplifier with a larger voltage range.

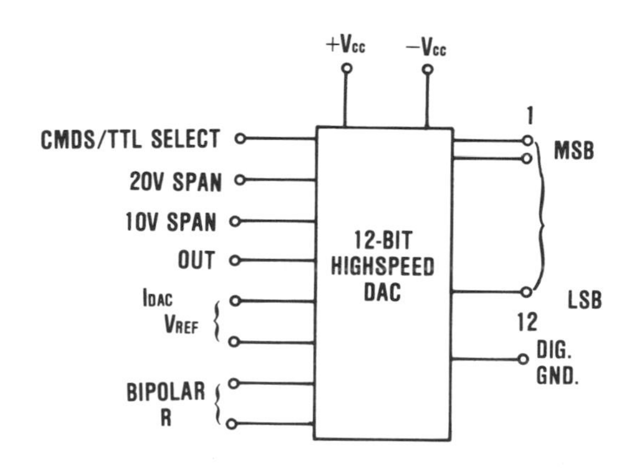

FIGURE I.2.3

KEY PARAMETERS

a) *Relative accuracy.* ±0.05% of full scale reading.
b) *Differential linearity.* ±¼ of the least significant bit (LSB).
c) *Settling time.* 200 ns is typical, with 400 ns maximum.
d) *Slewing rate.* 6 mA/μs.
e) *Gain temperature coefficient.* ±2 PPM/° C.
f) *Power supply sensitivity.* ±0.5 PPM of the full scale reading per % of the power-supply voltage.
g) *Maximum power dissipation.* 1W.

APPLICATIONS

This DAC is useful for high-speed analog-to-digital converters, data acquisition systems, and high reliability applications. Because of its high speed, it can be used for CRT display generators, video signal reconstruction, and waveform synthesizers.

1.2.4 Digital-to-Analog Converter (DAC), 10-Bit, Buffered

DESCRIPTION

As illustrated in the block diagram of Figure I.2.4, this IC consists of a 10-bit DAC, driven by a 10-bit register which, in turn, is driven in parallel by two separate shift registers. One of these is a 2-bit and the other an 8-bit shift register, and their special feature is the ability of shifting serial as well as parallel data. This IC is particularly suitable for microprocessors having an 8-bit byte bus system with additional two bits of control or flag information. The 2-bit shift register contains the most significant bit and is therefore referred to as the "high" register, while the 8-bit shift register contains the least significant bit (LSB) and is referred to as the "low" register. A common serial/parallel control line controls the operation of both registers. Serial data is entered at the LSB end and the output of serial data appears at the MSB end. A separate high and low strobe or clock signal controls the serial shifting operation of the two registers, and the 8-bit control is actuated when only eight bits instead of a full 10-bit input is used. Information from these

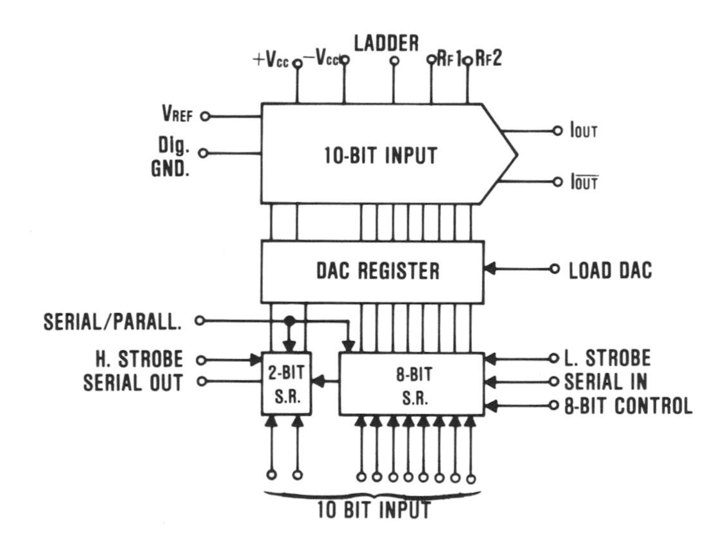

FIGURE I.2.4 Block Diagram—Buffered 10-Bit, D/A Converter

two shift registers is loaded into the DAC register only when the proper command is given to that register. The operation of the 10-bit DAC itself is conventional except for the availability of two separate feedback resistors, RF 1 and RF 2.

KEY PARAMETERS

a) *Relative accuracy.* ±0.5% of full scale reading.
b) *Differential linearity.* ±1% or ±2% of full scale reading (different models).
c) *Settling time.* 500 ns is typical.
d) *Nonlinearity temperature coefficient.* 2.0 PPM of full scale reading/°C.
e) *Gain temperature coefficient.* 10 PPM of full scale reading/°C.
f) *Maximum power dissipation.* 1W.

APPLICATIONS

Particularly suited for microprocessor interfacing to analog devices. A separate, external, op amp is required but, because the digital input can be loaded in either parallel or serial mode, and a number of external control and clock signals can be applied, this IC will be used mostly with microprocessors.

I.3.0 Level Shifter (Translator)

DESCRIPTION

When interfacing different IC families, it is often necessary that level shifters, translators, or special buffer circuits be used. Each of the major IC families includes such devices, usually arranged four or six per IC. In some instances the same voltages but different currents, and in other instances different voltages, must be applied to this device. Manufacturer's data should be consulted for buffer amplifiers intended for interfacing the two IC families.

Three representative examples of level shifters or translators are illustrated in Figure I.3.0. The CMOS-to-TTL level shifter illustrated in Figure I.3.0(a) is typical in its wide range of possible supply voltage. Its equivalent circuit consists of two complementary transistors in series, with a common, balanced, base input and emitter output. The main features of this CMOS-to-TTL level shifter are very low standby power, relatively high

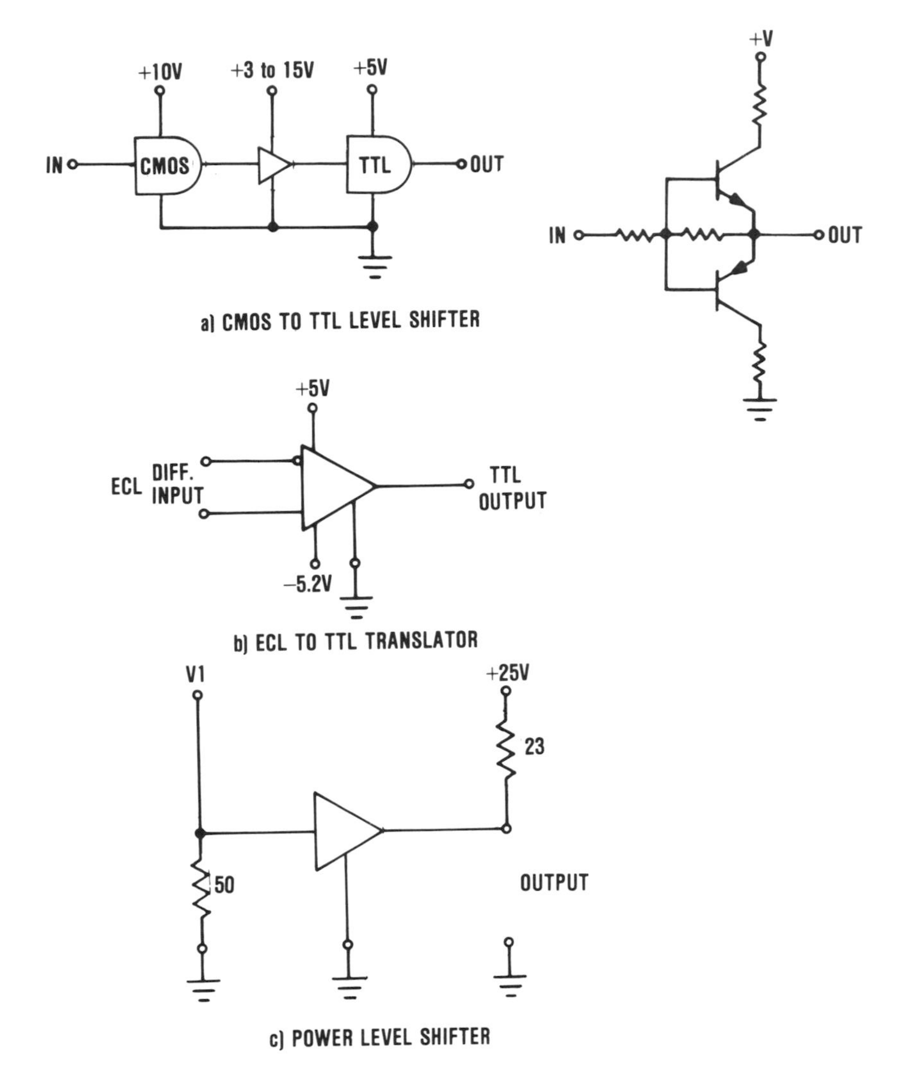

FIGURE I.3.0 (a) CMOS to TTL Level Shifter
(b) ECL to TTL Translator
(c) Power Level Shifter

current driving capability, and high-speed operation, as indicated by its key parameters.

The translator for emitter-coupled logic (ECL) to TTL logic is illustrated in Figure I.3.0(b) and features a differential input. This permits its use as inverting or noninverting translator or as differential line receiver, driving TTL ICs. ECL devices do not use saturated logic but TTL ICs do, and the change from the dual power supply, differential input device, to a single power TTL level can be accommodated by this IC.

The power level shifter illustrated in Figure I.3.0(c) can take its power supply from the output section and is capable of up to 15 W maximum power dissipation. It is provided with a heat dissipating stud and can be used for all sorts of high current devices. As indicated by its key parameters it can interface directly with almost all logic families and can provide relatively large outputs at high speeds.

KEY PARAMETERS

1) **CMOS to TTL level shifter:**

a) *Standby power.* The power required by the IC when no load is applied. 50 mW is typical.
b) *Supply voltage range.* This IC will operate with 3 to 15 V.
c) *Output loading.* This IC can drive up to ten standard TTL loads.
d) *Propagation delay,* **1** *input.* The time requried from input to output. 15 ns is typical with 50 pF capacity loading.
e) *Propagation delay,* **0** *input.* The time required from input to output. 30 ns is typical for 50 pF capacitive output.

2) **ECL to TTL translator:**

a) *Power dissipation.* 380 mW is typical when no loads are applied.
b) *Power-supply voltage.* +5.0 V and -5.2 V and a separate ground is required.
c) *Output loading.* This IC can drive the equivalent of ten standard TTL loads.
d) *Propagation delay time.* A minimum of 1.0 and a maximum of 6.0 ns is specified for this IC.
e) *Output pulse rise and fall time.* The time period between the 10% and 90% amplitudes of the rise and fall time of the outputs. 3.3 ns is typical.

3) **Power level shifter:**

a) *Maximum power dissipation.* 15 W is possible with a proper heat sink to keep the case temperature at 25°C.
b) *Output voltage.* Up to 18 V is possible with this IC. The output voltage is also the power-supply voltage.
c) *Input voltage.* The maximum for this IC is ±30 V.
d) *Input signal.* A minimum input current of less than 100 nA and a threshold voltage from 0.8 to 2.0 V permits direct interfaces to CMOS, TTL and other logic families.
e) *Switching or pulse characteristics.* This IC can pass a 50 ns wide pulse with an amplitude of 10 V at a repetition rate of 1.0 MHz over the circuit illustrated in Figure I.3.0(c).

APPLICATIONS

Level shifters or translators are used in the interface between different IC families and, sometimes, in the interface between an IC family and some peripheral equipment requiring considerably different voltage and current.

COMMENTS

For specific interface applications, the reader is referred to the manufacturer's literature describing the IC manufacturing method as refers to the output side of the interface.

I.4.0 Memory Interface Functions

DESCRIPTION

Microcomputers (MC) and microprocessors (MP) are generally used with arrays of internal memory ICs and with other memory devices, such as disc, tape, or magnetic bubble memories. Those memories that are located outside the housing that contains the MC or the MP are considered peripherals, and peripheral interface ICs are described in I.5.0. The memory ICs contained within the MC or MP housing also require some interface functions. The following five interface functions are typical of the many special purpose interface ICs used in micro- or minicomputer systems.

The memory address multiplexer, the direct memory access controller, and the memory expander are applicable to all types of IC memories. These functions are equally useful with read-only memories (ROM), programmable ROMs (PROM), EPROMs, EAROMs, and all kinds of random access memories (RAM). The refresh memory interface is required only where dynamic RAMs are used, but the sense amplifier is often applied to improve the signal-to-noise performance of memory data outputs.

KEY PARAMETERS

Both memories and MC or MP must be compatible in terms of logic levels, speed and bus specification, and the ICs interfacing between them must, of course, be equally compatible. In general, these ICs are part of the same digital IC family and only a few special features can be considered as key parameters, as described in the following pages for each specific IC.

APPLICATION

Memory interface function ICs are used in practically all mini- and microcomputers. Interface function ICs are usually installed whenever the memory capacity is increased.

I.4.1 Memory Address Multiplexer (MAM)

DESCRIPTION

Many large size memory ICs use multiplexed address inputs to accommodate all the addresses and still use a standard 16-pin IC package. A 16,384-word RAM, for example, would require 14 pins to accommodate the 14-bit address. Another seven pins are needed to perform the functions required to operate the RAM, such as the read, write, clocks, data in, etc.

Figure I.4.1 shows the functional block diagram of the memory address multiplexer (MAM). The address outputs of the microprocessor (MP) or microcomptuer (MC) are connected to the address input of MAM and the outputs of the MAM are connected ot the address inputs of the memory. If the row/col line is **1**, the odd number gates are activated, allowing the address A0 through A6 to be routed through the OR gates to the output of the IC. These outputs are stored in the row address register contained within the memory device. When the row/col line is **0**, the even-numbered AND gates are activated, connecting address inputs A7 through A13 to the outputs. The column address is then stored in the column address register inside the memory IC.

The timing signals required to store the row/column addresses in the selected row/column registers are synchronized with the row/col signals. This prevents an incorrect transfer. When both halves of the address word are stored in the memory register, the read or write functions are activated.

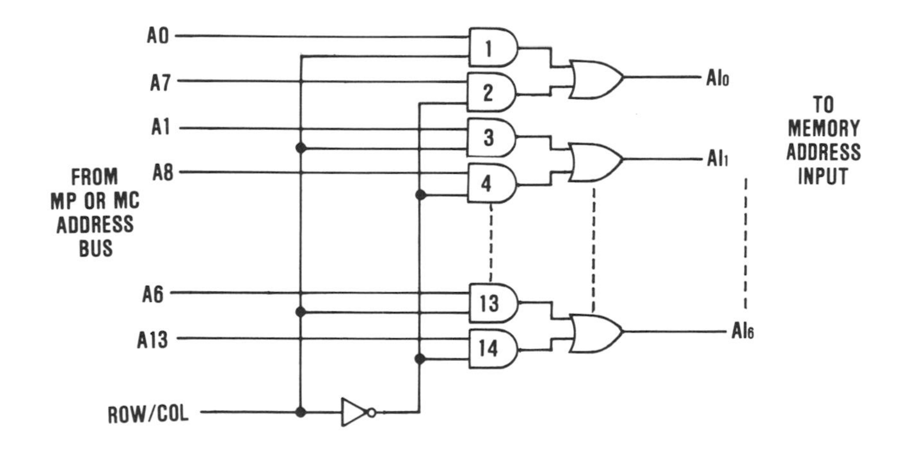

FIGURE I.4.1 Memory Address Multiplexer

KEY PARAMETERS

a) *Number of bits per column/row address.* The number of bits in the column/row address word defines the size of the memory the MAM is able to drive. Six bits are used to drive 4 K RAMs and seven bits for 16 K RAMs.
b) *Address input-to-output delay.* The time it takes for a change on the inputs to be detected at the outputs. This delay varies from 6 to 12 ns.
c) *Row/col-to-output delay.* The time it takes for a change in the row/col input to be detected at the output. This ranges between 6 and 30 ns.

APPLICATIONS

The MAM is used wherever multiplexed address memory devices are used.

COMMENTS

Some versions of this IC contain a refresh counter to retain the data in the dynamic RAMs.

I.4.2 Direct Memory Access Controller (DMAC)

DESCRIPTION

Direct memory access (DMA) is a technique used in computer systems which allows the transfer of binary information between memory and peripheral devices without going through the microprocessor (MP) or microcomputer (MC). It is often used in applications where the data transfer rates exceed the capability of the MP or MC. When a data transfer occurs, the MP or MC is placed in the "hold" mode. Control of the address, data and control bus is turned over to the DMA controller. The address and data bus is set up for the data transfer and the read or write signal is activated. Once the transfer is completed, control is returned to the MP or MC.

Figure I.4.2 shows the funcitonal block diagram of the DMA controller, which operates in the following manner:

Address and Word Counter: There are "n" channels in the DMAC, assigned to various peripheral devices. Channel 1 could be assigned to a floppy disc, channel 2 to a printer, etc. Each channel consists of an address counter and a word counter. The address counter determines the memory location where the block of data will be stored. The address counter is preset to a starting value by the MP or MC. After each data transfer the counter is

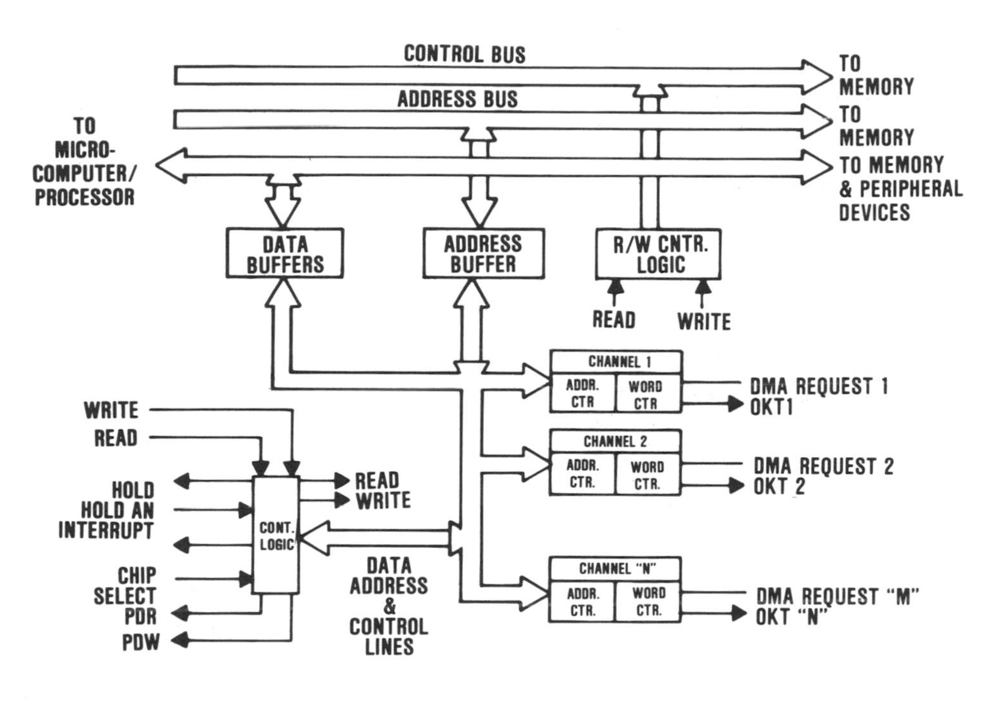

FIGURE I.4.2. DMA Controller

incremented to the next address. The word counter is used to record the total number of words transferred in a given block. Each time a data transfer occurs it is incremented by one. When the counter reaches the maximum count, the interrupt signal is sent to the MP or MC, indicating that a block of data words was transferred between the memory and peripheral device.

The DMA request signal from the peripheral device informs the DMAC IC that a transfer is requested. If another channel is not active, the control logic sends the hold signal to the MP or MC. This signal causes the MP or MC to disconnect itself from the address, data and control buses. The Hold A signal then informs the DMA controller that the MP or MC is disconnected. This allows the CMAC to take control of the bus. The OKT (OK to transfer) signal is then sent to the peripheral device. Concurrent with the generation of the OKT signal, the DMAC connects the address counter of the selected channel through the common address bus to the memory. When the address bus stabilizes to the value of the address counter, the control logic generates the read and PDR signals. The read or write signals are connected to the memory which executes the selected function. The PDR and PDW are command signals to the selected peripheral devices, which, when detected, will cause the specified operation. Once the data transfer cycle is ended, the selected address counter is incremented. The counter now points to the memory location that will be used in the next data transfer cycle.

R/W Control Logic: This logic function connects the necessary control signals to the control bus when the control logic generates the read or write commands. The MP or MC has already disconnected itself from the control bus upon receipt of the hold signal. The output signals are:

MEM Enable. This signal enables the memory in order to perform a read or write operation.

Read. When the read signal is activated on the control bus, the binary value stored in the selected memory location is read from the data bus to the selected peripheral device.

Write. When activated, the data value on the data bus, from the selected peripheral device, is written into the selected memory location. The memory location is determined by the value on the address bus.

Data Address Buffers. These two buffers connect the appropriate DMAC internal bus to the external address and data bus. They operate on command signal from the control logic. The data and address buffers are bidirectional, allowing data and address transfers to occur in both directions.

Control Logic. The control logic executes all commands from the MP or MC. In addition, it generates the necessary signals for a DMA transfer. The MP or MC must first enable the DMAC IC, and this is accomplished by activating the chip select line.

KEY PARAMETERS

a) *Number of channels.* The number of channels in the DMAC IC varies from 1 to 4.
b) *Clock period.* The speed of the basic clock cycle for the DMAC IC. This value varies from 0.25 to 5.0 μs.

APPLICATIONS

Used in computer systems which interface with high-speed peripheral devices such as discs, magnetic bubble memories, etc.

I.4.3 Memory Expander

DESCRIPTION

This IC is used whenever a number of memory banks are addressed by a microprocessor (MP) or microcomputer (MC). Memory banks can be of any size and the memory expander is used to select one bank out of the entire group. The specific address location for a particular block of data is sent to all banks, but only the one enabled by the memory expander can act on it. As illustrated in Figure I.4.3, a typical memory expander might have eight 4-input NAND gates, controlled by a single strobe gate and addressed by the true and complement levels of the three adress inputs A0, A1 and A2. In effect, the memory expander operates like a 1 of 8 decoder, at fairly high speed.

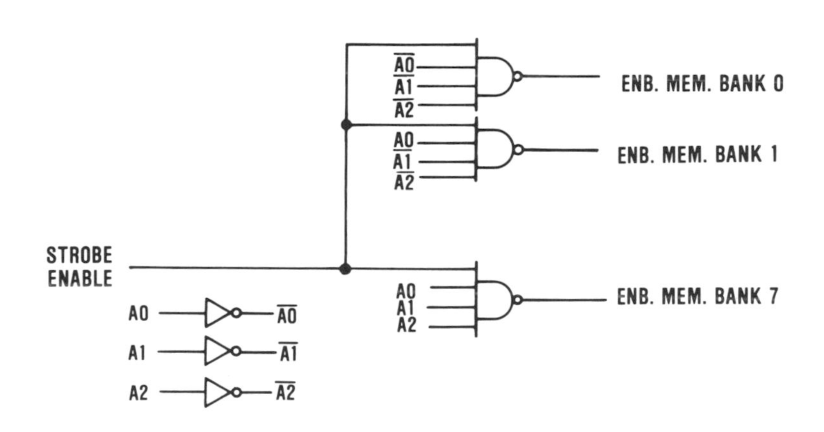

FIGURE I.4.3. Memory Expander

KEY PARAMETERS

The logic levels and other interface parameters must be compatible with the particule microprocessor or microcomputer family.

a) *Number of outputs*. Either four or eight.
b) *Propagation delay time*. The time required from the input to the output of the selected memory bank enable signal. Up to 18 ns is typical for CMOS ICs.

APPLICATIONS

Used in address decode logic functions where memory banks must be activated individually.

COMMENTS

If the speed requirement for the memory design is not critical, regular 1 of 8 decoders can be used.

I.4.4 Refresh Memory Interface (RMI)

DESCRIPTION

Dynamic random access memories (RAM) offer higher bit density, lower cost, and lower power requirements than static RAMs. The disadvantage of the dynamic RAM is the requirement for periodic updating or "refreshing" of the address inputs to prevent loss of data. The refresh memory interface (RMI) illustrated in Figure I.4.4 supplies all the necessary timing signals and gating required to perform the periodic updating of the RAM. It contains a 2-input address multiplexer, a refresh counter and the timing control logic, and operates in the following manner:

The address multiplexer connects either the address lines from the microprocessor (MP) or from the refresh counter to the address inputs of the dynamic RAM. The timing control logic controls the multiplexer through the select line. Then the select line is **0**, the dynamic RAM is under the control of the refresh counter, and when the select line goes to **1**, the MP drives the address bus. When the refresh counter is connected to the RAM, the timing control logic prevents the MP from addressing the RAM. The control signals are defined as follows:

Mem-Req:	The MP requests control of the RAM.
Req.-Granted:	The MP is given control of the RAM.
Mem-Active:	The memory is performing a read or write function.
Mem-Busy.	The memory cannot be used.

When the MP wants control of the RAM, the timing control logic examines the mem-busy line and determines if the RAM is in the refresh cycle. If the RAM is being refreshed, the request granted line to the MP is disabled. If the mem-busy is disabled and the system is not in the refresh cycle, the request granted line is enabled, and the address multiplexer selects the MP address bus. If the mem-req and the mem-busy lines are disabled, the refresh counter is connected to the memory.

KEY PARAMETERS

a) *Address input-to-output delay.* The delay from the time a change occurs at the address inputs to the time the output changes. This delay varies between five and 15 ns.

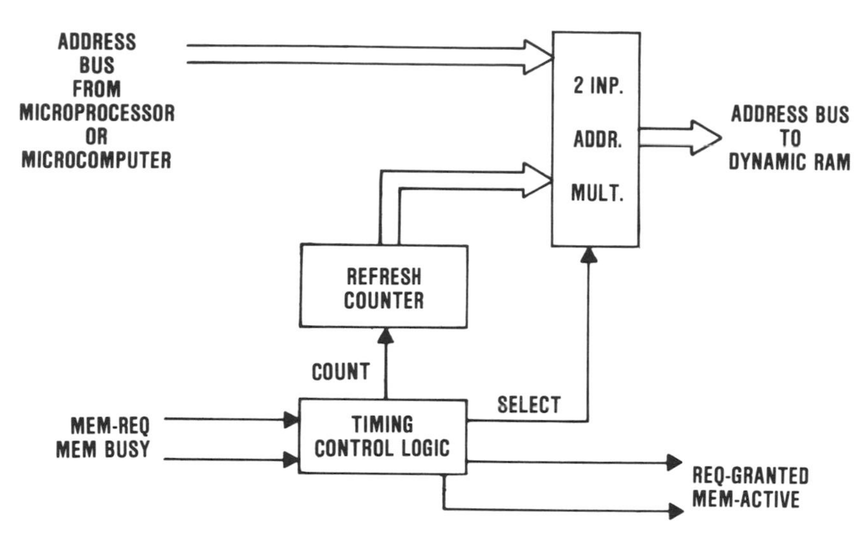

FIGURE I.4.4. Refresh Memory Interface

APPLICATIONS

This IC is for refreshing dynamic RAM ICs.

I.4.5 Sense Amplifier

DESCRIPTION

The sense amplifier detects low-level differential signals from the memory and converts them to levels compatible with the rest of the system.

Figure I.4.5 shows the logic diagram of a typical sense amplifier. The reference amplifier provides the threshold level for the comparator. This threshold level is usually set to a value slightly below the amplitude of the signal expected from the memory. If the differential input exceeds the threshold, the comparator is activated. Only when a strobe signal occurs, indicating that the memory is in the read cycle, can the amplifier output signal reach the output data line. Differential inputs are used to insure noise immunity at low signal levels.

KEY PARAMETERS

a) *Drive logic.* The sense amplifier is usually designed to drive TTL or DTL logic.

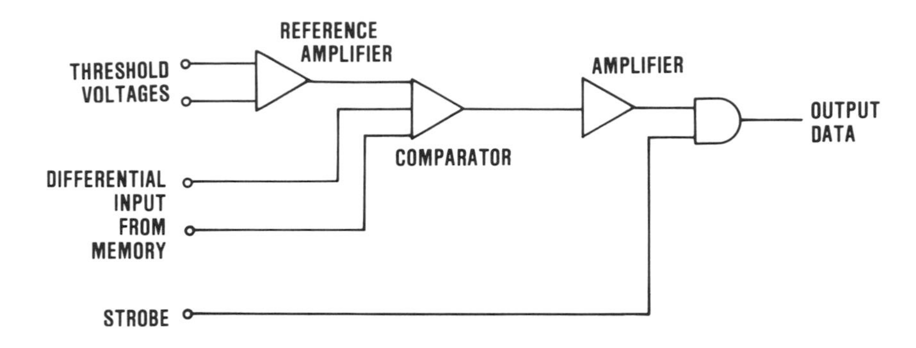

FIGURE I.4.5. Sense Amplifier

b) *Output types.* The output data line can be gated or latched.
c) *Operating threshold range.* The sense amplifiers will operate satisfactorily when the memory output signal is between ±10 to 15 millivolts.
d) *Power supply.* Sense amplifiers usually require both positive and negative voltages.

APPLICATIONS

Sense amplifiers are used whenever circuits such as memory elements have extremely small signal outputs.

I.5.0 System Interface Functions

DESCRIPTION

In order to connect the micro- or minicomputer with its peripherals, special interface functions are usually required to accommodate the difference in operating speed, data format, or signal level that applies to the specific device. A cathode ray tube (CRT) terminal with a keyboard, for example, needs some interface function to format the data from the computer so that it appears as alphanumeric characters on the screen. Another interface function is required to transfer the information entered into the keyboard by the human operator to the computer, in the computer's format and at the computer's speed.

A whole host of interface ICs is on the market, with many unique or custom-designed types for interfacing one particular peripheral device with one particular type of micro- or minicomputer. The ten system interface ICs described in the following pages have been chosen as typical examples, because each of them is available for every major microcomputer family and they interface with the most frequently used peripherals. CRT terminals, floppy disc drives, magnetic tape drives, keyboards and printers are used with almost all computer systems. The IEEE-488 instrument interface standard applies to the majority of computer controlled instrumentation systems and the needs of data communication systems are met by the three communciations interface ICs that are included here. The last interface IC, the contact debouncer, is a fairly simple but extremely useful device for single or multiple switch inputs.

KEY PARAMETERS

Interface ICs must be compatible with both sides of the interface, the computer and the peripheral. In practice this means that a particular IC will be part of the same microcomputer family and that the interface with the peripheral will be programmable, by the computer, to work with a number of different modes of this peripheral device. The key parameters are therefore generally the same as those of the microcomputer family. Unique parameters, peculiar to each type of IC, are listed with that IC's description in the following pages.

APPLICATION

All of the ICs listed in I.5.1 through I.5.10 are used in small computer systems.

COMMENT

In addition to consulting the manufacturer's data for a particular interface IC, a study of the application's notes for the complete system is also recommended. Only by carefully following the detailed specifications can troublefree interface operation be assured.

I.5.1 CRT Controller Interface

DESCRIPTION

Cathode ray tube (CRT) displays use a rectangular raster, just like a TV receiver. The screen displays a pattern of horizontal lines, each of which can be divided into a series of individual dots. A typical display can consist of 192 lines with each line containing 640 dots. In data CRT terminals the lines

COLUMN
1 2 3 4 5 6 7 8
LINE 1
LINE 2
LINE 3
LINE 4
LINE 5
LINE 6
LINE 7

a) DISPLAYED CHARACTER FROM DOT MATRIX GENERATOR

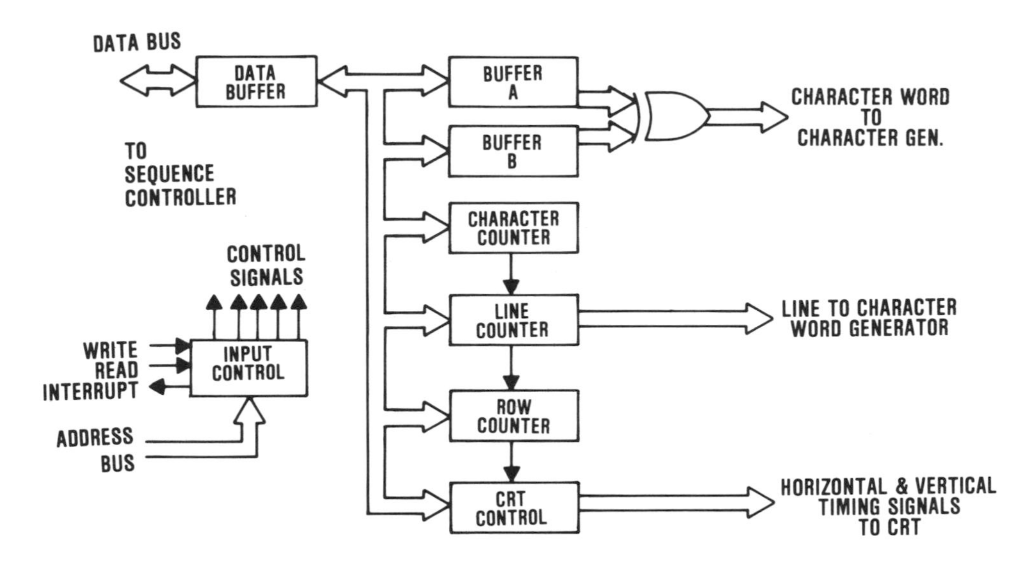

b) CRT CONTROLLER INTERFACE

FIGURE I.5.1 (a) Displayed Character from Dot Matrix Generator
(b) CRT Controller Interface

are subdivided into character rows, each consisting of seven lines with eight dots per character. The CRT display presented in this discussion would have 24 rows of characters with 80 characters per row. One character consists of seven lines and eight dots [see Figure I.5.1(a)], in order to duplicate all the necessary vertical, diagonal, and horizontal strokes for the selected alphanumeric or special symbol.

Vertical and horizontal timing circuits are used to sweep the beam across the screen. Starting from the upper left-hand corner, the beam sweeps to the right side of the screen and then retraces to the left again. The video signal, synchronized with the sweep signals, is connected to the CRT gun. A dot matrix generator generates the correct video signal for each successive line. When the beam reaches the end of the bottom line, it again moves to the upper left corner of the screen. Thirty complete "frames" or refresh cycles are displayed every second.

Figure I.5.1(b) shows the block diagram of the CRT controller. The character, line and row counters and the CRT control perform the line and refresh cycle. Buffers A and B are used to supply the characters to be presented on the CRT screen. The input control logic is used to interface the CRT controller with the microprocessor (MP). Most of the logic functions are programmable, allowing this IC to be used with most CRT terminals. A short description of each logic function follows:

Input Control: This logic function acts as the control link between the MP and the CRT controller. When the MP enables the CRT controller, the address and data bus are activated. Then the write command is enabled, causing the word on the data bus to be transferred to selected logic functions. The value on the address bus determines which logic function is being implemented. This cycle is repeated until all the logic functions are enabled. If the CRT controller requires information from the MP, the interrupt signal is generated and sent to the MP. The read signal is used to read information from the CRT controller to the MP over the data bus.

Buffers A and B: These buffers are used to store the characters that will be displayed on the next two rows. The numbers of characters stored in the buffers are programmable and both buffers are continuously activated. While one is shifting information to the CRT, the other buffer receives data from the MP for the next row.

Character Counter: The character counter is used to keep track of the number of characters displayed on the CRT. Once the last character is read out from the assigned buffer the other buffer is switched in. If the CRT terminal is designed to display 60 characters in one row, the character counter is programmed for a binary value of 60. The character counter also informs the row counter when the last character is displayed. This advances the row counter by one count, causing the raster timing signals to move down to the next line of the CRT.

Line Counter: The line counter is used to keep track of the number of lines per character row. If a dot matrix generator provides characters eight lines high, the line counter is programmed for eight. The line counter is updated for each horizontal line by one. When the counter reaches a count of eight, it resets to zero and updates the row counter. The cycle is repeated for the next character row.

Row Counter: This programmable counter is used to keep track of the number of rows that were scanned. If the CRT terminal is designed to display 24 rows of characters, the binary value of 24 is programmed into the row counter and is updated each time the line counter goes through a complete cycle. The information from the row counter is used to control the raster scan signals in order to position the sweep properly on the CRT screen.

CRT Control: The CRT control logic generates all the necessary horizontal and vertical timing signals for the CRT screen.

KEY PARAMETERS

a) *Character counter.* Is usually programmable up to 80 characters.
b) *Line counter.* This programmable counter can be set up to 16 lines per character row.
c) *Row counter.* Determines the number of character rows that can be displayed on the CRT screen. It is programmable up to 64 rows.

APPLICATIONS

The CRT controller can be used for interfacing the MP to a variety of CRT terminals.

COMMENTS

Some CRT controllers include a direct memory access (DMA) feature which allows rapid updating of the data buffers. A light pen register feature might also be included in this IC.

I.5.2 Floppy Disc Controller (FDC)

DESCRIPTION

Floppy discs are used as peripheral memories in many types of computers and are essentially a magnetic recording of digital data on a rotating, flexible disc. The standard size floppy disc is approximately eight inches in diameter, while the "minifloppy" is only five and a half inches in diameter. Most users of standard floppy discs use the IBM Flexible Diskette Data Exchange format. The disc contains 77 tracks, numbered from 00 to 76. Track 00 is located at the outmost track of the disc and is called the System Label Track. This track defines the data contents of the disc. Tracks 01 through 74 are used for data, and tracks 75 through 76 are reserved for alternate tracks. They can replace any two of the other 74 data tracks which might become defective.

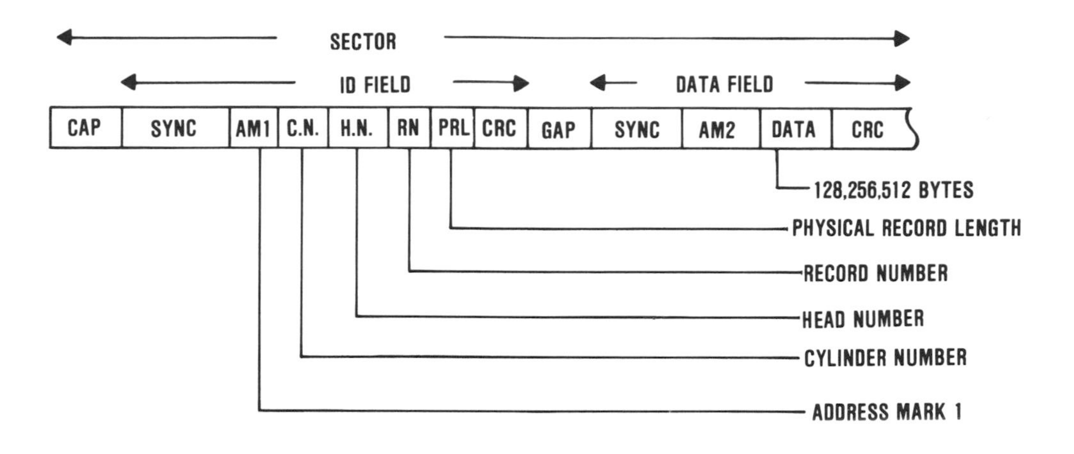

a) SECTOR FORMAT

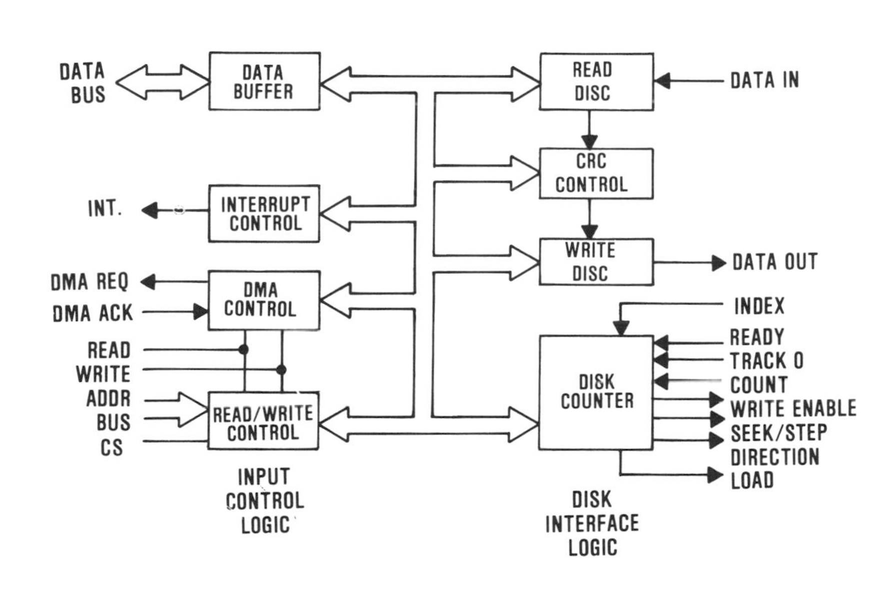

b) FLOPPY DISC CONTROLLER IC

FIGURE I.5.2. (a) Sector Format (b) Floppy Disc Controller IC

To mark the beginning of a track, a small hole near the center hole of the disc serves as index. A pulse is generated each time the read/write head passes over the index mark, and this pulse allows the synchronization of the read/write driver electronics to the disc. Each track can store 4096 bytes of information, which can be divided into sectors. These sectors can be

programmed for 128, 256, 512 bytes. The number of sectors per track is a function of the number of bytes per sector. A track can be programmed for 26, 15, 8 sectors. Each sector consists of an ID and data field as illustrated in Figure I.5.2(a).

The ID field contains information pertaining to the characteristics of the data field. This information is defined by the user. The data field contains the user's data. It can be 128, 256, 512 bytes long.

The floppy disc controller IC is used to provide the interface between the floppy disc and the processor and controls all read and write operations. The functional block diagram of the floppy disc controller (FDC) is shown in Figure I.5.2(b) and contains two parts: the input control and the disc interface logic.

The input control logic provides the interface between the processor and the disc interface logic and consists of the following functions:

DMA Control: The DMA control logic is used for direct data transfers between the internal memory and the floppy disc. This frees the processor from executing the data transfers. The DMA control logic provides signals which cause the DMA controller, in the computer, to transfer a data word to or from the memory. For each data word transferred between the floppy disc and the internal memory, the DMA req signal is activated. This signal is sent to the DMA controller, requesting a data transfer. If the memory is not busy, the DMA ACK signal is generated, together with the read or write signal.

Interrupt Control: The interrupt control logic is used by the FDC to inform the processor that it requires service.

Read/Write Control: This function is used by the processor to activate and initialize the FDC and determine whether data will move to or from the floppy disc.

The Disc Interface Logic shown in Figure I.5.2(b) is used to:

1. Format and store the track files onto the standard and minifloppy disc. In addition the FDC calculates the CRC bits and appends them onto the track files.
2. Read and process the track files from the disc. This also includes analyzing the CRC field of the file to determine if an error occurred in notifying the processor.
3. Control and maintain the location of the R/W head in order to read or write the track files onto or from the selected track on the disc.

Read Disc Logic: The read disc logic reads the information from the disc into the FDC. As the information is read into the FDC, it is routed to the appropriate section of the FDC for further processing. For example, the CRC field of the track file is routed to the CRC control logic for analysis. Which track of data is read into the FDC is determined by the disc control logic. Once the read/write head is positioned, the write enable signal is disabled. This causes a read operation.

Write Disc Logic: The write disc logic stores the information from the processor onto the disc. As the information is being written onto the disc, the CRC logic generates the necessary CRC bits to append them onto the track file.

CRC Logic: The CRC logic generates the cycle redundancy check bits as the data is written onto the diskette. The cycle redundancy check bits provide a form of error control for the track files. As the track file is read by the disc, the CRC is generated. If the resultant CRC into the CRC control are not the same as the CRC in the track file, an error has occurred. The processor is notified that the data block contains errors.

Disc Control: The disc control logic provides the supervisory protocol on the writing and reading of track files onto or from the diskette. In addition, it controls the position of the read/write head. In order to position the R/W head to any one of the 77 tracks, the track 0, seek/step, count and direction signals are used. A brief description of these signals follows:

Seek/Step. This signal is used to move the R/W head from track to track. When this signal is in the seek mode the count signal is monitored. Each time the R/W head goes to a different track, the count signal from the disc drive is activated. The number of count pulses generated indicates which track the R/W head has reached. The seek/step signal, in the step mode, causes the R/W head to move inward or outward one track.

Direction. This signal to the floppy disc determines the direction the read/write head will move when commanded by the seek/step signal.

Count. This signal from the floppy disc is used by the FDC to monitor the R/W head as it moves from track to track. Each time the R/W head is positioned over one of the 77 tracks a count pulse is generated.

Track 0. Whenever the R/W head is positioned over Track 0 (the outermost track on the disk), this signal to the FDC is activated. Using this signal as a reference point and tracking the count and seek/step signals, the position of the R/W head is always known.

Load. This signal is used to control the R/W head in the floppy disc. When activated, the R/W head is lowered against the disc. This operation occurs only after the disc control logic generates the necessary signals to position the R/W head over the selected track.

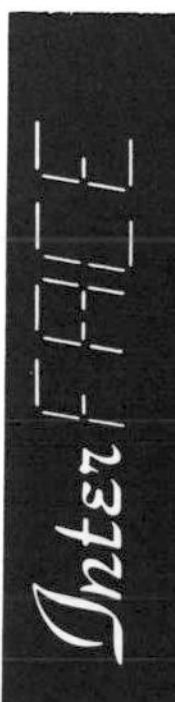

Write Enable. This signal, activated by the FDC, enables the write drive circuits at the disc.

Ready. This signal to the FDC is activated when the floppy disc is ready for operation.

Index. This signal from the disc drive indicates the start of a track. It synchronizes the FDC to the disc.

KEY PARAMETERS

a) *Number of discs*. The number of floppy discs that can be controlled by one FDC is programmable up to four.
b) *Types of discs*. A typical FDC can be programmed to control a standard floppy disc or a "minifloppy."

APPLICATIONS

The floppy disc controller is used as interface with floppy disc drivers.

I.5.3 High-Level Data Link Controller/Synchronous Data Link Control (HDLC/SDLC)

DESCRIPTION

This IC is specifically designed to provide an interface for data communications using the high-level data link, as well as the IBM developed, high-speed synchronous data link. These data links use a particular data format and signaling scheme, and there are several different data link types in use. The one described here is used for international data communications and employs the data format illustrated in Figure I.5.3(a). A brief description of this format follows.

Frame. The serial data stream is transmitted in frames and each frame consists of six fields:

Opening Flag. This field is used to flag the start of the frame.

Address Flag. This field, consisting of eight bits, identifies the station that will receive and store the frame of information.

Control Field. This field is used to define the type of information transmitted.

Information Field. This field is used to transmit the data. It is a variable length block.

Frame Check Sequence. This field is used for error control. It allows the receive station to determine if the received data frame is in error.

Closing Flag. The closing flag is used to indicate the end of the frame.

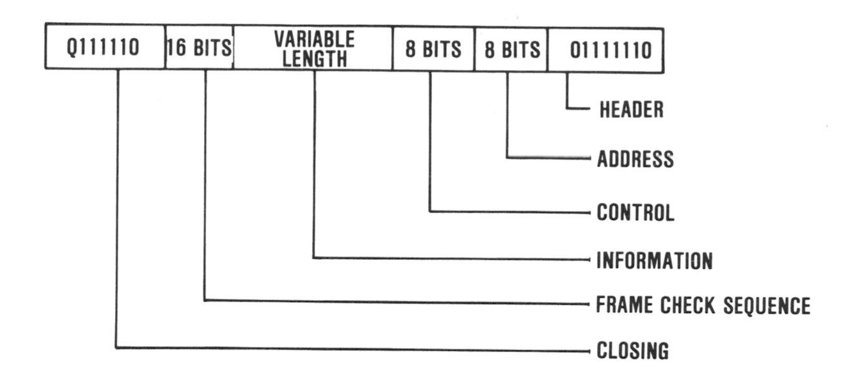

a) HDLC/SDLC FRAME FORMAT

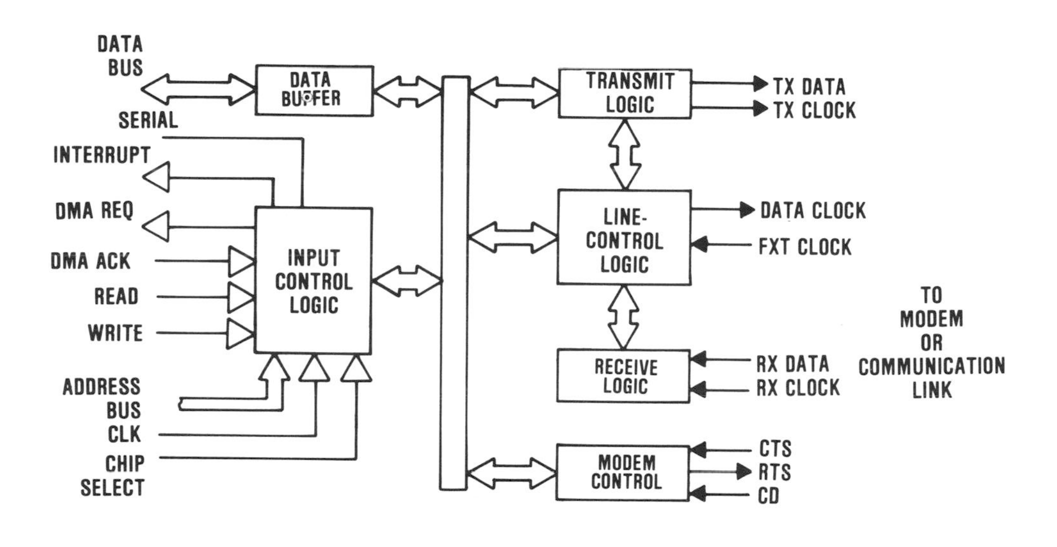

b) HDLC/SDLC INTERFACE IC

FIGURE I.5.3 (a) HDLC/SDLC Frame Format
(b) HDLC/SDLC Interface IC

The functional block diagram of Figure I.5.3.(b) illustrates the operation of the HDLC/SDLC. This IC can be programmed to serve either the HDL or the SDL communication protocol No. 1 and can operate either in the full-duplex or half-duplex mode. In the full-duplex mode the receive and transmit sections operate simultaneously and independently, and in the half-duplex mode either transmission or reception takes place. In either method of operation the DMA feature is activated, resulting in data transfers between the computer memory and the HDLC/SDLC IC. A brief description of the various functions follows.

Input Control Logic: This logic function controls the communications between the computer's processor and the HDLC/SDLC interface IC. The IC is always under the control of the processor. Command words are sent over the data bus and executed by the input control logic. The data bus is also used to transmit and receive information between the modem and the processor.

DMA: Direct memory access (DMA) is used when the data transfer between the modem and the HDLC/SDLC IS is at a rate faster than the processor can handle. Each time the data transfer occurs, the DMA request (DMA Req) signal is activated. This signal causes the processor to suspend operation. The DMA ACK is then enabled, causing the HDLC/SDLC IC to accept or send a data or command word over the data bus. Either of two modes of operation can be used, polling or interrupt.

Polled. A data or command transfer occurs between the processor and the IC when the address bus, write or read signals are activated. The processor initiates all read and write cycles.

Interrupt. This signal is activated when the HDLC/SDLC requires service from the processor.

Line Control Logic: The line control logic supervises all data transmissons between the HDLC/SDLC IC and the output interface. This interface can be a modem or another communications system. In addition, the line control logic is linked to the input control logic executing the command or data words from the processor.

The line control logic controls the receive and transmit logic. As the frames are transmitted or received, the line control logic automatically analyzes the data stream for possible errors. This is accomplished by using a hard wire algorithm for the frame check sequence. If the HDLC/SDLC is transmitting information to the modem, the error control bits are automatically generated and then appended at the end of the frame. When the IC receives data, the frame sequence check field is analyzed to determine if an error occurred. If there was an error, the processor is notified.

Transmit Logic: The transmit logic contains registers and control functions which enable the HDLC/SDLC IC to transmit the frame data bits over the TX data line. For each register of data bits transmitted, the transmit logic signals the line control logic to activate the DMA for another word from the memory.

Receive Logic: The receive logic contains registers and control functions which process the serial stream data from the modem. The receive logic, working with the line control logic, receives and stores the data bits as they are serially shifted over the RX data line to the HDLC/SDLC IC. When one of the registers is filled, the DMA is activated. During the DMA cycle, the second register in the receive logic is used to store the incoming information.

Modem Control: The modem control logic performs the necessary control functions in order to receive or transmit data over the communications link. In addition to the TX and RX data to the modem, the following signals are used:

Request to Send (RTS). When data are to be transmitted to the modem, the request to send line is activated. This signals the modem to begin conditioning the communication line.

Clear to Send (CTS). When the modem is ready to receive data from the HDLC/SDLC IC, the clear to send line is activated.

Carrier Detect (CD). This signal is used by the modem to indicate to the HDLC/SDLC IC that data is being received by the modem.

TX Clock. In order to maintain data synchronization between systems the transmitted data is serially shifted out of the HDLC/SDLC IC on the transition of the TX Clock.

RX Clock. The HDLC/SDLC uses the receive clock from the modem to synchronize itself to the receive data.

KEY PARAMETERS

The key parameters for this IC are uniquely determined by the HDL and the SDL communications protocol.

APPLICATIONS

The HDLC/SDLC IC is used as interface in high-speed data communication systems, such as satellite networks and packet-switching networks.

I.5.4 IEEE-488 Controller Interface

DESCRIPTION

The IEEE general purpose interface bus (GPIB) is used industry-wide for computer programmable instrumentation and is based on the interconnection standard No. 488. As many as 16 instruments can be connected to the same data bus, and all data movements can be controlled by a microprocessor (MP) or microcomputer (MC), or by a special sequence controller.

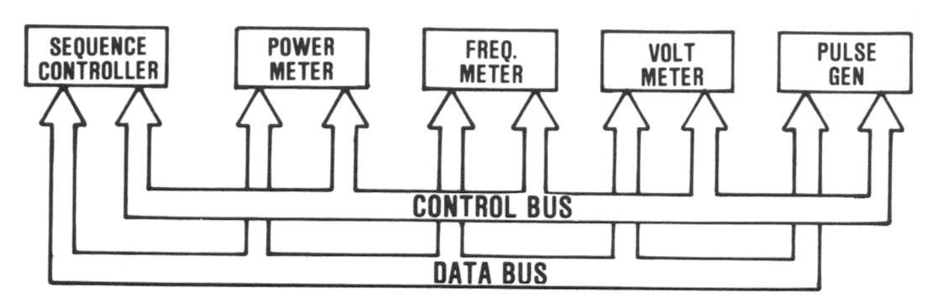

a) INSTRUMENT BUS STRUCTURE

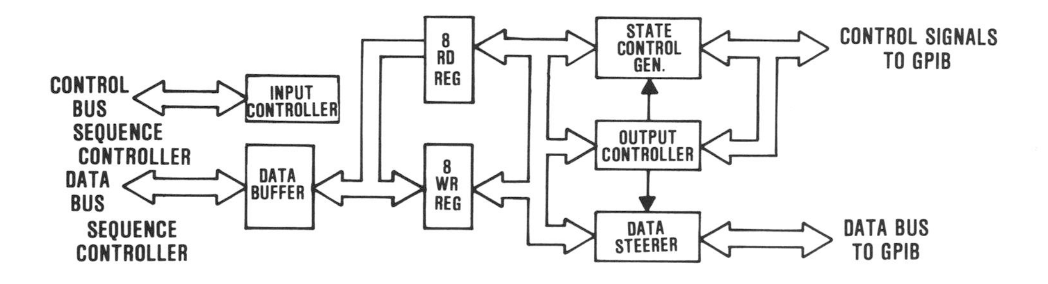

b) BLOCK DIAGRAM 488-IC

FIGURE I.5.4 (a) Instrument Bus Structure
(b) Block Diagram 488-IC

In the example of Figure I.5.4(a), a frequency meter, power meter, voltmeter, pulse generator and sequence controller (MP or MC) can be connected to the same data and control bus. Each of these units interfaces through a 488 controller. The block diagram of the 488 controller is shown in Figure I.5.4(b) and contains the following functions:

Input Control: Generates and monitors all the necessary signals between the sequencer controller and the internal logic elements of the 488-IC. When the sequence controller is ready to begin a data transfer to the GPIB, the write, chip select, and address steering signals will be activated on the control bus to the 488-IC. When these signals are detected by the input controller of the 488-IC, the data word on the data bus will be transferred to the selected register. This data word can be a command word indicating certain functions should begin, or it can be actual data words containing a message that might go to one of the instruments on the data bus. When data or commands enter the 488-IC from the GPIB, the input controller passes them to the sequence controller. In this case the input controller generates an interrupt signal over one of the lines on the control bus. This, in turn, causes the sequence controller to service the 488-IC, resulting in a read operation, transferring the contents of one of eight read registers through the data buffers onto the data bus.

16 Read/Write Registers: These registers are used by the sequence controller as a means to command the 488-IC to read or write data over the GPIB. The registers are used as follows:

R/W Data Registers. These two registers are used to move data between the GPIB and sequence controller.

Interrupt Registers. Two registers are used to generate an interrupt to the sequence controller. Two additional registers are used as interrupt mask registers, preventing certain occurrences from generating an interrupt. The sequence controller programs the interrupt mask registers accordingly.

R/W Serial Poll Registers. These registers are used to indicate the status and mode of the 488-IC. The mode could be "request for transmissions," "waiting for data," etc. The sequence controller can read these registers to determine what the 488 IC is doing.

Address Registers. These registers are used to set up the communications link in any one of "n" address modes. Mode A, for example, indicates an 8-bit address assigned to an instrument.

Command Register. This register is used whenever the GPIB wants to send a special command word to the sequence controller.

Auxiliary Registers. These registers are used to send special commands to the 488-IC from the sequence controller.

End of Sequence Register. This register is used to store and transfer its contents at the end of each message between the GPIB and the sequence controller.

Output Controller: This logic function in the 488-IC generates and monitors the various control signals on the GPIB. It determines if the 488-IC is a talker or listener. It sets up the data steering logic to the proper direction for each data block and commands the state control generator to begin operation.

State Control Generator: The state control generator is used to activate all the necessary control signals required in a data transfer. The GPIB control bus contains eight control signals which are used for the different modes a device can be in. These modes include request for transmission, waiting to receive data, awaiting service, address enable, end of transmision, attention, etc. A short description of the control signals follows:

DAV: Data is valid. (Information available for reading.)
NRFD: Not ready for the data. (Device cannot read the data at this time.)
NDAC: No data accepted. (Indicates whether or not the data has been accepted by the device.)
IFC: Interface clear. (Resets the 488-IC to a starting state.)
ATN: Attention. (Indicates how the data is interrupted—command or data.)

SRQ:	Service request. (Used to get the attention of the devices on line.)
REN:	Remote enable. (Indicates which of the devices on the GPIB is in the remote or manual mode.)
EOI:	End or identify. (Indicates the end of a block transmission.)

KEY PARAMETERS

Since the 488-IC is used in an Industry Standard there is little variation between manufacturers.

APPLICATIONS

This IC is found wherever the IEEE-488 data interface is used as a method of communication.

COMMENTS

Some 488 ICs have a direct memory access (DMA) feature which bypasses the CPU for data transfers until the block transmissions are complete.

I.5.5 Keyboard Interface (KBI)

DESCRIPTION

The keyboard interface (KBI) is designed to convert keyboard inputs to binary words. These binary words are uniquely coded for each key and sent to the microprocessor (MP) or microcomputer (MC) for further processing. The basic keyboard operation is illustrated in Figure I.5.5(a). Scan signals are sent out by the KBI to a 4-to-16 decoder which converts them to 16 outputs. Each output line is connected to a bank of eight keys. When a key contact is closed, one of the eight data lines is activated. The eight data lines are encoded into a 3-bit word. The coincidence of the scan lines and the data lines indicate which of the 128 keys were depressed, and the data lines are sent to the keyboard interface IC for further processing.

The block diagram of the keyboard interface is shown in Figure I.5.5(b). The scan counter continuously scans the keyboard to see if the operator has depressed a key. When a key is depressed, the key activate (KA) signal is generated and detected by the keyboard debouncer/detector. This circuit waits a fixed period of time, then samples the KA once again. A short delay is necessary to allow the keyboard to stabilize. The 4-bit scan word and the

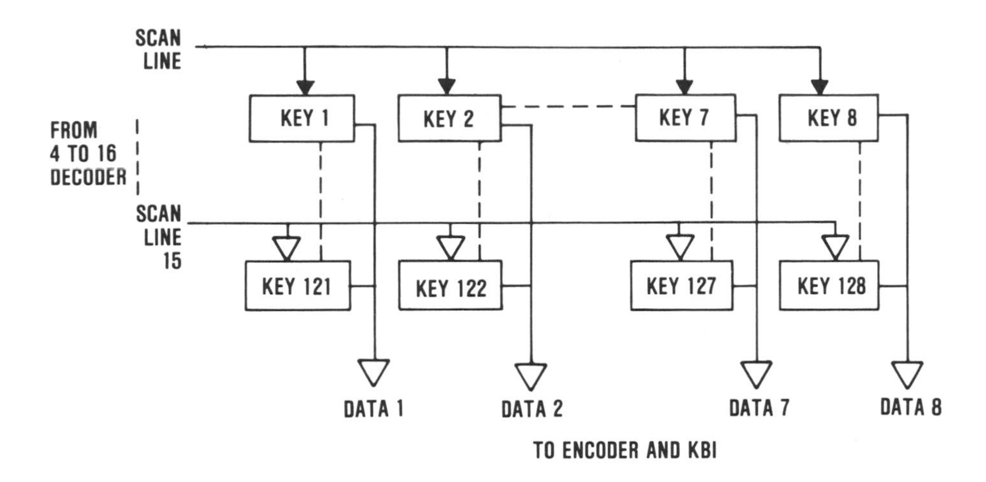

a) KEYBOARD OPERATION

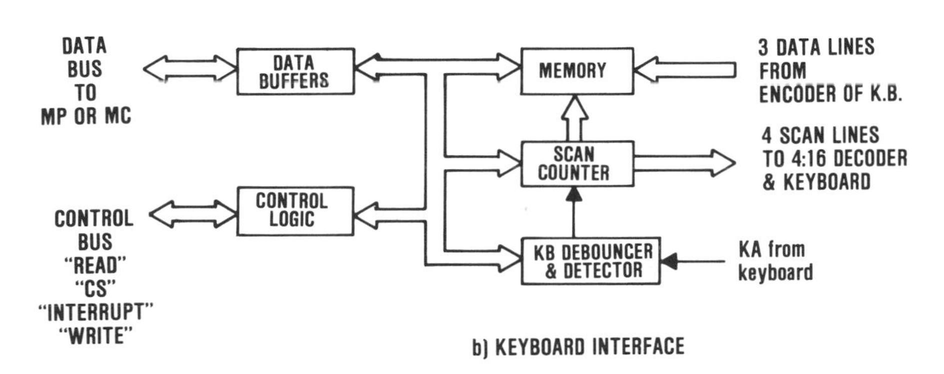

b) KEYBOARD INTERFACE

FIGURE I.5.5 (a) Keyboard Operation
(b) Keyboard Interface

status of the three data lines are stored in the memory. The 7-bit word stored in memory represents the alphanumeric symbol selected by the operator. If the operator depressed the key for the letter "A," the code from the scan counter might be "1010" and the code for the three data lines could be "100." The binary word stored in the memory therefore will be "1010100." This code is used to represent only one particular alphanumeric symbol. The memory is designed to store several 7-bit words from the keyboard and gives the MP or MC time to service the KBI. When a 7-bit word is stored in the memory, the control logic generates an interrupt signal to the MP or MC. This signal indicates that a data character is available for processing. If the MP or MC honors the interrupt, the necessary read and chip select signals are generated, to read the words stored in the memory. The data is read out of the memory through the data buffers to the MP or MC. Once the memory is empty, the interrupt request is disabled.

KEY PARAMETERS

a) *Size of keyboard.* The number of contact closures (keys) the KBI can process depends on the number of scan lines going out and the number of returned data lines. Most KBI ICs are designed either for 4 scan lines/3 return lines (128 characters) or 3 scan lines/3 return lines (64 characters).

APPLICATION

This IC is used for all size keyboards. The number of scan lines used for a given application is programmable.

I.5.6 Printer Interface

DESCRIPTION

Most printers used in computer systems print characters without the conventional ribbon, on special paper that darkens under pressure, or when heated. Instead of a unique key for each alphanumeric symbol, a universal print head, consisting of individual rods, is used. Figure I.5.6(a) shows the impact dot matrix used in this type of printer.

The universal print head consists of seven rods, assigned to the seven rows in the dot matrix. The print head starts at the left and moves across the paper towards the right. The seven rods are connected to seven solenoid drivers, which, when activated by the printer interface, cause the selected rods to strike the paper. In the example of Figure I.5.6(a) rods 3, 4, and 5 strike the paper in the first column. As the printer head moves to the right, rods 2 and 6 strike the paper in column 2. This is repeated until the complete character is printed. The head continues to move to the next character position until the printer detects a carriage return and starts a new line at the left side of the paper. It should be noted that the print process can occur as the head is moving from left to right or right to left. This doubles the speed of the print operation. The data must be formatted to accommodate a dual direction printing scheme.

Figure I.5.6(b) shows the functional block diagram of the impact printer interface. This IC generates all the necessary drive and timing signals. The format and the contents of the data are determined by the information received from the computer. In most systems one line of data or control

a) DOT MATRIX CHARACTER

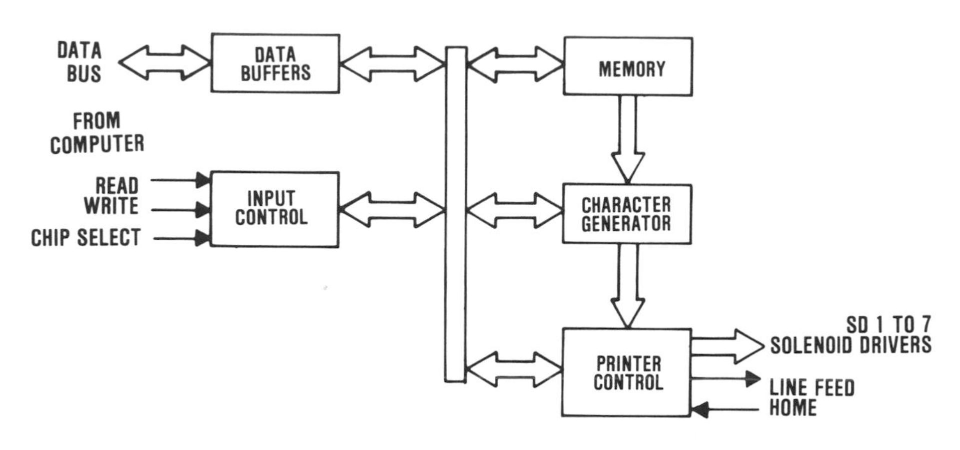

b) PRINTER INTERFACE IC

COLUMN 1 2 3 4 5 6 7
ROW 1 2 3 4 5 6 7

FIGURE I.5.6 (a) Dot Matrix Character
(b) Printer Interface IC

characters is sent over the data bus to the printer interface at a time. The input control logic stores the information in the memory and when the memory is filled, the printer control logic is activated. The information is then read out of the memory, one character at a time, to the character generator. If the information is a data character, the appropriate solenoid drivers are activated at the right time. If the information is a control character, the appropriate command signals are generated. A brief description of each logic function is presented below.

Input Control: The input control function interfaces with the computer for all data transfers. When the computer wants to implement a data transfer to the printer interface, the data is placed on the data bus and the write and chip select lines are activated. The chip select signal enables this IC. The write signal causes the input control to store the data in the

memory or excutes the command defined by the control character word on the data bus. The read signal is used to read information from the printer interface to the computer. The following types of characters are generally used:

Data characters: All the letter, numbers and special symbol characters are part of the text.

Control characters: These characters control the printer and include:

Line feed—generates a line feed to the printer.
Tab—positions the printer head to specify sections of the paper.
Home (carriage return)—sets the printer head to the "home" position on the page.

Memory: The memory logic function stores the characters that will be used with the printer. This memory unit frees the computer from continuously servicing the printer. The memory is filled up with one line of characters at a time.

Character Generation: The character generator converts each data word stored in memory to the format required by the universal print head. This data word must be converted to a 7-bit solenoid word which must be read out sequentially to activate the selected rods. Furthermore, the solenoid word must be synchronized to the print head position.

Printer Control: This logic function generates the necessary timing signals to drive the printer and converts the solenoid words to the proper voltage levels required to activate the rods. Another function is to monitor the printer for the home and line feed signals.

KEY PARAMETERS

a) *Memory size*. The number of characters that can be stored in the memory. The memory size varies from 40 to 120 characters.
b) *Dot matrix*. The resolution of the character generator. The more rows and columns are available, the better the definition of the character. A 7 × 7 dot matrix is typical.

APPLICATIONS

Printer interface ICs are used with impact printers.

COMMENTS

The input control logic can have a DMA, serial, or polling interface, depending on the operation specified by the computer.

I.5.7 Synchronous Communication Controller (SCC)

DESCRIPTION

Of the two forms of digital communication, synchronous transmission is generally faster than asynchronous transmission. In synchronous transmission systems the information is sent out in the form of data blocks, which are organized as shown in Figure I.5.7(a). The "header" characters are used by the receiver to synchronize its circuits to the data block. One header is used for each message and the data characters make up the message. The parity bits are used by the receiver to determine if an error has occurred in the message.

Figure I.5.7(b) shows the functional block diagram for the SCC. Control words are sent from the microprocessor (MP) or microcomputer (MC) over the data bus to start the SCC. They are used by control logic to set up the baud rate. A brief description of the SCC function follows.

Receive Mode: Data from the modem is connected to the receive data input and is stored in register A. Register B is permanently connected to the MP or MC. After each data character is shifted into register A, it is transferred to register B and the interrupt signal is generated. The interrupt causes the MP or MC to generate a read signal and the address value defining register B. The data character is then read out of the register over the data bus to the MP or MC.

Transmit Mode: When a block of data is transmitted from the MP or MC to the modem, the SCC is set to the transmit mode. The MP or MC sends a block of control characters to the transmit logic which sets up the character length, modem baud rate, etc. There are two data registers in the transmit logic: (1) Register A is used to transmit data to the modem and when it is empty, it is loaded with another data character from register B. (2) Register B is continuously updated with succeeding characters from the MP or MC. The interrupt signal is used to signal the MP or MC when to update the register.

Modem Control: When transmitting or receiving data over the telephone lines, a modem is required. In the transmit mode, the modem is used to convert the digital information into analog signals suitable for the telephone lines. The process is reversed in the receive mode. The modem logic provides the necessary control signals to inform the modem that the data transfer is about to occur. These include such signals as request to send (RTS) and terminal on (TON). The modem logic also monitors control signals from the modem, such as data set ready (DSR) and clear to send (CTS).

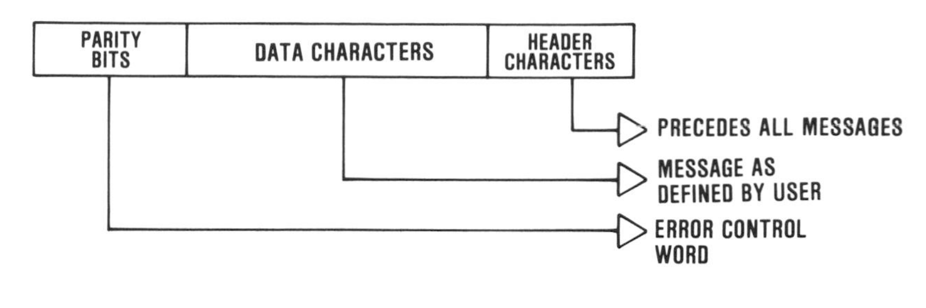

a) DATA BLOCK

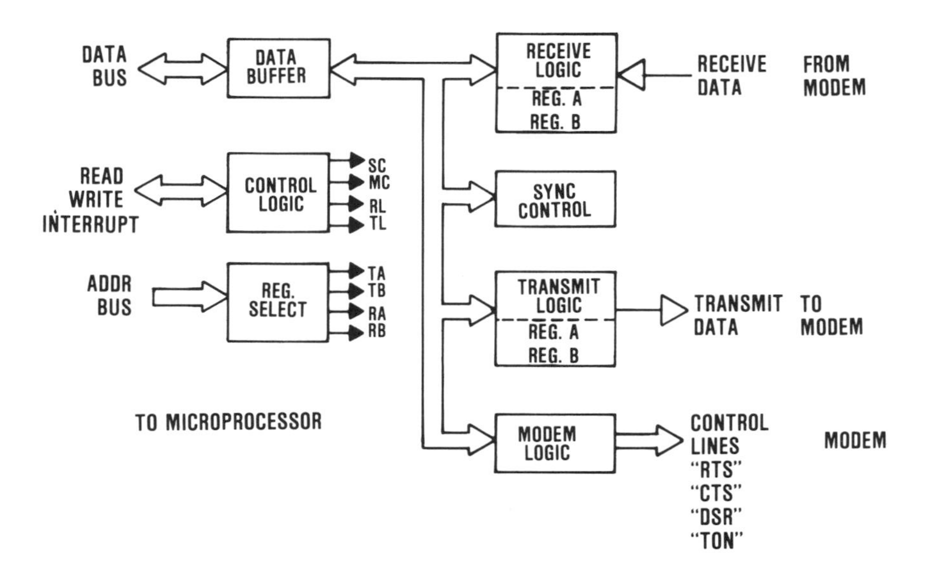

b) SYNCHRONOUS COMMUNICATIONS CONTROLLER

FIGURE I.5.7 (a) Data Block
(b) Synchronous Communications Controller

Sync Control: The sync control logic is used to generate and detect sync characters. In the transmit mode, the sync control logic is used to fill the message with sync characters whenever data characters are not available from the MP or MC. This is done in order to maintain synchronization between the transmit and receive stations. In the receive mode, the SCC looks for the beginning of the message from the transmitter. When the sync characters are detected, the SCC informs the MP or MC that a message is being received from the modem.

KEY PARAMETERS

a) *Baud rate.* The frequency at which the serial bits are transmitted over the telephone lines. It is programmable from DC to 64 K bits/s.
b) *Character length.* The SCC is programmable for 5 to 8-bit data characters
c) *Duplex mode.* The method of transmission is programmable for either full duplex (date transfers occur in both directions) or half duplex (data transfers occur in one direction at a time).

APPLICATION

The synchronous communications controller (SCC) is used whenever high-speed digital communication is required.

COMMENT

Some ICs contain both the synchronous and asynchronous functions (see UART, I.5.9).

I.5.8 Tape Data Controller (TDC)

DESCRIPTION

The tape data controller (TDC) provides the interface between a computer and a magnetic tape cassette or cartridge drive. Figure I.5.8(a) shows the block diagram of a typical tape system in which the tape data controller reads data from the tape drive to the processor and writes data onto the tape drive from the processor. The processor controls the tape drive by providing it with the necessary control signals. The processor, for example, commands the cassette/cartridge via the TDC to begin moving the tape. When the tape is at the proper speed, the processor may command the TDC to begin a data transfer in either direction. An explanation of the control signals between the processor and the tape drive follows:

Processor to Tape Drive

1. F/R: This signal determines the direction of the moving tape. If the F/R is **1**, the tape is commanded to move forward. When it is **0**, the tape moves in reverse.
2. F/S: This signal determines the rate at which the tape moves. In one state the tape moves fast; in the other state the tape moves slow.
3. G/S: This signal commands the tape drive to go or stop.
4. Write: This signal commands the tape drive to perform a write operation.
5. Select: This signal determines which tape drive is selected for operation.

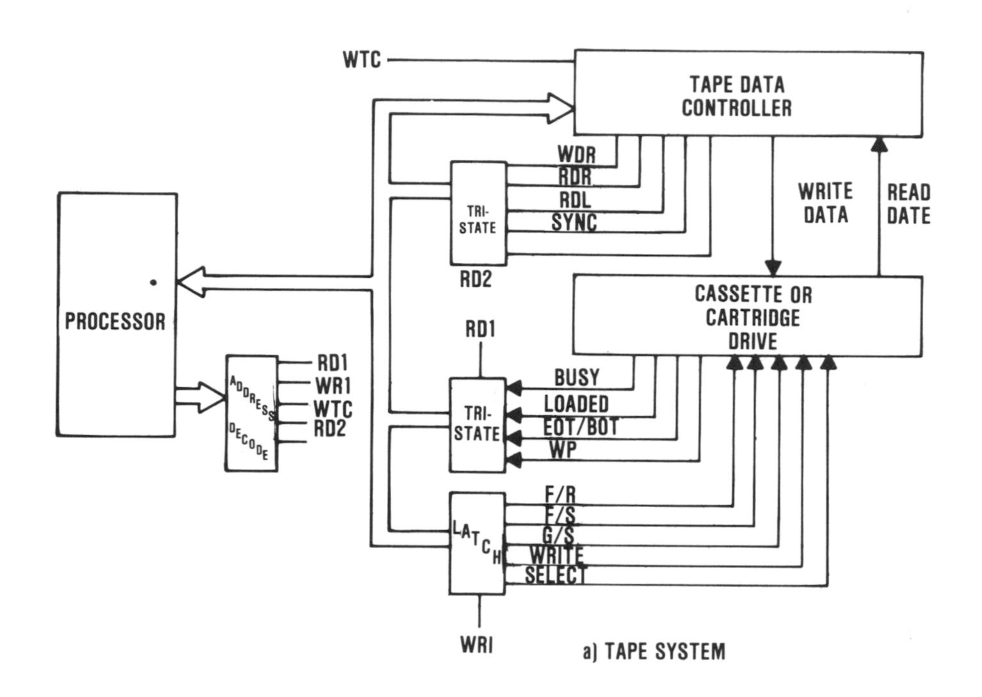

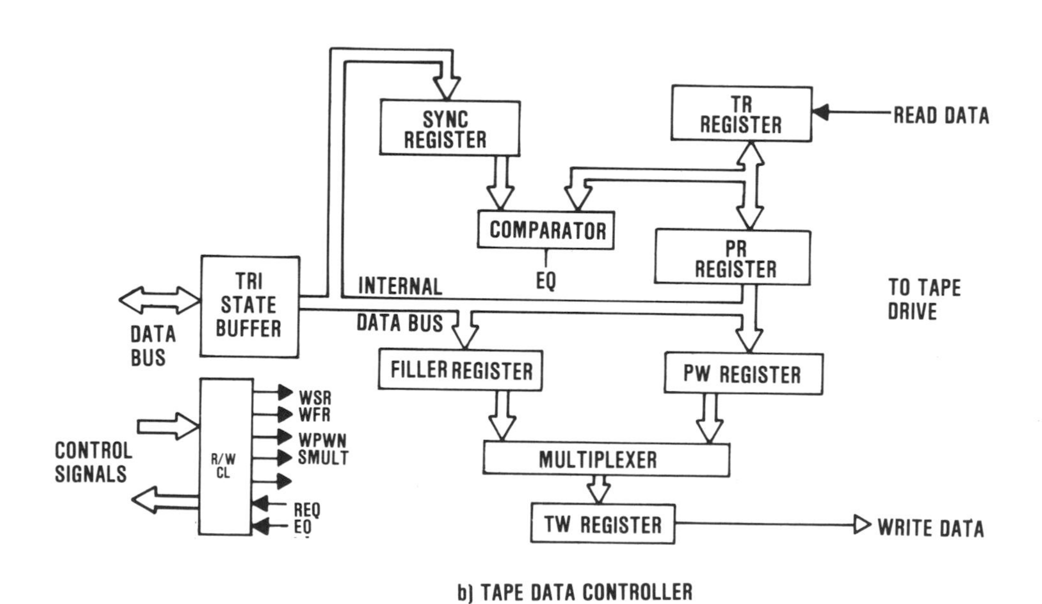

FIGURE I.5.8 (a) Tape System
(b) Tape Data Controller

Tape Drive to Processor

1. Busy: This signal indicates to the processor that the tape drive is busy.
2. Loaded: This signal tells the processor the tape drive is loaded with a tape and ready for operation.
3. EOT/BOT: This signal indicates the position of the end or beginning of the tape.
4. Write Protect: The processor uses this signal to determine if data can be written onto the tape.

Figure I.5.8(b) shows the block diagram of a tape data controller (TDC). Its two modes of operation are described below.

Write Mode: The processor first starts the tape drive to get the tape moving at the correct speed. This is accomplished by transferring a control word to the tape drive. Once the tape drive is set up for recording, the processor sets the TDC to the write mode.

The TDC uses the filler, processor write (PW), and tape write (TW) registers to perform a write operation. The R/W control logic, under the control of the processor, loads up the registers and generates all the necessary control signals to write information onto the tape. The TDC write mode operates as follows:

Once the tape drive is at operating speed, the processor sets up the filler and PW registers with the appropriate 8-bit characters. This is accomplished via the data and control bus to the TDC. The filler register is used to complete a block being recorded on the tape. Once the PW register contains a data character, it is transferred through the multiplexer to the TW register. The PW register then generates a request (Req) signal to the processor for the next data character. While the processor services the TDC, the TW register serially shifts out the information over the write data line to the read/write electronics of the tape drive where it is recorded onto the tape. The above cycle is repeated until the buffer of data from the processor is empty. If the tape file still has room and there is no processor data, the TDC automatically connects the multiplexer to the filler register. The filler character stored in the register is continuously transferred to the TW register for recording on the tape until the tape file is complete.

Read Mode: Information from the tape is serially shifted into the TDC, and the information is then processed by the TDC. A request (Req) signal is generated to the processor, indicating data is available. The processor then generates the necessary control signals to the R/W Control logic to read the data over the data bus. The TDC read mode operates as follows:

Once the processor gets the tape drive to operating speed, the sync register is loaded with an 8-bit sync character. The data bus and the WSR signal are activated in order to set up the sync register. As data is read off the tape, it is serially shifted over the read data line into the tape read (TR) register. Each character shifted into the TR register is compared with the

contents of the sync register. When a match occurs, the equate (EQ) signal is activated and sent back to the processor. A match signifies that the beginning of a tape file has been located. The TDC then connects the TR to the processor read (PR) register. Once the TDC reaches this stage all data read off the tape are transferred to the processor. The request signal and data bus are used for this operation.

KEY PARAMETERS

a) *Baud rate.* The frequency at which the data is transferred between the TDC and the tape drive. Most TDCs can operate up to 250 K bits per second.
b) *Mode.* TDCs can operate either in the full- or half-duplex mode. When in the half duplex mode, the IC is in either the read or the write mode but never in both. In full duplex the read and write modes are performed simultaneously.
c) *Tape driver type.* TCDs can operate with standard and mini-cassettes or with cartridges.

APPLICATIONS

The TDC is used in cassette and cartridge tape control operations.

COMMENTS

Some TDCs control only the data to and from the tape. They do not control the tape drive itself. Additional logic is then necessary to perform those functions. Data can be stored synchronously or asynchronously on tape. If stored synchronously there are two tracks. One track is reserved for the clock while the other is used to store the data. In the asynchronous mode the clock is embedded in the data bits. External logic is required to separate the clock from the data for the read mode and to combine the clock with the data for the write mode.

I.5.9 Universal Asynchronous Receiver Transmitter (UART)

DESCRIPTION

The UART shown in Figure I.5.9(a) interfaces microprocessors (MP) for synchronous serial data transmissions using the ASCII or ECSIDIC format of Figure I.5.9(b). The eight data bits can represent alphanumeric characters or control signals between systems. Each data character is preceded by its own start bit and is followed by one parity bit which is used

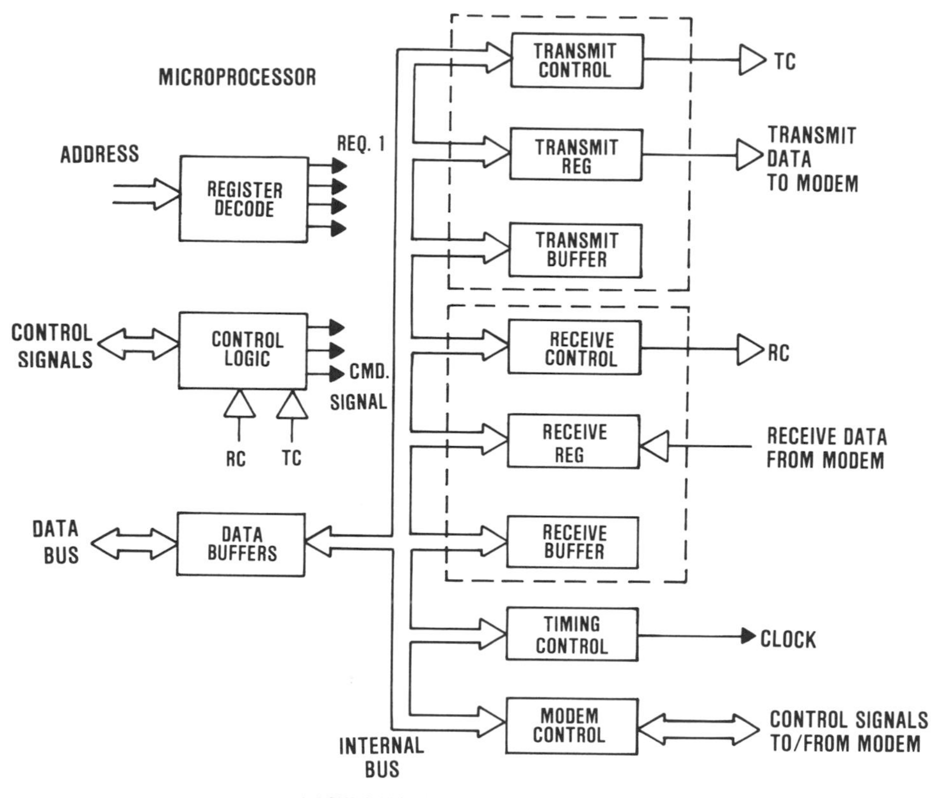

a) ASYNCHRONOUS COMMUNICATIONS CONTROLLER (UART)

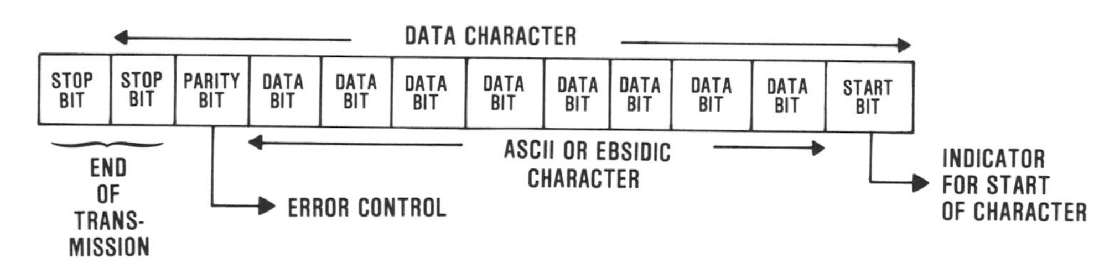

b) TYPICAL DATA CHARACTER

FIGURE I.5.9 (a) Asynchronous Communications Controller (UART)
(b) Typical Data Character

for error control. Two stop bits are assigned to signal the end of each data character.

Control words from the MP are sent over the data bus to set up the UART, and the address on the address bus determines which function receives the control words. Control signals are processed by the control logic to generate the command signal to the internal logic elements. The timing counter, for example, is loaded with a control/data word which defines the baud rate for the UART. The baud rate establishes the frequency at which the data bits are sent over the transmit or receive data lines. The modem control logic sets up the external modem for a communications cycle. A modem is a sepcialized interface device which converts the digital information from the UART into the analog signal suitable for the telephone lines. The modem also converts analog signals from the telephone lines to digital signals. A brief description of the transmit and receive operation follows.

Transmit Data Mode: The UART is set to the transmit mode when data is to be transferred from the MP to the modem. Next, a control word is sent over the data bus and loaded into the transmit control register. This control word formats the data character so that it can direct the UART to transmit a 6-, 7- or 8-bit data character and enables the odd or even parity mode. The data character itself has been stored in the transmit buffer. Next, the UART transfers the character from the transmit buffer to the transmit register which then transmits the data character serially to the transmit data line. When the transmit buffer is empty, the TC signal is generated and this causes an interrupt signal to go to the MP. This interrupt signal informs the MP that the UART can accept another data character.

Receive Data Cycle: The receive logic in the UART is used to receive data characters from the modem. All "overhead" bits are removed from the data character, including the start and stop and parity bits. If the parity is in error the MP is informed. After these operations are completed by the UART, an interrupt signal is sent to the MP, causing it to execute a program reading the characters from the UART. The receive control register is set up by the MP concerning the format of the expected data character, including the number of data bits (6, 7 or 8), the number of stop bits (1 or 2), and whether the expected character will have odd or even parity. The incoming data character on the receive data line is then shifted serially into the receive data register. During this shift operation, the overhead bits are stripped away. Once the register is full, its contents are transferred to the receive buffer. This also causes an interrupt signal to the MP which responds by sending a read signal and the register select value on the address bus. This causes the contents of the receive buffer to be transferred to the MP over the data bus.

KEY PARAMETERS

a) *Baud rate.* This parameter established the frequency at which the data characters are serially transferred between the UART and the mode. Baud rates up to 64 KHz are typical.

b) *Duplex mode*. Some UARTs have the capability of transmitting and receiving data simultaneously (full duplex). Other ICs can transmit or receive data only at any given time (half duplex).
c) *Error detection*. Various error detection schemes are available, but the vast majority of UARTs use the parity method.

APPLICATION

The UART is widely used in data communications as interface between a microprocessor (MP) or microcomputer (MC) and a modem or other communications encoder.

I.5.10 Contact Debouncer

DESCRIPTION

One of the important inputs to computer systems is a relay or switch contact, and there are two types of contacts: Form A, which has one input, and Form C, which has two inputs. Whenever the switch contact opens or closes, contact bounce occurs, and this is prevented by the contact debouncer IC.

Figure I.5.10(a) shows the block diagram for the Form A or C contact debouncer IC. When the contact is open, the signals contact close in and contact close out are both **1** (high). The contact close out signal, when high, indicates that the switch is open. Both **1**'s to the Exclusive OR gate cause continuous loading of **1**'s into the register. When the contact close in line goes to **0**, indicating that the contact is closed, the register switches to the shift mode. This causes **0**'s to be shifted into the FF's register. This shifting continues for each pulse. If contact bouncing occurs (input switches back to the **1** state) the register is switched to the load mode, causing **1** to be loaded into the FFs. This process is repeated when the contact opens.

Figure I.5.10(b) shows a contact debouncer for the Form C contact. Elimination of contact bounce for the Form C contact requires less logic than for the Form A contact because the Form C has two outputs rather than one. The contact output signal CON-CH is normally at the **1** state while the CON-CL signal is **0**. This causes the Q output of the cross-coupled latch to be in the **0** state. If the strobe enable is activated, the contact status line goes to **0**, indicating that the switch is disabled. When the switch or contact is activated, the CON-CL goes to **1**. The cross-coupled latch does not switch to the **1** state until the CON-CH signal goes to **0**.

If contact bounce occurs, the latch will not switch back to the disabled state.

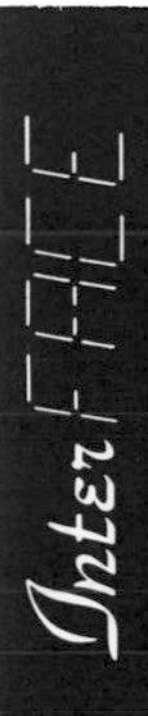

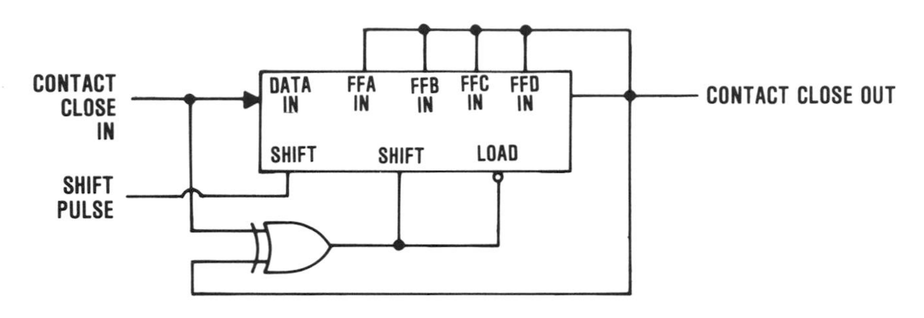

a) FORM A OR C CONTACT DEBOUNCER

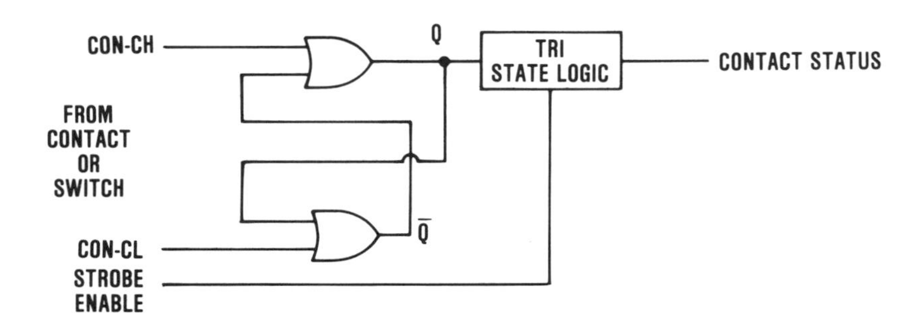

c) TRUTH TABLE

b) FORM C CONTACT DEBOUNCER

FIGURE I.5.10 (a) Form A or C Contact Debouncer
(b) Form C Contact Debouncer

KEY PARAMETERS

a) *Input voltage.* From 0 to +15 volts is typical.
b) *Number of contact filters.* The number of filters in a single IC varies from two to six.

APPLICATION

This IC is used whenever switches and relay contacts are part of the input.

APPENDIX

A. Dimensional Outlines*

CERAMIC DUAL-IN-LINE PACKAGES

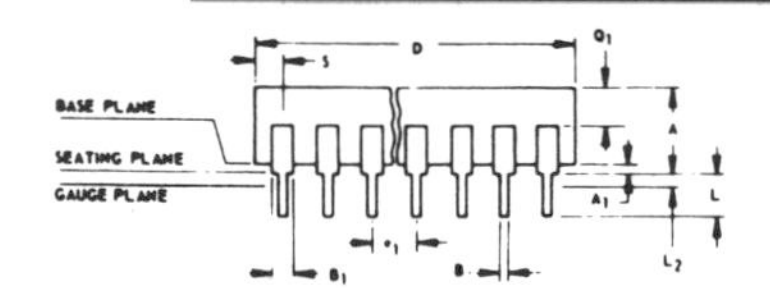

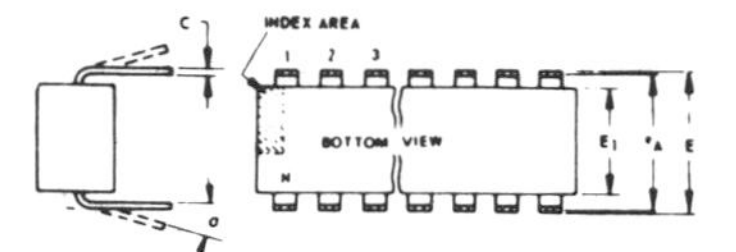

NOTES:
Refer to Rules for Dimensioning (JEDEC Publication No. 95) for Axial Lead Product Outlines.

1. When this device is supplied solder-dipped, the maximum lead thickness (narrow portion) will not exceed 0.013".
2. Leads within 0.005" (0.12 mm) radius of True Position (TP) at gauge plane with maximum material condition and unit installed.
3. e_A applies in zone L_2 when unit installed.
4. α applies to spread leads prior to installation.
5. N is the maximum quantity of lead positions.
6. N_1 is the quantity of allowable missing leads.

(D) Suffix (JEDEC MO-001-AD) 14-Lead

SYMBOL	INCHES MIN.	INCHES MAX.	NOTE	MILLIMETERS MIN.	MILLIMETERS MAX.
A	0.120	0.160		3.05	4.06
A_1	0.020	0.065		0.51	1.65
B	0.014	0.020		0.356	0.508
B_1	0.050	0.065		1.27	1.65
C	0.008	0.012	1	0.204	0.304
D	0.745	0.770		18.93	19.55
E	0.300	0.325		7.62	8.25
E_1	0.240	0.260		6.10	6.60
e_1	0.100 TP		2	2.54 TP	
e_A	0.300 TP		2, 3	7.62 TP	
L	0.125	0.150		3.18	3.81
L_2	0.000	0.030		0.000	0.76
α	0°	15°	4	0°	15°
N	14		5	14	
N_1	0		6	0	
Q_1	0.050	0.085		1.27	2.15
S	0.065	0.090		1.66	2.28

92SS-4411R2

(D) Suffix (JEDEC MO-001-AE) 16-Lead

SYMBOL	INCHES MIN.	INCHES MAX.	NOTE	MILLIMETERS MIN.	MILLIMETERS MAX.
A	0.120	0.160		3.05	4.06
A_1	0.020	0.065		0.51	1.65
B	0.014	0.020		0.356	0.508
B_1	0.035	0.065		0.89	1.65
C	0.008	0.012	1	0.204	0.304
D	0.745	0.785		18.93	19.93
E	0.300	0.325		7.62	8.25
E_1	0.240	0.260		6.10	6.60
e_1	0.100 TP		2	2.54 TP	
e_A	0.300 TP		2, 3	7.62 TP	
L	0.125	0.150		3.18	3.81
L_2	0.000	0.030		0.000	0.76
α	0°	15°	4	0°	15°
N	16		5	16	
N_1	0		6	0	
Q_1	0.050	0.085		1.27	2.15
S	0.015	0.060		0.39	1.52

92SS-4286R5

DUAL-IN-LINE PLASTIC AND FRIT-SEAL CERAMIC PACKAGES

(E) and (G) Suffixes (JEDEC MO-001-AN) 8-Lead Plastic (Mini-DIP)

SYMBOL	INCHES MIN.	INCHES MAX.	NOTE	MILLIMETERS MIN.	MILLIMETERS MAX.
A	0.155	0.200		3.94	5.08
A_1	0.020	0.050		0.508	1.27
B	0.014	0.020		0.356	0.508
B_1	0.035	0.065		0.889	1.65
C	0.008	0.012	1	0.203	0.304
D	0.370	0.400		9.40	10.16
E	0.300	0.325		7.62	8.25
E_1	0.240	0.260		6.10	6.60
e_1	0.100 TP		2	2.54 TP	
e_A	0.300 TP		2, 3	7.62 TP	
L	0.125	0.150		3.18	3.81
L_2	0.000	0.030		0.000	0.762
α	0	15	4	0	15°
N	8		5	8	
N_1	0		6	0	
Q_1	0.040	0.075		1.02	1.90
S	0.015	0.060		0.381	1.52

92CS-24026R1

(E, (F) and (G) Suffixes (JEDEC MO-001-AB) 14-Lead

SYMBOL	INCHES MIN.	INCHES MAX.	NOTE	MILLIMETERS MIN.	MILLIMETERS MAX.
A	0.155	0.200		3.94	5.08
A_1	0.020	0.050		0.51	1.27
B	0.014	0.020		0.356	0.508
B_1	0.050	0.065		1.27	1.65
C	0.008	0.012	1	0.204	0.304
D	0.745	0.770		18.93	19.55
E	0.300	0.325		7.62	8.25
E_1	0.240	0.260		6.10	6.60
e_1	0.100 TP		2	2.54 TP	
e_A	0.300 TP		2, 3	7.62 TP	
L	0.125	0.150		3.18	3.81
L_2	0.000	0.030		0.000	0.76
α	0°	15°	4	0°	15°
N	14		5	14	
N_1	0		6	0	
Q_1	0.040	0.075		1.02	1.90
S	0.065	0.090		1.66	2.28

92SS-4296R3

(E), (F), and (G) Suffixes (JEDEC MO-001-AC) 16-Lead

SYMBOL	INCHES MIN.	INCHES MAX.	NOTE	MILLIMETERS MIN.	MILLIMETERS MAX.
A	0.155	0.200		3.94	5.08
A_1	0.020	0.050		0.51	1.27
B	0.014	0.020		0.356	0.508
B_1	0.035	0.065		0.89	1.65
C	0.008	0.012	1	0.204	0.304
D	0.745	0.785		18.93	19.93
E	0.300	0.325		7.62	8.25
E_1	0.240	0.260		6.10	6.60
e_1	0.100 TP		2	2.54 TP	
e_A	0.300 TP		2, 3	7.62 TP	
L	0.125	0.150		3.18	3.81
L_2	0.000	0.030		0.000	0.76
α	0°	15°	4	0°	15°
N	16		5	16	
N_1	0		6	0	
Q_1	0.040	0.075		1.02	1.90
S	0.015	0.060		0.39	1.52

92CM-15967R4

CERAMIC FLAT PACKS

NOTES:

1. Refer to Rules for Dimensioning (JEDEC Publication No. 95) for Axial Lead Product Outlines.
2. Leads within .005" (.12 mm) radius of True Position (TP) at maximum material condition.
3. N is the maximum quantity of lead positions.
4. Z and Z_1 determine a zone within which all body and lead irregularities lie.

(K) Suffix (JEDEC MO-004-AF) 14-Lead

SYMBOL	INCHES MIN.	INCHES MAX.	NOTE	MILLIMETERS MIN.	MILLIMETERS MAX.
A	0.008	0.100		0.21	2.54
B	0.015	0.019	1	0.381	0.482
C	0.003	0.006	1	0.077	0.152
e	0.050 TP		2	1.27 TP	
E	0.200	0.300		5.1	7.6
H	0.600	1.000		15.3	25.4
L	0.150	0.350		3.9	8.8
N	14		3	14	
Q	0.005	0.050		0.13	1.27
S	0.000	0.050		0.00	1.27
Z	0.300		4	7.62	
Z_1	0.400		4	10.16	

92SS-4300R3

*Courtesy RCA

TO–5 STYLE PACKAGES

(T) Suffix (JEDEC MO-002-AL) 8-Lead TO-5 Style

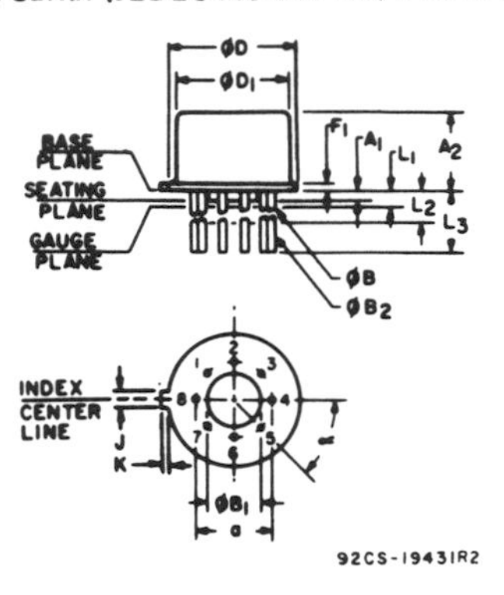

SYMBOL	INCHES		NOTE	MILLIMETERS	
	MIN.	MAX.		MIN.	MAX.
a	0.200 TP		2	5.88 TP	
A_1	0.010	0.050		0.26	1.27
A_2	0.165	0.185		4.20	4.69
ϕB	0.016	0.019	3	0.407	0.482
ϕB_1	0.125	0.160		3.18	4.06
ϕB_2	0.016	0.021	3	0.407	0.533
ϕD	0.335	0.370		8.51	9.39
ϕD_1	0.305	0.335		7.75	8.50
F_1	0.020	0.040		0.51	1.01
j	0.028	0.034		0.712	0.863
k	0.029	0.045	4	0.74	1.14
L_1	0.000	0.050	3	0.00	1.27
L_2	0.250	0.500	3	6.4	12.7
L_3	0.500	0.562	3	12 7	14.27
α	45° TP			45° TP	
N	8		6	8	
N_1	3		5	3	

(S) Suffix 8-Lead TO-5 Style with Dual-In-Line Formed Leads (DIL-CAN)

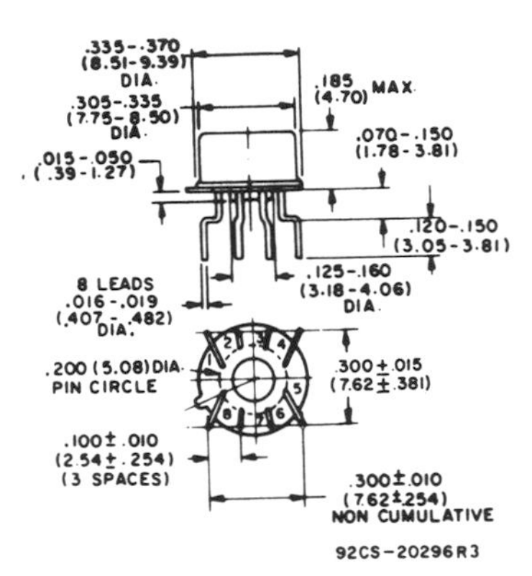

NOTES

1. Refer to JEDEC Publication No. 95 for Rules for Dimensioning Axial Lead Product Outlines.
2. Leads at gauge plane within 0.007" (0.178 mm) radium of True Position (TP) at maximum material condition.
3. ϕB applies between L_1 and L_2. ϕB_2 applies between L_2 and 0.500" (12.70 mm) from seating plane. Diameter is uncontrolled in L_1 and beyond 0.500" (12.70 mm).
4. Measure from Max. ϕD.
5. N_1 is the quantity of allowable missing leads.
6. N is the maximum quantity of lead positions.

(T) Suffix (JEDEC MO-006-AF) 10-Lead TO-5 Style

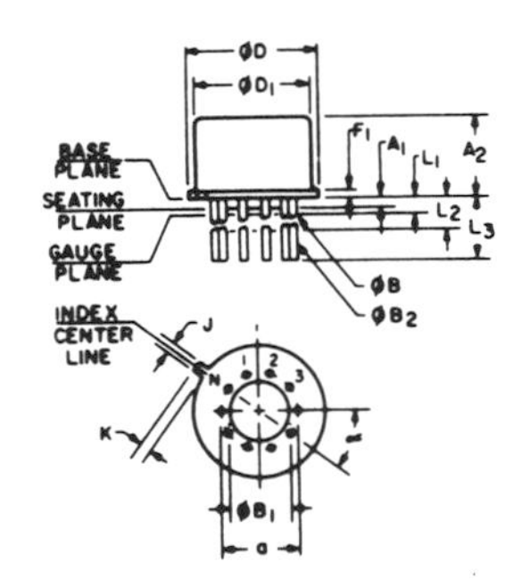

SYMBOL	INCHES		NOTE	MILLIMETERS	
	MIN.	MAX.		MIN.	MAX.
a	0.230 TP		2	5.84 TP	
A_1	0	0		0	0
A_2	0.165	0.185		4.19	4.70
ϕB	0.016	0.019	3	0.407	0.482
ϕB_1	0	0		0	0
ϕB_2	0.016	0.021	3	0.407	0.533
ϕD	0.335	0.370		8.51	9.39
ϕD_1	0.305	0.335		7.75	8.50
F_1	0.020	0.040		0.51	1.01
j	0.028	0.034		0.712	0.863
k	0.029	0.045	4	0.74	1.14
L_1	0.000	0.050	3	0.00	1.27
L_2	0.250	0.500	3	6.4	12.7
L_3	0.500	0.562	3	12.7	14.27
α	36° TP			36° TP	
N	10		6	10	
N_1	1		5	1	

(V) Suffix 10 Formed Leads Radially Arranged TO-5 Type

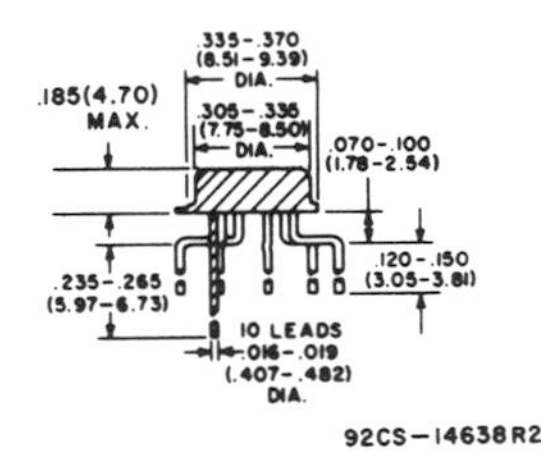

NOTES:

1. Refer to Rules for Dimensioning Axial Lead Product Outlines.
2. Leads at gauge plane within 0.007" (0.178 mm) radius of True Position (TP) at maximum material condition.
3. ϕB applies between L_1 and L_2. ϕB_2 applies between L_2 and 0.500" (12.70 mm) from seating plane. Diameter is uncontrolled in L_1 and beyond 0.500" (12.70 mm).
4. Measure from Max. ϕD.
5. N_1 is the quantity of allowable missing leads.
6. N is the maximum quantity of lead positions.

92CS-15835

TO–5 STYLE PACKAGE (Cont'd)

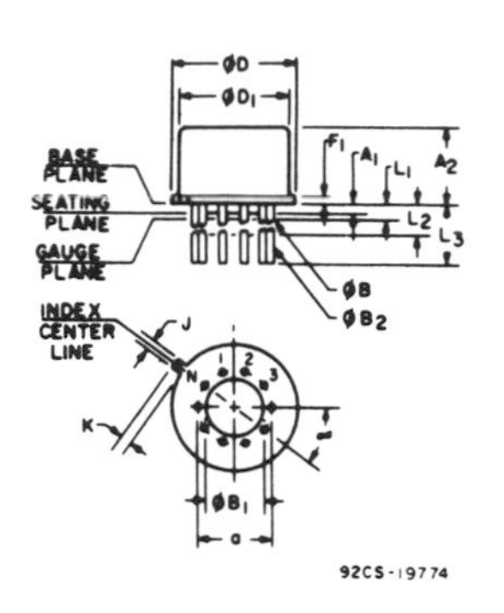

(T) Suffix (JEDEC MO-006-AG) 12-Lead TO-5 Style

SYMBOL	INCHES MIN.	INCHES MAX.	NOTE	MILLIMETERS MIN.	MILLIMETERS MAX.
a	0.230		2	5.84 TP	
A_1	0	0		0	0
A_2	0.165	0.185		4.19	4.70
ϕB	0.016	0.019	3	0.407	0.482
ϕB_1	0	0		0	0
ϕB_2	0.016	0.021	3	0.407	0.533
ϕD	0.335	0.370		8.51	9.39
ϕD_1	0.305	0.335		7.75	8.50
F_1	0.020	0.040		0.51	1.01
j	0.028	0.034		0.712	0.863
k	0.029	0.045	4	0.74	1.14
L_1	0.000	0.050	3	0.00	1.27
L_2	0.250	0.500	3	6.4	12.7
L_3	0.500	0.562	3	12.7	14.27
α	30° TP			30 TP	
N	12		6	12	
N_1	1		5	1	

NOTES:

1. Refer to Rules for Dimensioning Axial Lead Product Outlines.
2. Leads at gauge plane within 0.007" (0.178 mm) radius of True Position (TP) at maximum material condition.
3. ϕB applies between L_1 and L_2. ϕB_2 applies between L_2 and 0.500" (12.70 mm) from seating plane. Diameter is uncontrolled in L_1 and beyond 0.500" (12.70 mm).
4. Measure from Max. ϕD.
5. N_1 is the quantity of allowable missing leads.
6. N is the maximum quantity of lead positions.

JEDEC TO-72 PACKAGE

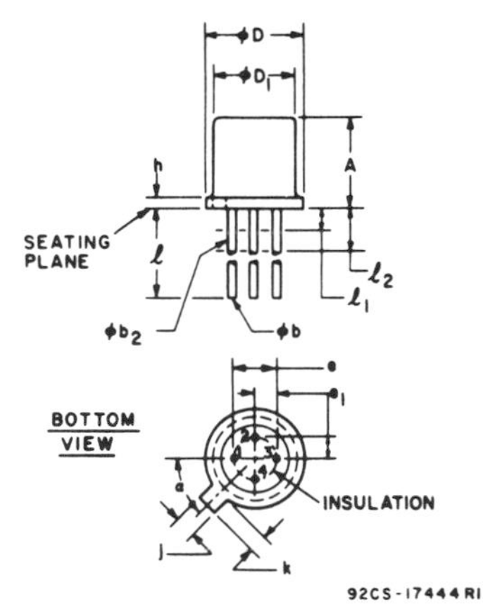

SYMBOL	INCHES MIN	INCHES MAX	MILLIMETERS MIN	MILLIMETERS MAX	NOTES
A	0.170	0.210	4.32	5.33	
ϕb	0.016	0.021	0.406	0.533	2
ϕb_2	0.016	0.019	0.406	0.483	2
ϕD	0.209	0.230	5.31	5.84	
ϕD_1	0.178	0.195	4.52	4.95	
e	0.100 T.P.		2.54 T.P.		4
e_1	0.050 T.P.		1.27 T.P.		4
h		0.030		0.762	
j	0.036	0.046	0.914	1.17	
k	0.028	0.048	0.711	1.22	3
l	0.500		12.70		2
l_1		0.050		1.27	2
l_2	0.250		6.35		2
α	45° T.P.		45° T.P.		4, 6

Note 1: (Four leads) Maximum number leads omitted in this outline. "none" (0). The number and position of leads actually present are indicated in the product registration. Outline designation determined by the location and mimimum angular or linear spacing of any two adjacent leads.

Note 2: (All leads) ϕb_2 applies between l_1 and l_2. ϕb applies between l_2 and 500" (12.70 mm) from seating plane. Diameter is uncontrolled in 1_1 and beyond 500" (12 70 mm) from seating plane.

Note 3: Measured from maximum diameter of the product.

Note 4: Leads having maximum diameter 019" (483 mm) measured in gaging plane 054" (1.37 mm) + 001" (025 mm) - 000" (000 mm) below the seating plane of the product shall be within 007" (.178 mm) of their true position relative to a maximum width tab.

Note 5: The product may be measured by direct methods or by gage.

Note 6: Tab centering.

QUAD-IN-LINE PLASTIC PACKAGES

(Q) Suffix 14-Lead Staggered

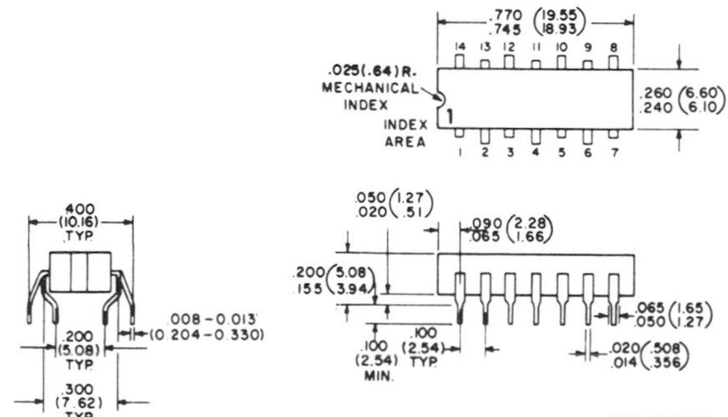

1. Body width is measured 0.040" (1.02mm) from top surface.
2. Seating plane defined as the junction of the angle with the narrow portion of the lead.

Dimensions in parentheses are millimeter equivalents of the basic inch dimensions.

Recommended Mounting – Hole Dimensions and Spacing

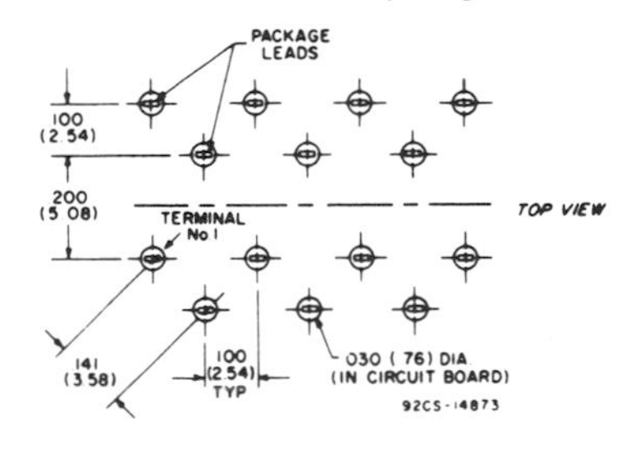

QUAD IN-LINE PACKAGES (Cont'd)

(W) Suffix 16-Lead Staggered

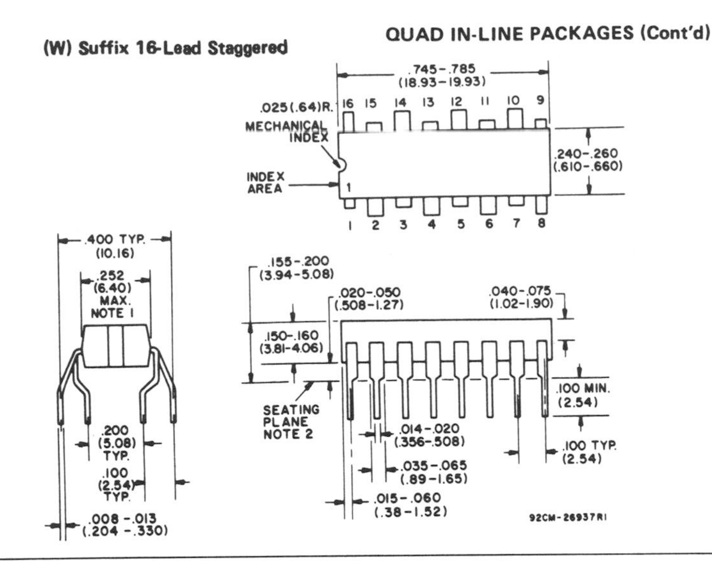

Recommended Mounting – Hole Dimensions and Spacing

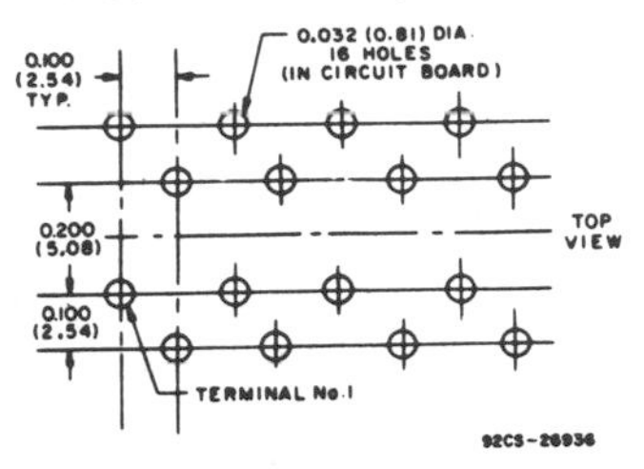

Notes:

1. Body width is measured 0.040" (1.02 mm) from top surface.
2. Seating plane defined as the junction of the angle with the narrow portion of the lead.

Dimensions in parentheses are millimeter equivalents of the basic inch dimensions.

(Q) Suffix 16-Lead

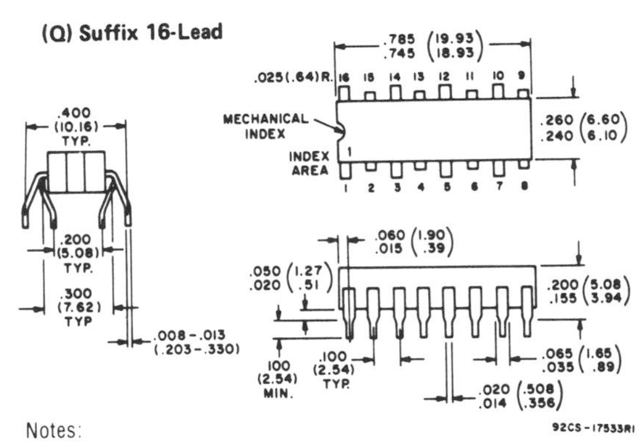

Recommended Mounting – Hole Dimensions and Spacing

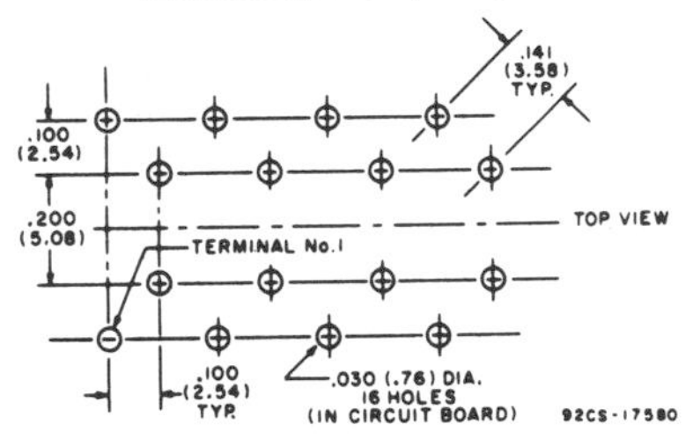

Notes:

1. Body width is measured 0.040" (1.02 mm) from top surface.
2. Seating plane defined as the junction of the angle with the narrow portion of the lead.

Dimensions in parentheses are millimeter equivalents of the basic inch dimensions.

20-Lead Shielded

1.010 (25.65)
.990 (25.15)
.790 (20.06)
.770 (19.56)
.030-.040
(.76-1.01)
.100-.120
(2.54-3.04)
.260 (6.60)
.240 (6.10)
INDEX NOTCH
.050 (1.27)
.020 (.51)
.045-.065
(1.14-1.65)
.400
(10.16)
TYP.
.200 (5.08)
.155 (3.94)
.100
(2.54)
MIN.
.100
(2.54)
TYP.
.075 (1.90)
.035 (.89)
.020 (.508)
.014 (.356)
.200
(5.08)
TYP.
.300
(7.62)
TYP.
.008-.013
(.204-.330)

NOTE: TERMINALS 11 AND 20 ARE OMITTED.

92CS-17587R1

Recommended Mounting – Hole Dimensions and Spacing

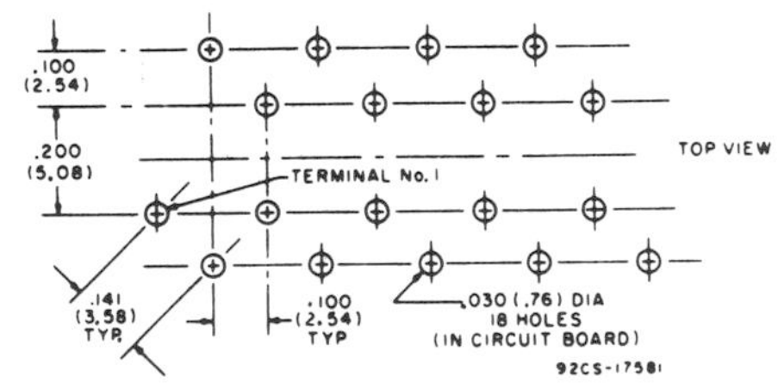

Notes:

1. Body width is measured 0.040" (1.02 mm) from top surface.
2. Seating plane defined as the junction of the angle with the narrow portion of the lead.

Dimensions in parentheses are millimeter equivalents of the basic inch dimensions.

E) Suffix
6-Lead "Power-Stud" Package

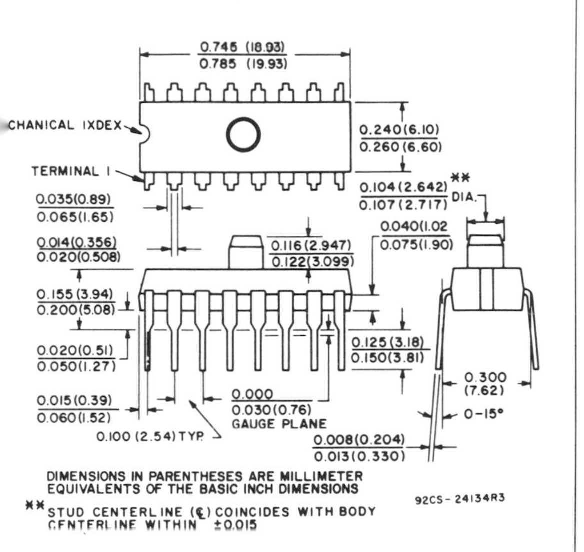

(EM) Suffix
Modified 16-Lead with Integral Heat Sink

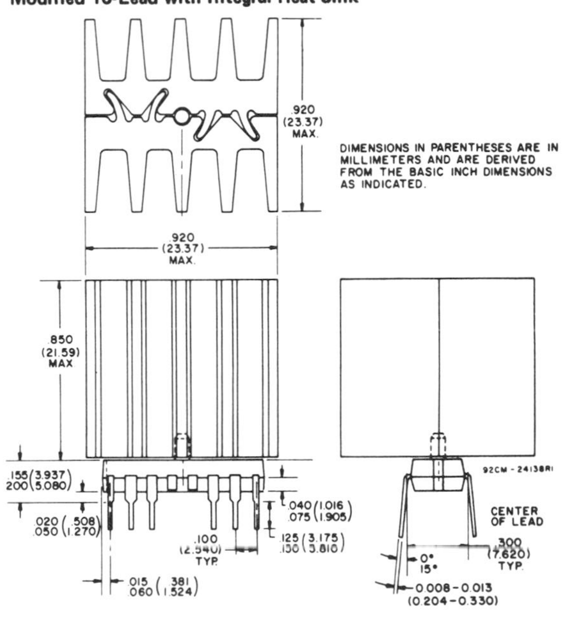

Q) Suffix
odified 16-Lead with Integral Bent Down Wing-Tab Heat Sink

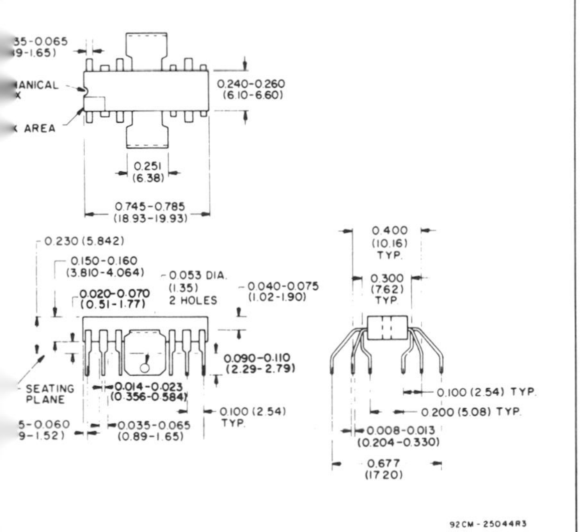

ENSIONS IN PARENTHESES ARE MILLIMETER
VALENTS OF THE BASIC INCH DIMENSIONS

(QM) Suffix
Modified 16-Lead with Integral Flat Wing-Tab Heat Sink

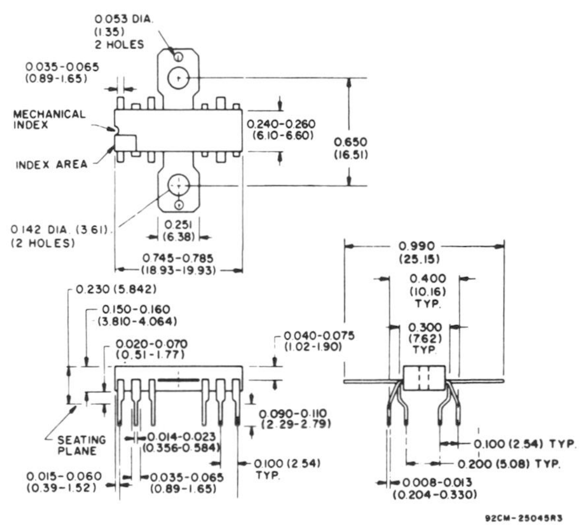

DIMENSIONS IN PARENTHESES ARE MILLIMETER
EQUIVALENTS OF THE BASIC INCH DIMENSIONS

DUAL-IN-LINE AND QUAD-IN-LINE PLASTIC PACKAGES
(Power Stud and Heat-Sink Types)

(EM) Suffix
16-Lead with Integral Strap Heat Sink

0.745-0.785 (18.93-19.93)
END ALIGNMENT OF HEAT SINK TO CENTER OF PLASTIC BODY
PLASTIC BODY 0.240-0.260 (6.10-6.60)
0.093-0.157 (2.36-3.99)
0.250 (6.35)
HEAT SINK
TERMINAL 1
0.537-0.587 (13.64-14.91)
0.280(7.11) MAX. IN DEBOSSING AREAS
0.030-0.036 (0.76-0.91)
INDEX AREA
1.125(28.58)
PLASTIC BODY
0.125 MAX. (3.18)
SILVER CONDUCTIVE EPOXY JOINT
EPOXY (2 DABS)
0.155-0.200 (3.94-5.08)
0.350 (8.89)
0°
15°
0.020-0.050 (0.51-1.27)
0.125-0.150 (3.18-3.81)
0.100 (2.54) TYP.
0.300 (7.62) TYP. CENTER OF LEAD
0.015-0.060 (0.39-1.52)
0.008-0.013 (0.204-0.330)
92CM-25649R2

(QM) Suffix
16-Lead Staggered with Integral Strap Heat Sink

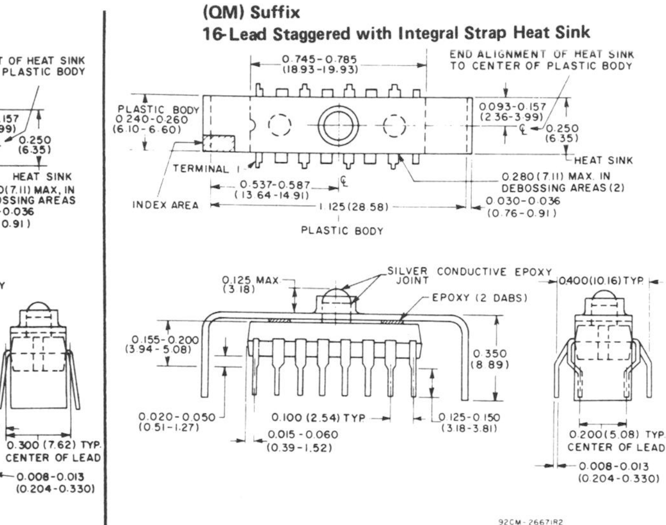

TO-220-STYLE (VERSA-V) PLASTIC PACKAGE

VERTICAL MOUNT

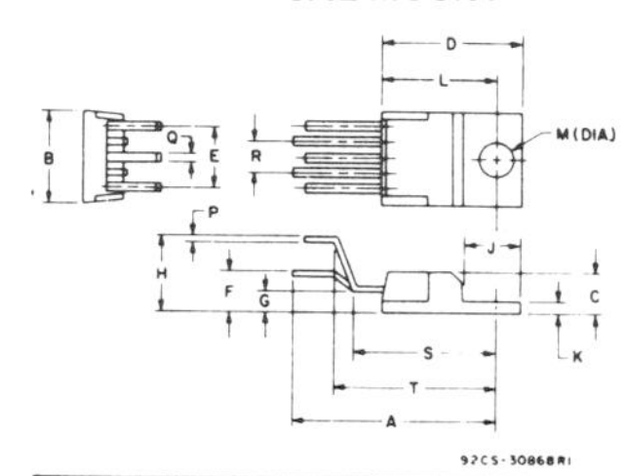

SYMBOL	INCHES		MILLIMETERS	
	MIN.	MAX.	MIN.	MAX.
A	0.876	0.896	22.25	22.75
B	0.396	0.408	10.06	10.36
C	0.173	0.182	4.395	4.622
D	0.604	0.619	15.35	15.72
E	0.263	0.273	6.681	6.934
F	0.168	0.188	4.268	4.775
G	0.100	0.104	2.540	2.641
H	0.320	0.340	8.128	8.638
J	0.246	0.254	6.249	6.451
K	0.046	0.054	1.169	1.371
L	0.496	0.508	12.60	12.90
M	0.140	0.150	3.556	3.810
N	5		5	
P	0.015	0.020	0.381	0.406
Q	0.033	0.040	0.839	1.016
R	0 129	0.139	3.277	3.530
S	ʃ.600	0.630	15.24	16.00
T	0.680	0.710	17.27	18.03

HORIZONTAL MOUNT (M Suffix)

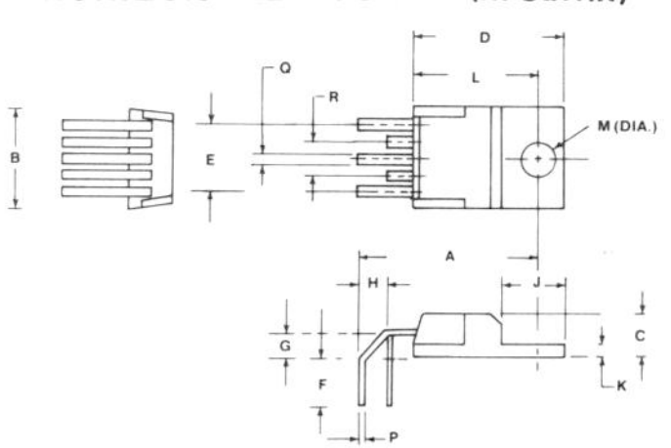

SYMBOL	INCHES		MILLIMETERS	
	MIN.	MAX.	MIN.	MAX.
A	0.726	0.746	18.44	18.94
B	0.396	0.408	10.06	10.36
C	0.173	0.182	4.395	4.622
D	0.604	0.619	15.35	15.72
E	0.263	0.273	6.681	6.934
F	0.221	0.251	5.614	6.375
G	0.100	0.104	2.540	2.641
H	0.143	0.163	3.633	4.140
J	0.246	0.254	6.249	6.451
K	0.046	0.054	1.169	1.371
L	0.496	0.508	12.60	12.90
M	0.140	0.150	3.556	3.810
N	5		5	
P	0.015	0.020	0.381	0.406
Q	0.033	0.040	0.839	1.016
R	0.129	0.139	3.277	3.530

B. List of Major IC Manufacturers

Advanced Micro Devices
901 Thompson Pl.
Sunnyvale, Ca. 94086
(408-732-2400)

American Micro Systems
3800 Homestead Road
Santa Clara, Ca. 95051
(408-246-0330)

Analog Devices, Inc.
P.O. Box 280
Norwood, Ma. 02062
(617-329-4700)

Datel Systems, Inc.
1020 Turnpike St.
Canton, Ma. 02021
(617-828-8000)

EXAR Integrated Systems, Inc.
750 Palomai Avenue
Sunnyvale, Ca. 94086
(408-733-7700)

Fairchild Semiconductor
464 Ellis St.
Mountainview, Ca. 94042
(415-962-5011)

Ferranti Electric, Inc.
E. Bethpage Road
Plainview, N.Y. 11803
(516-293-8383)

Harris Semiconductor
P.O. Box 883
Melbourne, Fl. 32901
(305-724-7407)

Hughes Aircraft Co., Microelectronics Prod. Division
500 Superior Avenue
Newport Beach, Ca. 92675
(714-548-0671)

Intel Corp.
3065 Bowers Avenue
Santa Clara, Ca. 95051
(408-246-7501)

Intersil, Inc.
10900 Tantau Avenue
Santa Clara, Ca. 95014
(407-984-2170)

ITT Semiconductors
74 Commerce Way
Woburn, Ma. 01801
(617-935-7910)

MOS Technology, Inc.
950 Rittenhouse Road
Norristown, Pa. 19401
(215-666-7950)

Mostek Corp.
1215 W. Crosby Road
Carrollton, Tx. 75006
(214-242-0444)

Motorola Semiconductor Prod., Inc.
P.O. Box 20912
Phoenix, Az. 85036
(602-244-6900)

National Semiconductor Corp.
2900 Semiconductor Drive
Santa Clara, Ca. 95051
(408-732-5000)

NEC America Inc.
3070 Lawrence Expressway
Santa Clara, Ca. 95051
(408-738-2180)

Plessy Semicondcutors
1674 McGaw Avenue
Santa Ana, Ca. 92715
(714-540-9945)

Precision Monolithics
1500 Space Park Drive
Santa Clara, Ca. 95050
(408-246-9211)

Raytheon Co., Semiconductor Division
350 Ellis St.
Mountainview, Ca. 94042
(415-968-9211)

RCA Solid State Division
Somerville, N.J. 08876
(201-685-6000)

Rockwell International, Microelectr. Device Division
3310 Miraloma Avenue
Anaheim, Ca. 92803
(714-632-3729)

Signetics Corp.
811 E. Arques Avenue
Sunnyvale, Ca. 94086
(408-739-7700)

Silicon General, Inc.
7382 Bolsa Avenue
Westminster, Ca. 92683
(714-892-5531)

Siliconix, Inc.
2201 Laurelwood Road
Santa Clara, Ca. 95054
(408-246-8000)

Solid State Scientific, Inc.
Montgomeryville, Pa. 18936
(215-855-8400)

Solitron Devices, Inc.
8808 Balboa Avenue
San Diego, Ca. 92123
(714-278-8780)

Sprague Electric Co.
645 Marshall St.
N. Adams, Ma. 01247
(413-664-4411)

Standard Microsystems Corp.
35 Marcus Blvd.
Hauppauge, N.Y. 11787
(516-273-3100)

Teledyne Semiconductor
1300 Terra Bella Avenue
Mountainview, Ca. 94043
(415-968-9241)

Texas Instruments, Inc.
P.O. Box 5012, Mail Station 84
Dallas, Tx. 75222
(214-238-2011)

TRW Semiconductor
14520 Aviation Blvd.
Lawndale, Ca. 90260
(213-679-4561)

Varian Assoc. Inc., Electron Devices
611 HansenWay
Palo Alto, Ca. 94303
(415-592-1221)

Watkins-Johnson Co.
3333 Hillview Avenue
Palo Alto, Ca. 94304
(415-493-4141)

Zilog
10460 Bubb Road
Cupertino, Ca. 95014
(408-446-4666)

C. Operating and Handling Considerations for Linear ICs*

Solid state devices are being designed into an increasing variety of electronic equipment because of their high standards of reliability and performance. However, it is essential that equipment designers be mindful of good engineering practices in the use of these devices to achieve the desired performance.

This Note summarizes important operating recommendations and precautions which should be followed in the interest of maintaining the high standards of performance of linear integrated circuits and MOS field-effect transistors.

The ratings included in RCA data bulletins are based on the Absolute Maximum Rating System, which is defined by the following Industry Standard (JEDEC) statement:

Absolute-Maximum Ratings are limiting values of operating and environmental conditions applicable to any electron device of a specified type as defined by its published data, and should not be exceeded under the worst probable conditions.

The device manufacturer chooses these values to provide acceptable serviceability of the device, taking no responsibility for equipment variations, environmental variations, and the effects of changes in operating conditions due to variations in device characteristics.

The equipment manufacturer should design so that initially and throughout life no absolute-maximum value for the intended service is exceeded with any device under the worst probable operating conditions with respect to supply-voltage variation, equipment component variation, equipment control adjustment, load variation, signal variation, environmental conditions, and variations in device characteristics.

It is recommended that equipment manufacturers consult RCA whenever device applications involve unusual electrical, mechanical or environmental operating conditions.

GENERAL CONSIDERATIONS

The design flexibility provided by integrated circuits and MOS/FET's makes possible their use in a broad range of applications and under many different operating conditions. When incorporating these devices in equipment, designers should anticipate the rare possibility of device failure and make certain that no safety hazard would result from such an occurrence.

The small size of these devices provides obvious advantages to the designers of electronic equipment. However, it should be recognized that these compact devices usually provide only relatively small insulation area between adjacent leads and the metal envelope. When these devices are used

*Courtesy RCA.

in moist or contaminated atmospheres, therefore, supplemental protection must be provided to prevent the development of electrical conductive paths across the relatively small insulating surfaces.

Devices should not be connected into or disconnected from circuits with the power on because high transient voltages may cause permanent damage to the devices.

TESTING PRECAUTIONS

In common with many electronic components, solid-state devices should be operated and tested in circuits which have reasonable values of current limiting resistance, or other forms of effective current overload protection. Failure to observe these precautions can cause excessive internal heating of the device resulting in destruction and/or possible shattering of the enclosure.

MOUNTING

Integrated circuits are normally supplied with lead-tin plated leads to facilitate soldering into circuit boards. In those relatively few applications requiring welding of the device leads, rather than soldering, the devices may be obtained with gold or nickel plated Kovar leads.* It should be recognized that this type of plating will not provide complete protection against lead corrosion in the presence of high humidity and mechanical stress. The aluminum-foil-lined cardboard "sandwich pack" employed for static protection of the flat-pack also provides some additional protection against lead corrosion, and it is recommended that the devices be stored in this package until used.

When integrated circuits are welded onto printed circuit boards or equipment, the presence of moisture between the closely spaced terminals can result in conductive paths that may impair device performance in high-impedance applications. It is therefore recommended that conformal coatings or potting be provided as an added measure of protection against moisture penetration.

In any method of mounting integrated circuits which involves bending or forming of the device leads, it is extremely important that the lead be supported and clamped between the bend and the package seal, and that bending be done with care to avoid damage to lead plating. In no case should the radius of the bend be less than the diameter of the lead, or in the case of rectangular leads, such as those used in RCA 14-lead and 16-lead flat-packages, less than the lead thickness. It is also extremely important that the ends of the bent leads be straight to assure proper insertion through the holes in the printed-circuit board.

---

*MIL-38510A, paragraph 3.5.6.1(a), lead material.

MOS FIELD-EFFECT TRANSISTORS

Insulated-Gate Metal Oxide-Semiconductor Field-Effect Transistors (MOS FETs), like bipolar high-frequency transistors, are susceptible to gate insulation damage by the electrostatic discharge of energy through the devices. Electrostatic discharges can occur in an MOS FET if a type with an unprotected gate is picked up and the static charge, built in the handler's body capacitance, is discharged through the device. With proper handling and applications procedures, however, MOS transistors are currently being extensively used in production by numerous equipment manufacturers in military, industrial, and consumer applications, with virtually no problems of damage due to electrostatic discharge.

In some MOS FETs, diodes are electrically connected between each insulated gate and the transistor's source. These diodes offer protection against static discharge and in-circuit transients without the need for external shorting mechanisms. MOS FETs which do not include gate-protection diodes can be handled safely if the following basic precautions are taken:

1. Prior to assembly into a circuit, all leads should be kept shorted together either by the use of metal shorting springs attached to the device by the vendor, or by the insertion into conductive materials such as "ECCOSORB* LD26" or equivalent.

   (NOTE: Polystyrene *insulating* "SNOW" is not sufficiently conductive and should not be used.)
2. When devices are removed by hand from their carriers, the hand being used should be grounded by any suitable means, for example, with a metallic wristband.
3. Tips of soldering irons should be grounded.
4. Devices should never be inserted into or removed from circuits with power on.

SOLID STATE CHIPS

Solid state chips, unlike packaged devices, are non-hermetic devices, normally fragile and small in physical size, and therefore, require special handling considerations as follows:

1. Chips must be stored under proper conditions to insure that they are not subjected to a moise and/or contaminated atmosphere that could alter their electrical, physical, or mechanical characteristics.

---

*Trade Mark: Emerson and Cumming, Inc.

After the shipping container is opened, the chip must be stored under the following conditions:

A. Storage temperature, 40°C max.

B. Relative humidity, 50% max.

C. Clean, dust-free environment.

2. The user must exercise proper care when handling chips to prevent even the slightest physical damage to the chip.
3. During mounting and lead bonding of chips the user must use proper assembly techniques to obtain proper electrical, thermal, and mechanical performance.
4. After the chip has been mounted and bonded, any necessary procedure must be followed by the user to insure that these non-hermetic chips are not subjected to moist or contaminated atmosphere which might cause the development of electrical conductive paths across the relatively small insulating surfaces. In addition, proper consideration must be given to the protection of these devices from other harmful environments which could conceivably adversely affect their proper performance.

D. General Operating and Application Considerations for COS/MOS ICs*

GENERAL OPERATING AND APPLICATION CONSIDERATIONS

This section is intended as a guide to circuit and equipment designers in the operation and application of MOS integrated circuits. It covers general operating and handling considerations with respect to the following critical factors:

- Operating supply-voltage range
- Power dissipation and derating
- System noise considerations
- Power-source rules
- Gate-oxide protection networks
- Input signals and ratings
- Chip assembly and storage
- Device mounting
- Testing

More specific information is then given on significant features, special design and application requirements, and standard ratings and electrical characteristics for COS/MOS A- and B-series logic circuits, and on COS/MOS timekeeping and special circuits.

GENERAL OPERATING AND HANDLING CONSIDERATIONS

The following paragraphs discuss some key operating and handling considerations that must be taken into account to achieve maximum advantage of the COS/MOS technology. Additional information on the operation and handling of COS/MOS integrated circuits is given in ICAN-6525, "Guide to Better Handling and Operation of CMOS Integrated Circuits," included in the Application Notes Section of this DATA BOOK.

Operating Supply-Voltage Range

Because logic systems occasionally experience transient conditions on the power-supply line which, when added to the nominal power-bus voltage, could exceed the safe limits of circuits connected to the power bus, the recommended operating supply-voltage ranges are 3 to 12 volts for A-series devices and 3 to 18 volts for B-series devices. The recommended maximum power-supply limit is substantially below the minimum primary breakdown limit for the devices to allow for limited power-supply transient and regulation limits. For circuits that operate in a linear mode over a portion of the voltage range, such as RC or crystal oscillators, a minimum supply voltage of at least 4 volts is recommended.

Power Dissipation and Derating

The power dissipation of a COS/MOS integrated circuit is the sum of a dc (quiescent) component and an ac (dynamic) component. The dc component is the sum of the net integrated-circuit reverse diode-junction current and the surface leakage current times the supply voltage. In standard A- or B-series logic devices, the dc dissipation typically ranges, depending upon device complexity, from 100 to 400 milliwatts for a supply voltage of 10 volts. Worst-case dc dissipation is the product of the maximum quiescent current (given in the data sheet on each device) and the dc supply voltage V_{DD}.

The dynamic (ac) power dissipation is approximately equal to the product CV^2f, where C is the net output capacitive load being charged and discharged by the integrated circuit, V is the supply voltage, and f is the output switching frequency. The product CV^2f accounts for approximately 90 per cent of the dynamic dissipation. The remaining 10 per cent is contributed by the momentary flow of IC switching current through the p- and n-MOS transistors to ground.

---

*Reprinted from "RCA COS/MOS Integrated Circuits Databook," 78. Courtesy RCA Corporation.

All COS/MOS devices are rated at 200 mW per package at the maximum operating ambient temperature rating (T_A) for the package type (85°C for plastic packages and 125°C for ceramic packages). Power ratings for temperatures below the maximum operating temperature are shown in the standard COS-/MOS thermal derating chart in Fig. 1. This chart assumes that (a) the device is mounted and soldered (or placed in a socket) on a PC board; (b) there is natural convection cooling, with the PC board mounted horizontally; and (c) the pressure is standard (14.7 psia). In addition to the over-all package dissipation, device dissipation per output transistor is limited 100 mW maximum over the full package operating-temperature range.

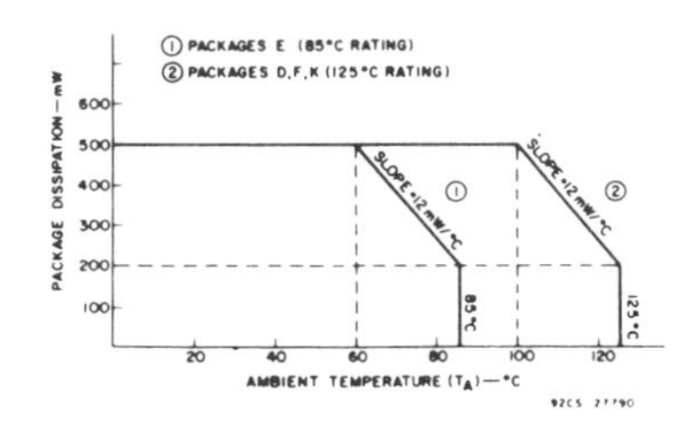

Fig. 1 — Standard COS/MOS thermal derating chart.

System Noise Considerations

In general, COS/MOS devices are much less sensitive to noise on power and ground lines than bipolar logic families (such as TTL or DTL). However, this sensitivity varies as a function of the power-supply voltage, and more importantly as a function of synchronism between noise spikes and input transitions. Good power distribution in digital systems requires that the power bus have a low dynamic impedance; for this purpose, discrete decoupling capacitors should be distributed across the power bus. A more detailed discussion of COS/MOS noise immunity is provided by ICAN-6587, "Noise Immunity of B-series COS/MOS Integrated Circuits," in the Application Notes Section.

Power-Source Rules

Fig. 2 shows the basic COS/MOS inverter and its gate-oxide protection network plus inherent diodes. The safe operating procedures listed below can be understood by reference to this inverter:

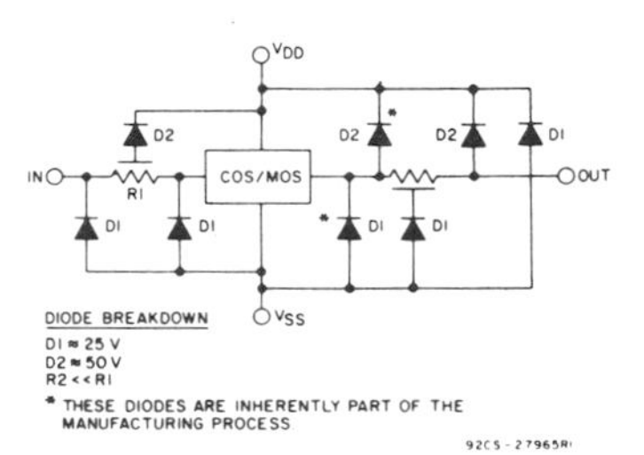

Fig. 2— Basic COS/MOS inverter with B-series types protection network.

1. When separate power supplies are used for the COS/MOS device and for the device inputs, the device power supply should always be turned on before the independent input signal sources, and the input signals should be turned off before the power supply is turned off ($V_{SS} \leqslant V_I \leqslant V_{DD}$ as a maximum limit). This rule will prevent overdissipation and possible damage to the D2 input-protection diode when the device power supply is grounded. When the device power supply is an open circuit, violation of this rule can result in undesired circuit operation although device damage should not result; ac inputs can be rectified by diode D2 to act as a power supply.
2. The power-supply operating voltage should be kept safely below the absolute maximum supply rating, as indicated previously.
3. The power supply polarity for COS/MOS circuits should not be reversed. The positive (V_{DD}) terminal should never be more than 0.5

volt negative with respect to the negative (VSS) terminal (VDD-VSS > -0.5 V). Reversal of polarities will forward-bias and short the structural and protection diode between VDD and VSS.

4. VDD should be equal to or greater than VCC for COS/MOS buffers which have two power supplies (in particular, for CD4009 and CD4010 COS/MOS-to-TTL "down"-conversion devices).
5. Power-source current capability should be limited to as low a value as reasonable to assure good logic operation.
6. Large values of resistors in series with VDD or VSS should be avoided; transient turn-on of input protection diodes can result from drops across such resistors during switching.

Gate-Oxide Protection Network

A problem occasionally encountered in handling and testing low-power semiconductor devices, including MOS and small-geometry bipolar devices, has been damage to gate oxide and/or p-n junctions. Fig. 3 shows the gate-oxide protection circuits used to protect COS/MOS devices from static electricity damage. ICAN-6572 gives further information on protection circuits. Although these circuits are included in all COS/MOS devices, the handling precautions in ICAN-6572 and ICAN-6525 should be observed.

Input Signals and Ratings

1. Input signals should be maintained within the power-supply voltage range, $V_{SS} \leqslant V_I \leqslant V_{DD}$. In applications such as astable and monostable multivibrators, input current can flow and, for optimal performance, should be limited to 100 microamperes by use of a resistor in series with the input terminal. The added protection assures proper circuit operation and prevents possible parasitic bipolar effects.

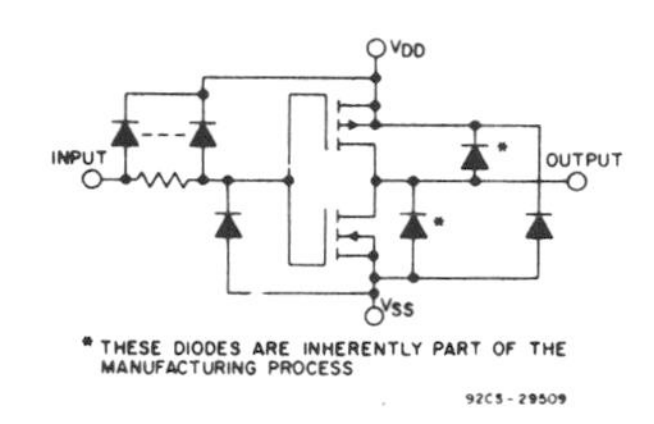

(a) **For standard A-series COS/MOS product.**

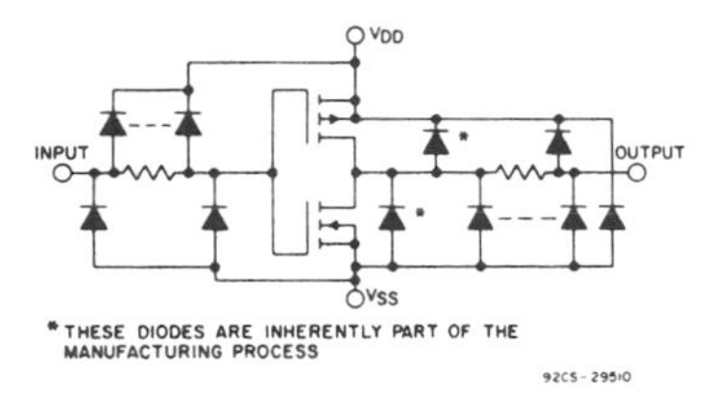

(b) **For improved B-series COS/MOS product.**

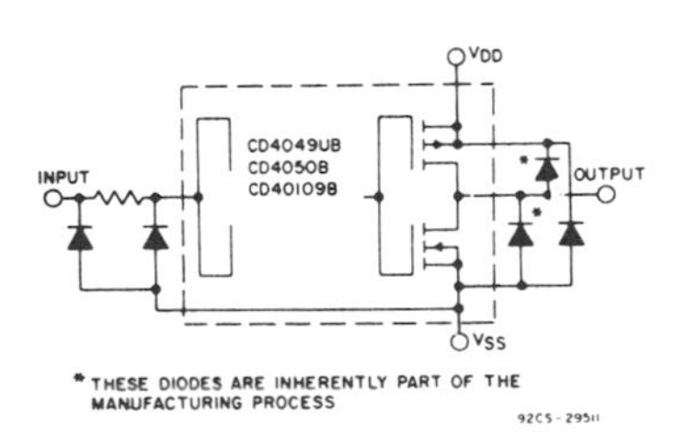

(c) **For CD4049UB, CD4050B, and CD40109B COS/MOS types.**

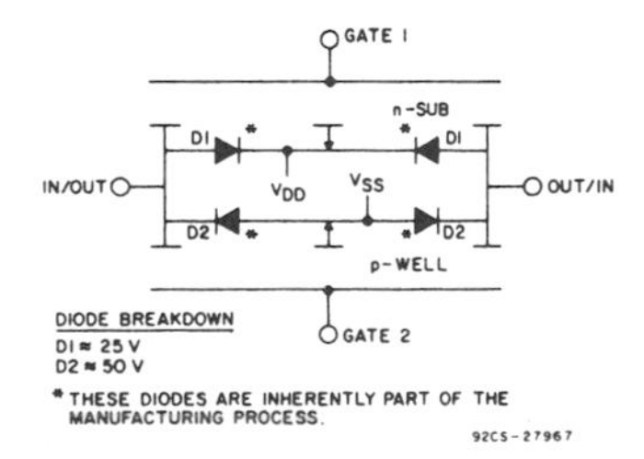

(d) **For COS/MOS transmission gates.**

Fig. 3— Gate-oxide protection networks used in RCA COS/MOS integrated circuits.

2. All COS/MOS inputs should be terminated. When COS/MOS inputs are wired to edge card connectors with COS/MOS drive coming from another PC board, a shunt resistor should be connected to V_{DD} or V_{SS} in case the inputs become unterminated with the power supply on.
3. When COS/MOS circuits are driven by TTL logic, a "pull-up" resistor should be connected from the COS/MOS input to 5 volts (further information is given in ICAN-6602).
4. Input signals, should be maintained within the recommended input-signal-swing range.

Output Rules

1. The power dissipation in a COS/MOS package should not exceed the rated value for the ambient temperature specified. The actual dissipation should be calculated when (a) shorting outputs directly to V_{DD} or V_{SS}, (b) driving low-impedance loads, or (c) directly driving the base of p-n-p or n-p-n bipolar transistor.
2. Output short circuits often result from testing errors or improper board assembly. Shorts on buffer outputs or across power supplies greater than 5 volts can damage COS/MOS devices.
3. COS/MOS, like active pull-up TTL, cannot be connected in the "wire-OR" configuration because an "on" PMOS and an "on" NMOS transistor could be directly shorted across the power-supply rails.
4. Paralleling inputs and outputs of gates is recommended only when the gates are within the same IC package.
5. Output loads should return to a voltage within the supply-voltage range (V_{DD} to V_{SS}).
6. Large capacitive loads (greater than 5000 pF) on COS/MOS buffers or high-current drivers act like short circuits and may over-dissipate output transistors.
7. Output transistors may be over-dissipated by operating buffers as linear amplifiers or using these types as one-shot or astable multivibrators.

Noise Immunity and Noise Margin

The complementary structure of the inverter, common to all COS/MOS logic devices, results in a near-ideal input-output transfer characteristic, with switching point midway (45% to 55%) between the 0 and 1 output logic levels. The result is high dc noise immunity.

Fig. 4 shows a typical transfer curve that may be used to define the noise immunity of COS/MOS integrated circuits. The noise-immunity voltage (V_{IL} or V_{IH}) is the noise voltage at any one input that does not propagate through the system. Minimum noise immunity for buffered B-series COS/MOS devices is 30, 30, and 27 percent, respectively for supply voltages V_{DD} of 5, 10, and 15 volts and 20 percent of V_{DD} for all unbuffered gates. The V_{IL} and V_{IH} specifications define the maximum permissible additive noise voltage at an input terminal when input signals are within 50 millivolts of the supply rails.

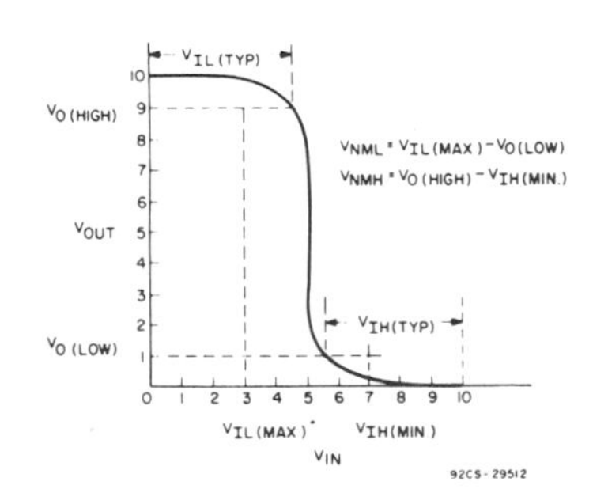

Fig. 4— Typical transfer curve for an inverting gate at V_{DD}=10 V.

Noise margin is the difference between the noise-immunity voltage (V_{IL} or V_{IH}) and the output voltage V_o. Noise-margin voltage is the maximum voltage that can be impressed upon an input voltage V_{IN} (where V_{IN} is the V_{OL} or V_{OH} voltage of the preceding state) at any (or all) logic I/O terminals without upsetting the logic or causing any output to exceed the output voltage (V_o) conditions specified for V_{IL} and V_{IH} ratings. Fig. 5 illustrates the noise-margin concept in a simple system. Noise margins for buffered B-series COS/MOS devices are 1, 2, and 2.5 volts, respectively, for supply voltages of 5, 10, and 15 volts.

Of the two noise-limitation specifications (noise immunity and noise margin), RCA considers noise immunity to be more practical for COS/MOS devices because COS/MOS outputs are normally within 50 millivolts of supply rails.

Noise immunity increases as the input pulse width becomes less than the propagation delay of the circuit. This condition is often described as ac noise immunity. (Further information on noise immunity is given in ICAN-6385.)

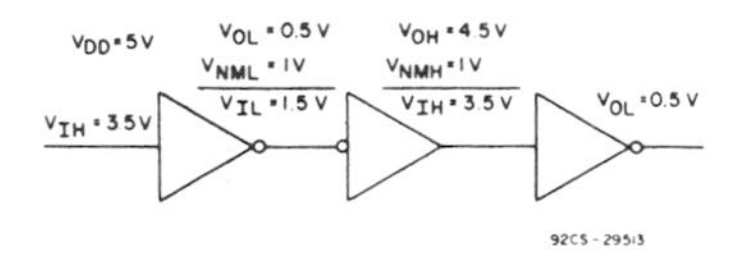

Fig. 5 — Noise margin example using inverters.

Clock Rise-and Fall-Time Requirements

Most COS/MOS clocked devices have maximum rise- and fall-time ratings (normally 5 to 15 microseconds). With longer rise or fall times, a device may not function properly because of data ripple-through, false triggering problems, etc. Whenever feasible, B-series COS/MOS counters have Schmitt-trigger shaping circuits built into the clock circuit thereby negating the restriction for input rise or fall times of 5 to 15 microseconds. Long rise and fall times on COS/MOS buffer-type inputs cause increased power dissipation which may exceed device capability for operating power-supply voltages greater than 5 volts.

Parallel Clocking

Process variations leading to differences in input threshold voltage among random device samples can cause loss of data between certain synchronously clocked sequential circuits, as shown in Fig. 6. This problem can be avoided if the clock rise time (t_rCL) is made less than the total of the fixed propagation delay plus the output transition time of the first stage, as determined from the device data for the specific loading condition in effect. Schmitt trigger circuits such as the CD4093B are an ideal solution to applications requiring wave-shaping.

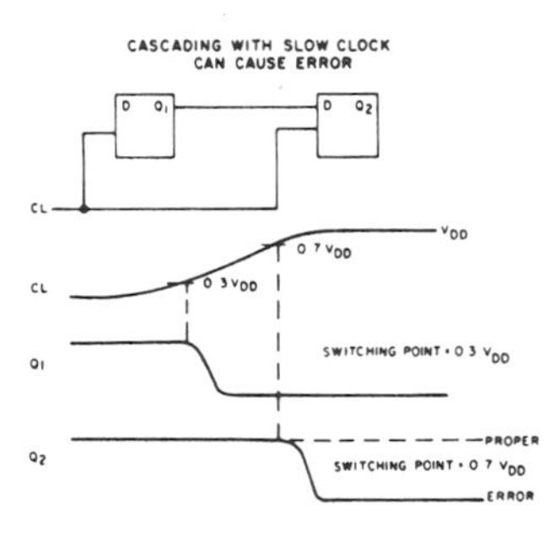

Fig. 6 — Error effect that results from a slow clock in cascaded circuits.

Three-State Logic

Three-state logic can be easily implemented by use of a transmission gate in the output circuit; this technique provides a solution to the wire-OR problem in many cases.

Chip Assembly and Storage

RCA COS/MOS integrated circuits are provided in chip form (H suffix) to allow customer design of special and complex circuits to suit individual needs. COS/MOS chips are electrically identical and offer the features of their counterparts sealed in ceramic and plastic packages. The following paragraphs describe mounting considerations, packaging, shipping and storage criteria, handling criteria, visual inspection criteria, testing criteria, and bonding pad layout and dimensions for each chip.

Mounting Considerations. All COS/MOS chips are non-gold backed and require the use of epoxy mounting. DuPont No. 5504A conductive silver paste or equivalent is recommended. In any case the manufacturer's recommendations for storage and use should be followed. If DuPont No. 5504A paste is used, the bond should be cured at temperatures between 185°C and 200° for 75 minutes.

In COS/MOS circuits MOS-transistor p-channel substrates (n-type bulk material) are connected to V_{DD}, therefore, when chips are mounted and a conductive paste is used care must be taken to keep the active substrate isolated from ground or other circuit elements.

Packing, Shipping, and Storage Criteria. Solid-state chips, unlike packaged devices, are non-hermetic devices, normally fragile and small in physical size, and therefore, require special handling considerations as follows:

1. Chips must be stored under proper conditions to insure that they are not subjected to a moist and/or contaminated atmosphere that could alter their electrical, physical, or mechanical characteristics. After the shipping container is opened, the chip must be stored under the following conditions:
   A. Storage temperature, 40°C max.
   B. Relative humidity, 50% max.
   C. Clean, dust-free environment.
2. The user must exercise proper care when handling chips to prevent ever the slightest physical damage to the chip.
3. During mounting and lead bonding of chips the user must use proper assembly techniques to obtain proper electrical, thermal, and mechanical performance.
4. After the chip has been mounted and bonded, any necessary procedure must be followed by the user to insure that these non-hermetic chips are not subjected to moist and contaminated atmosphere which might cause the development of electrical conductive paths across the relatively small insulating surfaces. In addition, proper consideration must be given to the protection of these devices from other harmful environments which could conceivably adversely affect their proper performances.

Handling Criteria. The user should find the following suggested precautions helpful in handling COS/MOS chips.

In any event, because of the extremely small size and fragile nature of chips, the equipment designer should exercise care in handling these devices.

For additional handling considerations for COS/MOS devices, refer to ICAN-6525, "Guide to Better Handling and Operation of CMOS Integrated Circuits."

1. Grounding

a. Bonders, pellet pick-up tools, table tops, trim and form tools, sealing equipment, and other equipment used in chip handling should be properly grounded.

b. The operator should be properly grounded.

2. In-Process handling

a. Assemblies or subassemblies of chips should be transported and stored in conductive carriers.

b. All external leads of the assemblies or subassemblies should be shorted together.

3. Bonding Sequence

a. Connect VDD first to external connections, for example, terminal 14 of the CD4001AH.

b. Remaining functions may be connected to their external connections in any sequence.

4. Testing

a. Transport all assemblies of chips in conductive carriers.

b. In testing chip assemblies or subassemblies, the operator should be properly grounded.

Visual Inspection Criteria. All standard commercial COS/MOS chips undergo a visual inspection which is patterned after MIL-STD-883, Method 2010.1, Condition B with modifications reflecting COS/MOS requirements.

Testing Criteria. COS/MOS chips are dc electrically tested 100% in accordance with the same standards prescribed for RCA devices in standard packages.

Device Testing

RCA COS/MOS circuits are 100-percent tested by circuit probe in the wafer stage and are 100-percent tested again after they have been packaged. DC tests of RCA devices are performed at 5, 10, 15, and 20 volts; functionality is checked at 3, 17, and 22 volts depending on family (i.e., A or B series). Sample testing is used to assure adherence to quality requirements and ac specifications.

Static tests, high-speed functional and dc parametric tests, are performed at wafer and package stages by means of a Teradyne J283 test set. A Teradyne S157CM test set and a Marcodata MD 154 test set are used in dynamic testing. Dynamic tests are performed with 15 and 50 picofarad loads. Testing at 15 picofarads is accomplished primarily by laboratory "bench-test" techniques; automatic testing at 15 picofarads is difficult because of the high input capacitance (approximately 20 to 35 picofarads) of most automatic ac test sets.

Users should follow the sequence below when testing COS/MOS devices:

1. Insert the device into the test socket.
2. Apply VDD.
3. Apply the input signal.
4. Perform the test.
5. On completion of test, remove the input signal.
6. Turn off the power supply (VDD).
7. Remove the device from the test socket and insert it into a conductive carrier. COS/MOS devices under test must not be exposed to electrostatic discharge or forward biasing of the intrinsic protective diodes shown in Fig. 3.

Detailed information on the techniques employed in the testing of RCA COS/MOS integrated circuits are described in ICAN-6532 included in the Application Notes section of this DATABOOK.

Device Mounting

Integrated circuits are normally supplied with lead-tin plated leads to facilitate soldering into circuit board. In those relatively few applications requiring welding of the device leads, rather than soldering, the devices may be obtained with gold or nickel plated Kovar leads.* It should be recognized that this type of plating will not pro-

*Mil-M-38510A, paragraph 3.5.6.1 (a), lead material.

vide complete protection against lead corrosion in the presence of high humidity and mechanical stress.

In any method of mounting integrated circuits which involves bending or forming of the device leads, it is extremely important that the lead be supported and clamped between the bend and the package seal, and that bending be done with care to avoid damage to lead plating. In no case should the radius of the bend be less than the diameter of the lead. It is also extremely important that the ends of bent leads be straight to assure proper insertion through the holes in the printed-circuit board.

A-SERIES COS/MOS INTEGRATED CIRCUITS

RCA CD4000A-series types have a maximum dc supply-voltage rating of -0.5 to 15 volts, and a recommended operating supply-voltage range of 3 to 12 volts. The major features of this series are as follows:

- Quiescent current specified to 15 volts
- 5-volt and 10-volt parametric ratings
- Maximum input leakage of 1 μA at 15 volts over the full package operating-temperature range
- 1-volt noise margin (full package temperature range)

Table I shows the maximum ratings and the recommended operating supply-voltage range for RCA A-series COS/MOS integrated circuits.

Static Electrical Characteristics

Table II shows the standard dc electrical characteristics for A-series types. The data sheet for each of these types contains the family characteristics shown in Table I plus additional dc characteristics that are type-dependent.

Dynamic Electrical Characteristics

A-series dynamic electrical characteristics are specified for individual types under the following conditions: V_{DD} = 5 V and 10 V; T_A = 25°C (temperature coefficient is typically 0.3%/°C); C_L = 15 pF; t_r and t_f of inputs = 20 ns.

HIGH-VOLTAGE B-SERIES COS/MOS INTEGRATED CIRCUITS

RCA-CD 4000B-series types have a maximum dc supply-voltage rating of -0.5 to 20 volts, and a recommended operating supply-voltage range of 3 to 18 volts. The major features of this series are as follows:

- High-voltage (20-V) ratings
- 100% tested for quiescent current at 20 V
- 5-V, 10-V, and 15-V parametric ratings
- Standardized, symmetrical output characteristics
- Maximum input current of 1 μA at 18 V over full package-temperature range; 100 nA at 18 V and 25°C
- Noise margin (full package-temperature range) =

  1 V at V_{DD} = 5 V
  2V at V_{DD} = 10 V
  2.5 V at V_{DD} = 15 V

- Meets all requirements of JEDEC Tentative Standard No. 13A, “Standard Specifications for Description of ‘B’ Series CMSO Devices.

JEDEC Minimum Standard

Under the sponsorship of the Joint Electron Devices Engineering Council (JEDEC) of the Electronic Industries Association (EIA), minimum industrial standards have been established for the maximum ratings and static electrical characteristics of B-series CMOS integrated circuits. The JEDEC standard (JEDEC Tentative Standard No. 13A) defines B-series CMOS inte-

Table I— Maximum Rating and Recommended Operating Conditions for A-Series COS/MOS Integrated Circuits

MAXIMUM RATINGS, *Absolute-Maximum Values:*

DC SUPPLY-VOLTAGE RANGE, V_{DD} (Voltages referenced to V_{SS} terminal) ..... -0.5 to +15V
INPUT VOLTAGE RANGE, ALL INPUTS ..... -0.5 to V_{DD} +0.5V
POWER DISSIPATION PER PACKAGE (P_D):
For T_A = -40 to +60° C (Package Type E) ..... 500 mW
For T_A = +60 to +85° C (Package Type E) ..... Derate Linearly to 200 mW
For T_A = −55 to +100°C (Package Types D, H) ..... 500 mW
For TA = +100 to +125° C (Package Types D, H) ..... Derate linearly to 100 mW
DEVICE DISSIPATION PER OUTPUT TRANSISTOR:
For T_A = Full package-temperature range (All package types) 100mW
OPERATING-TEMPERATURE RANGE (T_A):
Package Types D, H ..... −55 to +125°C
Package Type E ..... −44 to +85°C
STORAGE-TEMPERATURE RANGE (T_{STG}) ..... −65 to 150°C
LEAD TEMPERATURE (During Soldering):
At distance 1/16 ± 1/32 inch (1.59 ± 0.79 mm) from case for 10 s max. ..... +265° C

RECOMMENDED OPERATING CONDITIONS
For maximum reliability, nominal operating conditions should be selected so that operation is always within the following ranges:

CHARACTERISTIC	LIMITS		UNITS
	Min.	Max.	
Supply-Voltage Range (For T_A = Full Package-Temperature Range)	3	12	V

Table II— A-Series Static Electrical Characteristics (Full Package Temperature Range)

SYMBOL	PARAMETER	CONDITIONS				LIMITS			UNITS
		V_{IN}	V_O (volts)		V_{DD}				
		VOLTS	MIN.	MAX.	VOLTS	MIN.	TYP.	MAX.	
V_{OL}	Output Low Voltage	5 10	– –	– –	5 10	– –	0 0	0.05 0.05	V V
V_{OH}	Output High Voltage	0 0	– –	– –	5 10	4.95 9.95	5 10	– –	V V
V_{NL} (SSI Types)	Noise Voltage (Input Low)	– –	3.6 7.2	– –	5 10	1.5 3	2.25 4.5	– –	V V
V_{NH} (SSI Types)	Noise Voltage (Input High)	– –	– –	1.4 2.8	5 10	1.5 3	2.25 4.5	– –	V V
V_{NL} (MSI Types)	Noise Voltage (Input Low)	– –	4.2 9.0	– –	5 10	1.5 3	2.25 4.5	– –	V V
V_{NH} (MSI Types)	Noise Voltage (Input High)	– –	– –	0.8 1.0	5 10	1.5 3	2.25 4.5	– –	V V
V_{NML}	Noise Margin (Input Low)	– –	4.5 9.0	– –	5 10	1 1	– –	– –	V V
V_{NMH}	Noise Margin (Input High)	– –	– –	0.5 1.0	5 10	1 1	– –	– –	V V
I_{IL}, I_{IH}	Input Leakage Low	–	–	–	15	–	$\pm 10^{-5}$	±1	µA
I_L	Quiescent Device Leakage	–	–	–	5,10,15	See Data Sheets			µA
I_{DN}, I_{DP}	Output Source and Sink current	–	–	–	5,10	See Data Sheets			ma

Note: Logic Level Inversion Assumed in Table II.

grated circuits as a uniform family of both buffered and unbuffered types that have an absolute dc supply-voltage rating of at least 18 votes.

Buffered CMOS devices are types in which the output "on" impedance is independent of any and all valid input logic conditions, both preceding and present. All such CMOS product are designated by the suffix "B" following the basic type number.

Unbuffered CMOS devices are types that meet all B-series specifications except that the logical outputs are not buffered and the noise-immunity voltages, V_{IL} and V_{IH}, are specified as 20 and 80 per cent, respectively, of V_{DD} for operation from 5 or 10 volts and 17 and 83 per cent, respectively, of V_{DD} for operation time from 15 volts. All such CMOS products are designated by the suffix "UB."

The JEDEC minimum standard also includes in the B-series CMOS types that have analog inputs or outputs and, in addition, have maximum ratings and logical input and output parameters that conform to B-series specifications wherever applicable. These CMOS devices are also designated by the suffix "B."

All B-series CMOS devices can directly replace their A-series counterparts in most applications. The UB types are high-voltage versions of corresponding A-series (unbuffered) types.

Table III lists the JEDEC minimum standard established for the maximum ratings and recommended operation conditions for B-series CMOS integrated circuits.

Table IV shows the JEDEC standards for the static electrical characteristics of CMOS B-series integrated circuits.

Standardized RCA Ratings and Static Characteristics

RCA B-series COS/MOS integrated circuits meet or exceed the most stringent requirements of the JEDEC B-series specifications. Table V shows the standardized maximum ratings and recommended operating supply-voltage range for RCA B-series COS/MOS integrated circuits. The standardized static electrical characteristics for these devices are shown in Table VI. As with the JEDEC specifications, the RCA standardized characteristics classify the B-series devices into three leakage (quiescent-device-current) categories. Table VII lists the RCA types in each category and indi-

Table III— JEDEC Minimum Standards for Maximum Ratings and Recommended Operating Conditions for B -Series CMOS Integrated Circuits*

Absolute Maximum Ratings (Voltages referenced to V_{SS}):			
DC Supply Voltage	V_{DD}	-0.5 to +18	Vdc
Input Voltage	V_{IN}	-0.5 to V_{DD} +0.5	Vdc
DC Input Current (any one input)	I_{IN}	±10	mAdc
Storate Temperature Range	T_S	-65 to +150	°C
Recommended Operating Conditions:			
DC Supply Voltage	V_{DD}	+3 to +15	Vdc
Operating-Temperature Range:	T_A		
Military-Range Devices		-55 to +125	°C
Commercial-Range Devices		-40 to +85	°C

***Reprinted from JEDEC Tentative Standard No. 13-A, "Standard Specifications for Description of B-Series CMOS Devices."**

cates types that, although they are still B-series types, differ in one or more static characteristics.

Tables V and VI show that, in a number of important respects, RCA has established new performance standards for B-series COS/MOS logic circuits:

1. **Tight limits for all packages**
   RCA devices use the same set of limits for all package styles. The JEDEC standard establishes two sets of limits for most dc (static) parameters: a tight set for products having a full operating temperature range of -55°C to +125°C (normally used for ceramic packages), and a relaxed set for products having a limited temperature range of -40°C to +85°C (normally used for plastic packages). Because RCA supplies only one premium grade of B-series product in all package styles (i.e., fall-out chips are *not* used), all B-series COS/MOS devices are specified to the tight set of limits only.

2. **Improved voltage rating**
   All RCA B-series devices are tested to voltages as high as 22 volts and have an absolute maximum dc supply voltage rating of 20 volts. This higher rating permits greater derating for reliable 15-volt operation, permits greater 15-volt supply tolerance and peak transients, and permits system use to 18-volts with confidence.

3. **Wider operating range**
   All RCA B-series devices have a recommended maximum operating voltage of 18-volts. This higher limit permits 18-volt system supply operation, and also permits wider power-source tolerances and transients for supplies normally set up to 18 volts.

4. **Lower leakage current**
   The JEDEC standard establishes three sets of limits for quiescent device current (I_{DD}) intended to match chip complexity to device leakage current as realistically as possible. For all three levels of chip complexity, all RCA B-series devices (regardless of package) conform to the tighter set of limits established in the standard. In addition, a maximum rating is specified at 20V, as well as at 5V, 10V, and 15V. As a result:
   (a) In current-limited applications, COS/MOS users can depend on one tight leakage limit independent of package style selected.
   (b) Customer use of COS/MOS product up through 18 volts is protected by a published tight leakage current specification at 20 volts (as well as by an input leakage specification at 18 volts).

5. **Symmetrical output**
   All RCA B-series devices have balanced complementary output drive (i.e., the output high current I_{OH} rating is the same as the output low current I_{OL} rating) specified to the tighter set of limits established in the JEDEC standard. The balanced output provides uniform rise and fall time performance, improved system noise energy (dynamic) immunity, optimum device speed for both output switching low-to-high (tp_{LH}) and output switching high-to-low (tp_{HL}), and in general the identical high and low dc and ac characteristics normally associated with a good complementary output drive circuit. MOS system design, simulation, and performance are significantly enhanced by equal high and low dc and ac performance ratings and one tight specification limit for all package styles.

6. **Improved input current (leakage) ratings**
   All RCA B-series devices (regardless of package) have a maximum input leakage current (I_{IN}) rating of 100 nA specified at voltages up to 18V, and a maximum limit of 1 μA at the upper limit of the

Table IV— JEDEC Standard for Static Characteristic of B-Series CMOS Integrated Circuits▲

PARAMETER		TEMP. RANGE	V_{DD} (Vdc)	CONDITIONS	LIMITS T_{LOW}* Min	T_{LOW}* Max	+25°C Min	+25°C Typ	+25°C Max	T_{HIGH}* Min	T_{HIGH}* Max	Units
I_{DD}	Quiescent Device Current GATES	Mil	5 10 15	$V_{IN} = V_{SS}$ or V_{DD}		0.25 0.5 1.0			0.25 0.5 1.0		7.5 15 30	uAdc
		Comm	5 10 15	All valid input combinations		1.0 2.0 4.0			1.0 2.0 4.0		7.5 15 30	
	BUFFERS, FLIP-FLOPS	Mil	5 10 15	$V_{IN} = V_{SS}$ or V_{DD}		1.0 2.0 4.0			1.0 2.0 4.0		30 60 120	uAdc
		Comm	5 10 15	All valid input combinations		4 8 16			4.0 8.0 16.0		30 60 120	
	MSI	Mil	5 10 15	$V_{IN} = V_{SS}$ or V_{DD}		5 10 20			5 10 20		150 300 600	uAdc
		Comm	5 10 15	All valid input combinations		20 40 80			20 40 80		150 300 600	
V_{OL}	Low-Level Output Voltage	All	5 10 15	$V_{IN} = V_{SS}$ or V_{DD} $\|I_O\| < 1uA$		0.05 0.05 0.05			0.05 0.05 0.05		0.05 0.05 0.05	Vdc
V_{OH}	High-Level Output Voltage	All	5 10 15	$V_{IN} = V_{SS}$ or V_{DD} $\|I_O\| < 1uA$	4.95 9.95 14.95		4.95 9.95 14.95			4.95 9.95 14.95		Vdc
V_{IL}	Input Low Voltage	All	5 10 15	V_O = 0.5V or 4.5V V_O = 1.0V or 9.0V V_O = 1.5V or 13.5V $\|I_O\| < 1uA$		1.5 3.0 4.0			1.5 3.0 4.0		1.5 3.0 4.0	Vdc
V_{IH}	Input High Voltage	All	5 10 15	V_O = 0.5V or 4.5V V_O = 1.0V or 9.0V V_O = 1.5V or 13.5V $\|I_O\| < 1uA$	3.5 7.0 11.0		3.5 7.0 11.0			3.5 7.0 11.0		Vdc
I_{OL}	Output Low (Sink) Current	Mil	5 10 15	V_O = 0.4V, V_{IN} = 0 or 5V V_O = 0.5V, V_{IN} = 0 or 10V V_O = 1.5V, V_{IN} = 0 or 15V	0.64 1.6 4.2		0.51 1.3 3.4			0.36 0.9 2.4		mAdc
		Comm	5 10 15	V_O = 0.4V, V_{IN} = 0 or 5V V_O = 0.5V, V_{IN} = 0 or 10V V_O = 1.5V, V_{IN} = 0 or 15V	0.52 1.3 3.6		0.44 1.1 3.0			0.36 0.9 2.4		mAdc
I_{OH}	Output High (Source) Current	Mil	5 10 15	V_O = 4.6V, V_{IN} = 0 or 5V V_O = 9.5V, V_{IN} = 0 or 10V V_O = 13.5V, V_{IN} = 0 or 15V	−0.25 −0.62 −1.8		−0.2 −0.5 −1.5			−0.14 −0.35 −1.1		mAdc

Table IV—JEDEC Standard for Static Characteristics of B-Series CMOS Integrated Circuits▲ (cont'd)

PARAMETER		TEMP. RANGE	V_{DD} (Vdc)	CONDITIONS	LIMITS T_{LOW}* Min	T_{LOW}* Max	+25°C Min	+25°C Typ	+25°C Max	T_{HIGH}* Min	T_{HIGH}* Max	Units
I_{OH}	Output High (Source) Current (cont'd)	Comm	5	V_O = 4.6V, V_{IN} = 0 or 5V	-0.2		-0.16			-0.12		mAdc
			10	V_O = 9.5V, V_{IN} = 0 or 10V	-0.5		-0.4			-0.3		mAdc
			15	V_O = 13.5V, V_{IN} = 0 or 15V	-1.4		-1.2			-1.0		mAdc
I_{IN}	Input Current	Mil	15	V_{IN} = 0 or 15V		±0.1			±0.1		±1.0	uAdc
		Comm	15	V_{IN} = 0 or 15V		±0.3			±0.3		±1.0	uAdc
C_{IN}	Input Capacitance per Unit Load	All	–	Any Input					7.5			pF

*T_{LOW} = -55°C for Military Temp. Range device, -40°C for Commercial Temp. Range device
*T_{HIGH}= +125°C for Military Temp. Range device, +85°C for Commercial Temp. Range device
▲Reprinted from JEDEC Tentative Standard No. 13-A, "JEDEC Standard Specification for Description of B-Series CMOS Device."

Table V—RCA Standardized Maximum Ratings and Recommended Operating Conditions for B-Series COS/MOS Integrated Circuits

Maximum Ratings, Absolute-Maximum Values:

DC SUPPLY-VOLTAGE RANGE, (V_{DD})
(Voltages referenced to V_{SS} Terminal) .... -0.5 to +20 V
INPUT VOLTAGE RANGE, ALL INPUTS .... -0.5 to V_{DD} +0.5 V
DC INPUT CURRENT, ANY ONE INPUT .... ±10 mA
POWER DISSIPATION PER PACKAGE (P_D:
For T_A = -40 to +60°C (PACKAGE TYPE E) .... 500 mW
For T_A = +60 to +85°C (PACKAGE TYPE E) Derate Linearly at 12 mW/°C to 200 mW
For T_A = -55 to +100°C (PACKAGE TYPES D,F) .... 500 mW
For T_A = +100 to +125°C (PACKAGE TYPES D,F) Derate Linearly at 12 mW/°C to 200 mW
DEVICE DISSIPATION PER OUTPUT TRANSISTOR
For T_A = FULL PACKAGE-TEMPERATURE RANGE (All Package Types) 100 mW
OPERATING-TEMPERATURE RANGE (T_A):
PACKAGE TYPES, D, F, H .... -55 to +125°C
PACKAGE TYPE E .... -40 to +85°C
STORAGE TEMPERATURE RANGE (T_{stg}) .... -65 to +150°C
LEAD TEMPERATURE (DURING SOLDERING):
At distance 1/16 ± 1/32 inch (1.59 ± 0.79 mm) from case for 10s max ..... +265°C

Recommended Operating Conditions:

For maximum reliability, nominal operating conditions should be selected so that operation is always within the following ranges:

CHARACTERISTIC	LIMITS MIN.	LIMITS MAX.	UNITS
Supply-Voltage Range (For T_A = Full Package Temperature Range)	3	18	V

Table VI—RCA B-Series COS/MOS Standardized Electrical Characteristics

CHARACTERISTIC	CONDITIONS			LIMITS AT INDICATED TEMPERATURES (°C) Values at -55, +25, +125 Apply to D,F,H Packages Values at -40, +25, +85 Apply to E Package							UNITS
	V_O (V)	V_{IN} (V)	V_{DD} (V)	-55	-40	+85	+125	+25			
								Min.	Typ.	Max.	
Quiescent Device Current, I_{DD} Max. Gates, Inverters▲	–	0,5	5	0.25	0.25	7.5	7.5	–	0.01	0.25	μA
	–	0,10	10	0.5	0.5	15	15	–	0.01	0.5	
	–	0,15	15	1	1	30	30	–	0.01	1	
	–	0,20	20	5	5	150	150	–	0.02	5	
Buffers, Flip-Flops, Latches, Multi-Level Gates (MSI-1 Types)▲		0,5	5	1	1	30	30	–	0.02	1	
		0,10	10	2	2	60	60	–	0.02	2	
		0,15	15	4	4	120	120	–	0.02	4	
		0,20	20	20	20	600	600	–	0.04	20	
Complex Logic (MSI-2 Types)▲		0,5	5	5	5	150	150	–	0.04	5	
		0,10	10	10	10	300	300	–	0.04	10	
		0,15	15	20	20	600	600	–	0.04	20	
		0,20	20	100	100	3000	3000	–	0.08	100	
Output Low (Sink) Current I_{OL} Min.	0.4	0,5	5	0.64	0.61	0.42	0.36	0.51	1	–	mA
	0.5	0,10	10	1.6	1.5	1.1	0.9	1.3	2.6	–	
	1.5	0,15	15	4.2	4	2.8	2.4	3.4	6.8	–	
Output High (Source) Current, I_{OH} Min.	4.6	0,5	5	-0.64	-0.61	-0.42	-0.36	-0.51	-1	–	
	2.5	0,5	5	-2	-1.8	-1.3	-1.15	-1.6	-3.2	–	
	9.5	0,10	10	-1.6	-1.5	-1.1	-0.9	-1.3	-2.6	–	
	13.5	0,15	15	-4.2	-4	-2.8	-2.4	-3.4	-6.8	–	
Output Voltage: Low-Level, V_{OL} Max.	–	0,5	5	0.05				–	0	0.05	V
	–	0,10	10	0.05				–	0	0.05	
	–	0,15	15	0.05				–	0	0.05	
Output Voltage: High-Level V_{OH} Min.	–	0,5	5	4.95				4.95	5	–	
	–	0,10	10	9.95				9.95	10	–	
	–	0,15	15	14.95				14.95	15	–	
Input Low Voltage, V_{IL} Max. B Types	0.5, 4.5	–	5	1.5				–	–	1.5	V
	1,9	–	10	3				–	–	3	
	1.5, 13.5	–	15	4				–	–	4	
UB Types	0.5, 4.5	–	5	1				–	–	1	
	1, 9	–	10	2				–	–	2	
	1.5, 13.5	–	15	2.5				–	–	2.5	
Input High Voltage, V_{IH} Min. B Types	0.5, 4.5	–	5	3.5				3.5	–	–	
	1, 9	–	10	7				7	–	–	
	1.5, 13.5	–	15	11				11	–	–	
UB Types	0.5, 4.5	–	5	4				4	–	–	
	1, 9	–	10	8				8	–	–	
	1.5, 13.5	–	15	12.5				12.5	–	–	
Input Current I_{IN} Max.	–	0,18	18	±0.1	±0.1	±1	±1	–	$\pm 10^{-5}$	±0.1	μA
3-State Output Leakage Current I_{OUT} Max.	0, 18	0,18	18	±0.4	±0.4	±12	±12	–	$\pm 10^{-4}$	± 0.4	μA

▲Classifications of RCA COS/MOS B-Series Types are shown in Table VII.

package-temperature range. Actually, the 100 nA rating is a practical specification limited by the capability of commercial test equipment to measure lower currents. Laboratory tests show that input leakage currents of RCA B-series COS/MOS devices are significantly lower than this limit, typically ranging from 10 to 100 pA/.

7. **Buffered and unbuffered gates**
The new industry standard establishes a suffix "UB" for CMOS products that meet all B-series specifications except that the logical outputs of the devices are not buffered and the V_{ii} and V_{IH} specifications are relaxed. The suffix "B" defines only buffered-output devices in which the output "on" impedance is independent of any and all valid input logic conditions, both preceding and present.
RCA will supply both buffered ("B") and unbuffered ("UB") versions of the popular NOR and NAND gates to make available to designers the advantages of both. The chart below briefly compares the features of the two versions (a more detailed coverage of the special features of B- and UB-series COS/MOS gates is provided by ICAN-6558 in the Application-Notes section).

8. **Reliability**
RCA B-Series COS/MOS integrated circuits incorporate the latest improvements in processing technology and plastic and ceramic packaging techniques. Product quality is real-time control using accelerated-temperature group quality screening in which tight B-series limits for dc parameters are criticized as test points.

Figs. 7 through 10 show the standardized n- and p-channel drain characteristics for B-series COS/MOS devices, and Figs. 11 through 14 show the normalized variation of output source and sink currents with respect to temperature and voltage in these devices.

B-Series Dynamic Electrical Characteristics

B-series dynamic electrical characteristics are specified for individual types under the following conditions: V_{DD} = 5 V; 10 V, and 15 V; T_A = 25°C; C_L = 50 pF; R_L = 200 kΩ; t_r and t_f = 20 ns. Table VIII lists dynamic characteristics specified for RCA B-series COS/MOS integrated circuits. Figs. 15 through 18 show the variation of B-series dynamic parameters with temperature. Figs. 19 and 20 show the variation of output transition time with supply voltage. Fig. 21 shows the variation of the standardized output transition time with load capacitance.

Maximum propagation delay or transition times for values of C_L other than the specified 50 picofarads can be determined by use of the multiplication factor (usually 2) between the typical and maximum values given in the dynamic characteristics chart included in the technical data for each device applied to the typical curves, also shown in the device technical data.

Characteristic	Buffered Version ("B")	Unbuffered Version ("UB")
Propagation Delay (Speed)	Moderate	Fast
Noise Immunity/Margin	Excellent	Good
Output Impedance and Output Transition Time	Constant	Variable
AC Gain	High	Low
Output Oscillation for Slow Inputs	Yes	No
Input Capacitance	Low	High

Table VII — Classification of RCA B-Series COS/MOT Integrated Circuits

Gates/ Inverters		Buffers/Flip-Flop/ Latches/Multi-Level Gates (MSI-1)		Complex Logic (MSI-2)		
CD4000B	CD4023UB	CD4009UB■	CD4070B	CD4006B	CD4051B■	CD4527B
CD4000UB	CD4025B	CD4010B■	CD4077B	CD4008B	CD4052B■	CD4532B
CD4001B	CD4025UB	CD4013B	CD4085B	CD4014B	CD4053B■	CD4536B
CD4001UB	CD4048B	CD4019B	CD4086B	CD4015B	CD4054B■	CD4555B
CD4002B	CD4066B ■	CD4027B	CD4093B	CD4017B	CD4055B■	CD4556B
CD4002UB	CD4068B	CD4030B	CD4095B	CD4018B	CD4056B■	CD4585B
CD4007UB	CD4069UB	CD4041UB ■	CD4096B	CD4020B	CD4060B	CD4724B
CD4011B	CD4071B	CD4042B	CD4098B	CD4021B	CD4063B	CD40100B
CD4011UB	CD4072B	CD4043B	CD4502B ■	CD4022B	CD4067B■	CD40101B
CD4012B	CD4073B	CD4044B	CD40106B	CD4024B	CD4076B	CD40102B
CD4012UB	CD4075B	CD4047B	CD40107B■	CD4026B	CD4089B	CD40103B
CD4016B■	CD4078B	CD4049UB■	CD40109B ■	CD4028B	CD4094B	CD40105B
CD4023B	CD4081B	CD4050B■	CD40174B	CD4029B	CD4097B■	CD40108B
	CD4082B		CD40257B	CD4031B	CD4099B	CD40110B ■
				CD4032B	CD4508B	CD40114B
				CD4033B	CD4510B	CD40147B
				CD4034B	CD4511B■	CD40160B
				CD4035B	CD4512B	CD40161B
				CD4038B	CD4514B	CD40162B
				CD4040B	CD4515B	CD40163B
				CD4045B ■	CD4516B	CD40181B
				CD4046B■	CD4517B	CD40182B
					CD4518B	CD40192B
					CD4520B	CD40193B
						CD40208B

■Indicated types for which, because of special design requirements, one or more static characteristics differ from the standardized data. Refer to RCA data pages on these types for specific differences.

Fig. 7— Typical output low (sink) current characteristics.

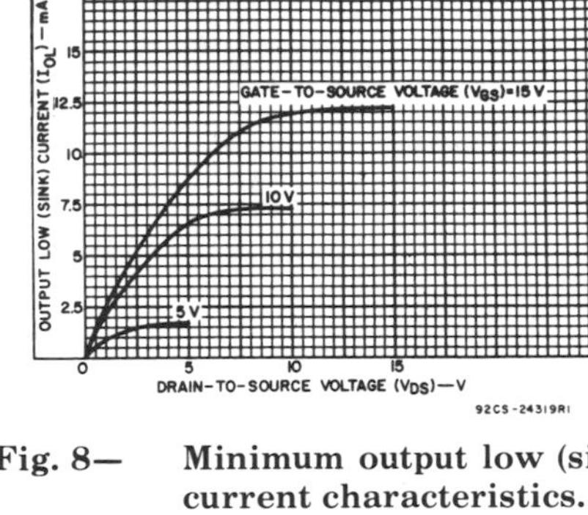

Fig. 8— Minimum output low (sink) current characteristics.

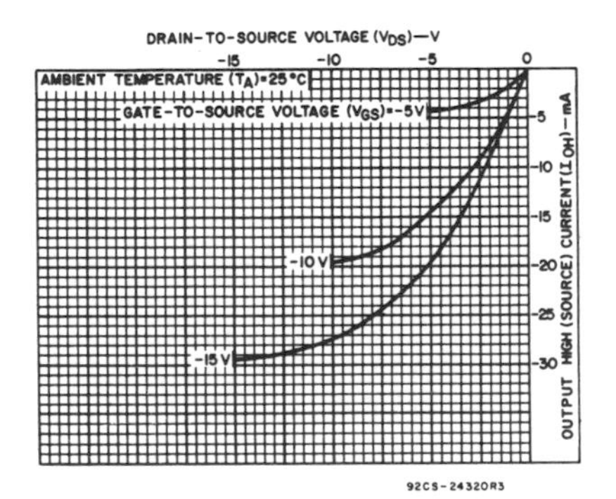

Fig. 9— Typical output high (source) current characteristics.

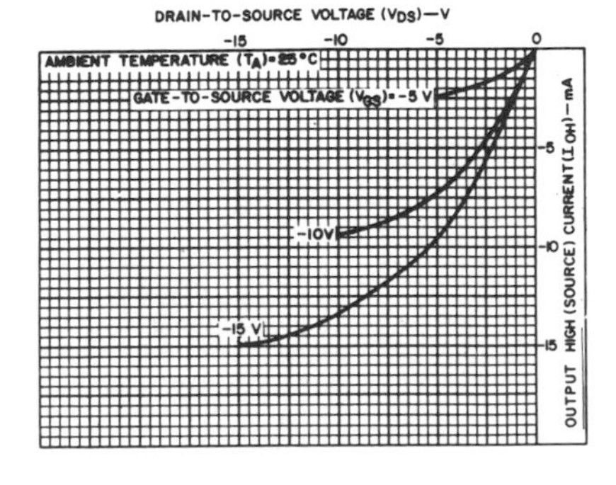

Fig. 10— Minimum output high (source) current characteristics.

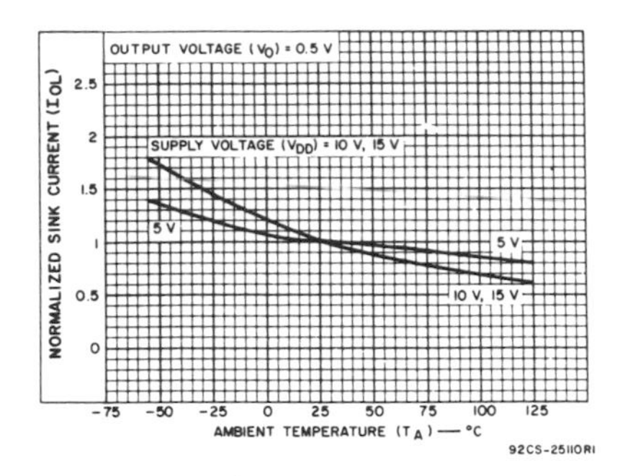

Fig. 11— Variation of normalized output low (sink) current (J_{OL}) with temperature.

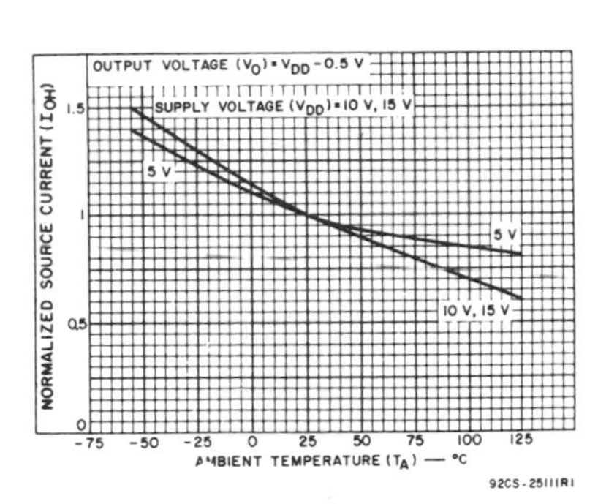

Fig. 12— Variation of normalized output high (source) current (I_{OH}) with temperature.

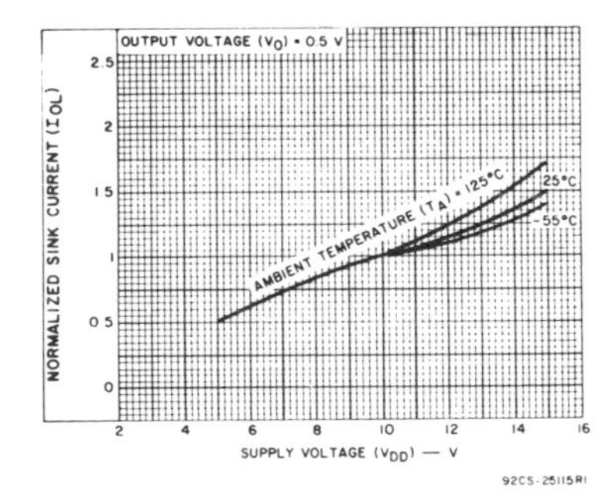

Fig. 13— Variation of normalized output low (sink) current (I_{OL}) with supply voltage.

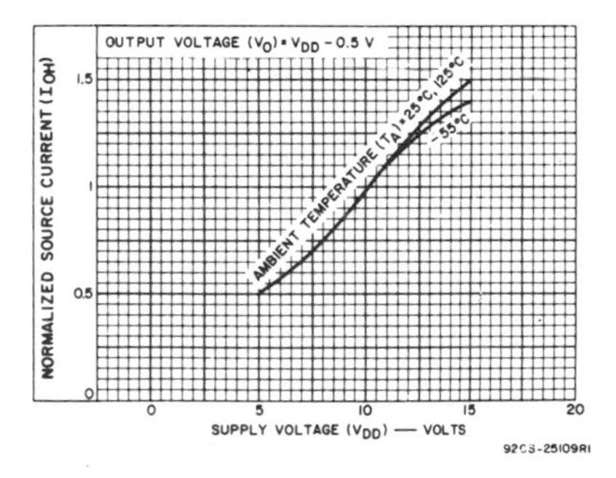

Fig. 14— Variation of normalized output high (source) current (I_{OH}) with supply voltage.

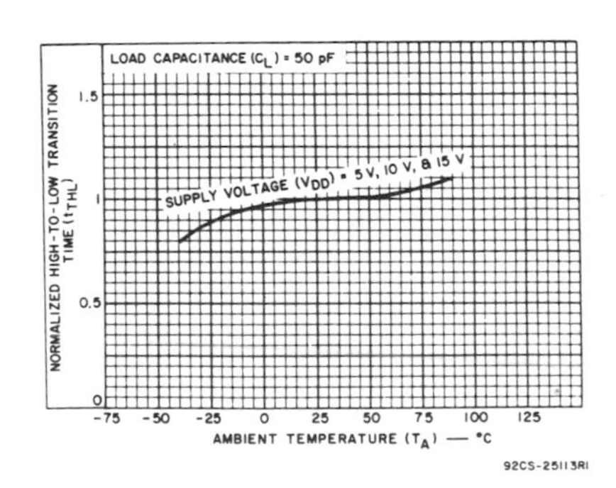

Fig. 15— Variation of high-to-low transition time (t_{THL}) with temperature.

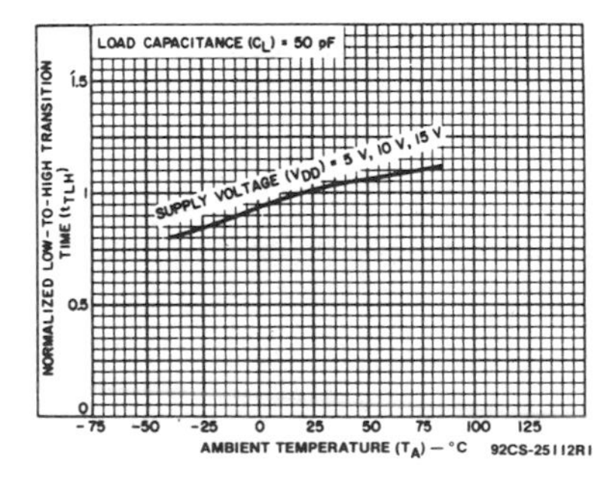

Fig. 16— Variation of low-to-high transition time (t_{TLH}) with temperature.

Table VIII—Dynamic Electrical Characteristics—Definitions

Characteristic	Symbol	Limits		Notes
		Max.	Min.	
Propagation Delay:				
Outputs going high to low	t_{PHL}	X		
Outputs going low to high	t_{PLH}	X		
Output Transition Time:				
Outputs going high to low	t_{THL}	X		
Outputs going low to high	t_{TLH}	X		
Pulse Width-Set, Reset, Preset Enable, Disable, Strobe, Clock	t_{WL} or t_{WH}		X	1
Clock Input Frequency	f_{CL}	X		1, 2
Clock Input Rise and Fall Time	t_{rCL}, t_{fCL}	X		
Set-Up Time	t_{SU}		X	1
Hold Time	t_H		X	1
Removal Time - Set, Reset, Preset-Enable	t_{REM}		X	1
Three State Disable Delay Times:				
High level to high impedance	t_{PHZ}	X		
High impedance to low level	t_{PZL}	X		
Low level to high impedance	t_{PLZ}	X		
High impedance to high level	t_{PZH}	X		

NOTE: (1) By placing a defining min. or max. in front of definition, the limits can change from min. to max., or vice versa.

(2) Clock input waveform should have a 50% duty cycle and be such as to cause the outputs to be switching from 10% V_{DD} to 90% V_{DD} in accordance with the divice truth table.

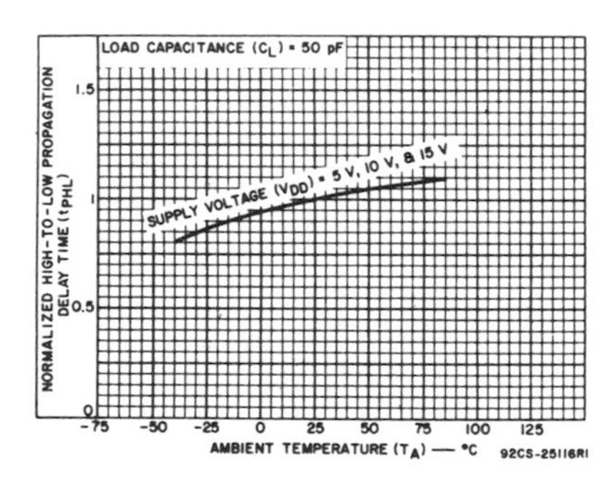

Fig. 17— Variation of high-to-low propagation delay time (t_{PHL}) with temperature.

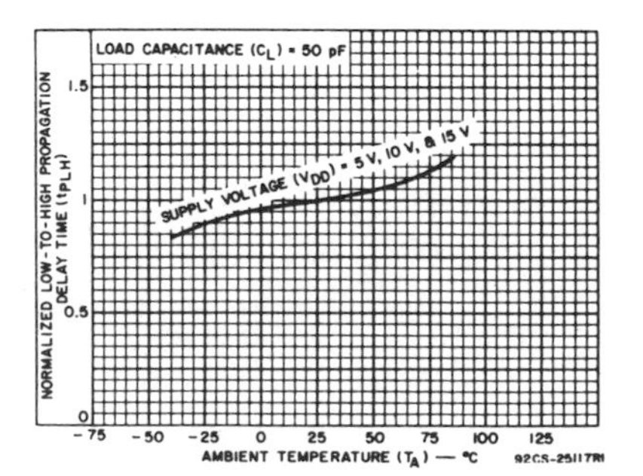

Fig. 18— Variation of low-to-high propagation delay time (t_{PLH}) with temperature.

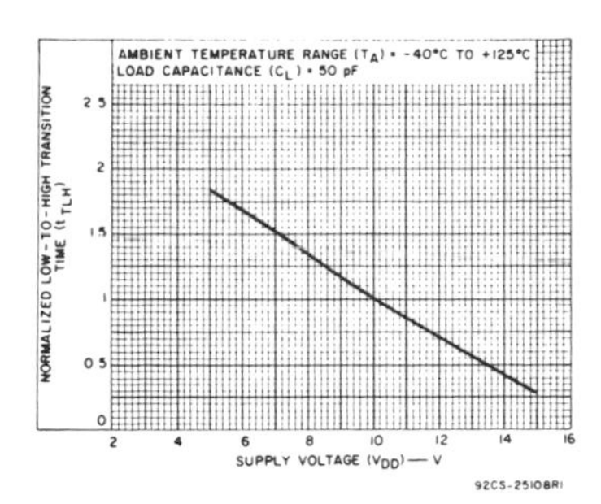

Fig. 19— Variation of low-to-high transition time (t_{TLH}) with supply voltage.

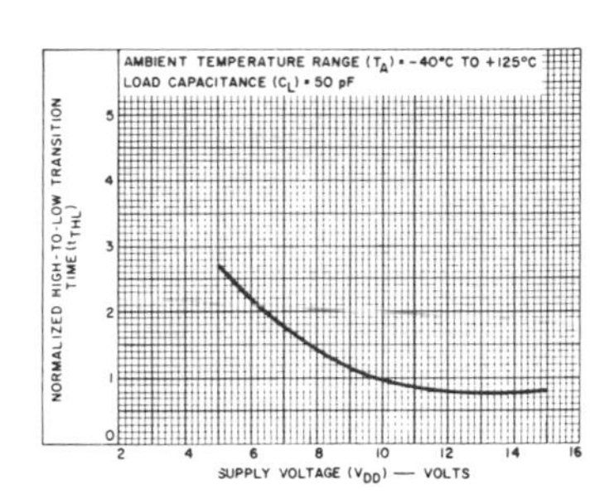

Fig. 20— Variation of high-to-low transition time (t_{THL}) with supply voltage.

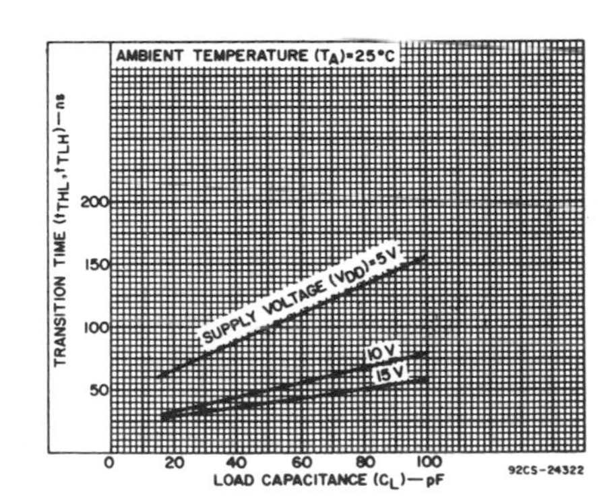

Fig. 21— Variation of transition time (t_{THL}) (t_{TLH}) with load capacitance of three levels of supply voltage.

B-Series Dynamic (AC) Switching Parameters

Table VIII defines major COS/MOS ac characteristics, with reference to the waveforms shown in Figs. 22 through 25. Test conditions of V_{DD}, low capacitance (C_L), and input conditions are given for individual types in the published data.

COS/MOS TIMEKEEPING AND SPECIAL PRODUCTS

RCA supplies a variety of COS/MOS timekeeping products and a group of special COS/MOS products that have operating supply-voltage ranges and other characteristics that differ from standardized data specified for A- and B-series COS/MOS integrated circuits.

COS/MOS Timekeeping Products

The RCA line of COS/MOS timekeeping circuits (CD220XX types) includes types that are designed for use as watch circuits, as wall- or auto-clock circuits, and as industrial times.

Watch Circuits: RAC standard commercial COS/MOS watch products include a line of liquid-crystal-display watch circuits that range from a 3½-digit, 2-function chip to a 6-digit, 6-function circuit with various additional features such as stopwatch and alarm circuitry. These circuits are

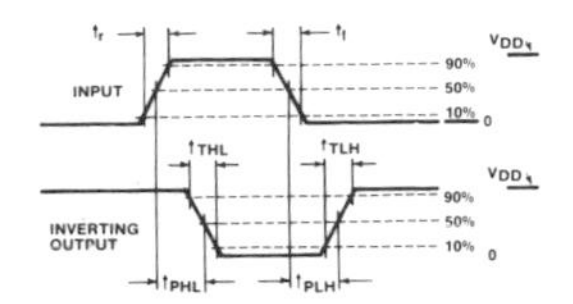

Fig. 22— Transition times and propagation delay times, combination logic.

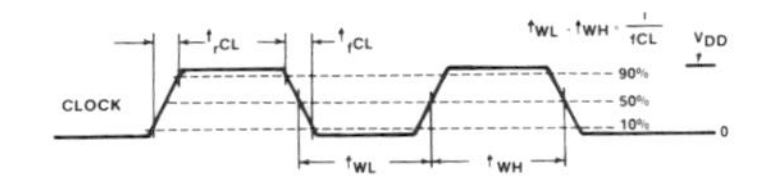

Fig. 23— Clock-pulse rise and fall times and pulse width.

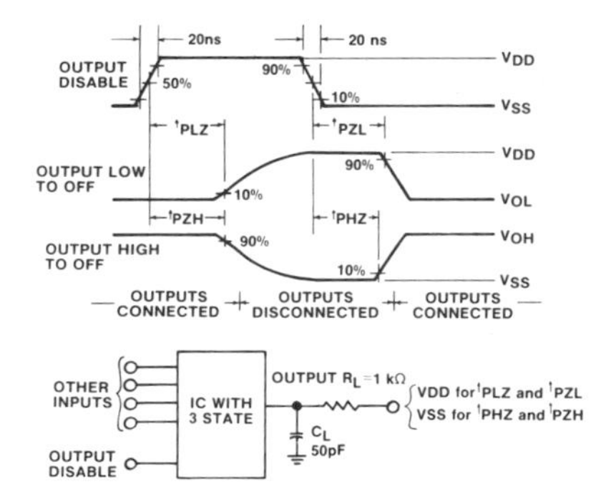

Fig. 24— Three-state propagation delay wave shapes and test circuit.

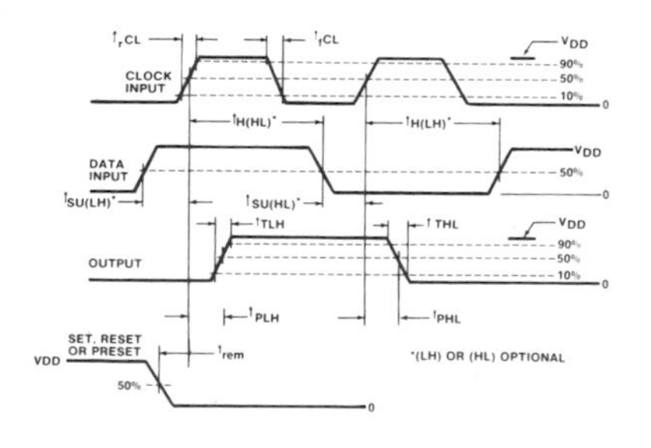

Fig. 25— Setup times, hold times, removal time, and propagation delay times for positive-edge triggered sequential logic circuits.

designed to operate from a single battery cell.

Several RCA watch-circuit chips are offered in both direct and mirror-image versions to allow mounting on either side of a module board.

Wall-Clock Circuits. RCA COS/MOS wall-clock circuits are intended for operation from one or two battery cells under crystal control and designed to drive stepping motors. Standard commercial types include a 32-kHz circuit that is designed to drive a pulse-width-controlled Portescap motor and a 4-MHz silicon-on-sapphire circuit that operates from a single 1.5-volt cell.

Auto-Clock Circuits. RCA standard commercial auto-clock circuits include a line of stepping-motor-drive clock circuits that operate with crystal frequencies of 2, 3, and 4 MHz. All types are designed to meet the voltages and other conditions associated with the automobile environment.

Industrial Timers. Although RCA timekeeping products are usually intended primarily for use in watches, wall clocks, and auto clocks, they can also be used in a variety of industrial timing applications. Several standard COS/MOS timekeeping products have been used in industrial applications as appliance and thermostat-control timers.

Special COS/MOS Products

RCA COS/MOS integrated circuits classified as special types include a group of crosspoint switches (CD221XX types) intended for use in telephone and PBX systems, in studio audio switching applications, and as multi-system bus interconnects.

E. Circuit Characteristics of Low-Power Schottky TTL ICs*

LSTTL circuit features are best understood by examining the LS00 2-input NAND gate *(Figure 1-1)*. The input/output circuits of all LSTTL are almost identical. Although the logic function and basic structure of LS circuits are the same as conventional TTL, there are also significant differences.

INPUT CONFIGURATION. With a few exceptions, LSTTL circuits do not use the multi-emitter input structure that originally gave TTL its name. Most LS elements use a DTL type input circuit with Schottky diodes to perform the AND function, as exemplified by D3 and D4 in *Figure 1-1*. Compared to the classical multi-emitter structure, this circuit is faster and it increases the input breakdown voltage. Inputs of this type are tested for leakage with an applied input voltage of 10 V and the input breakdown voltage is typically 15 V or more. *Figure 1-2* shows the V_{OUT}/V_{IN} transfer function of an LS00 gate. The input threshold is approximately 0.1 V lower than for standard TTL.

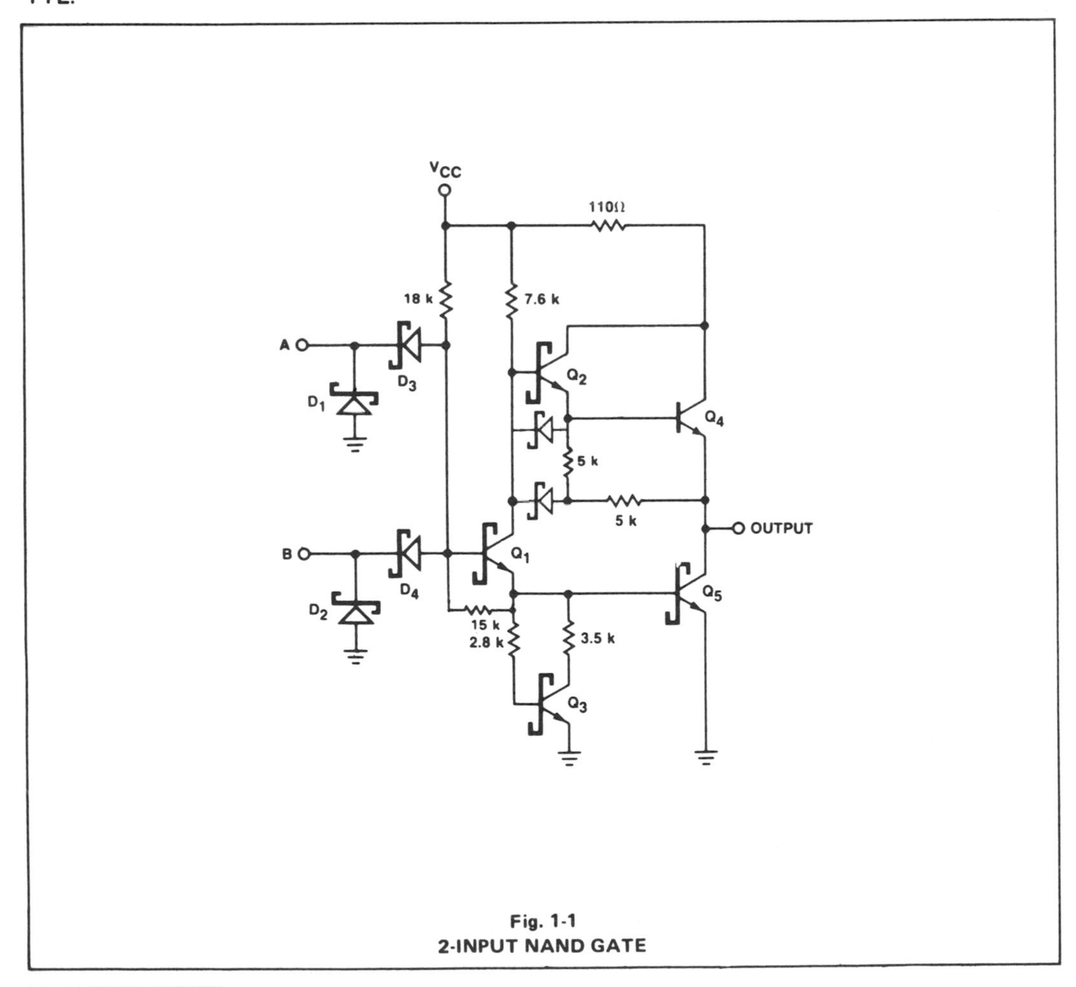

Fig. 1-1
2-INPUT NAND GATE

*Courtesy Motorola

Another input arrangement often used in MSI has three diodes connected as shown in *Figure 1-3.* This configuration gives a slightly higher input threshold than that of *Figure 1-1.*

A third input configuration that is sometimes used employs a vertical PNP transistor as shown in *Figure 1-4.* This arrangement also gives a higher input threshold and has the additional advantage of reducing the amount of current that the signal source must sink. Both the diode cluster arrangement and the PNP input configuration have breakdown voltage ratings greater than 10 V.

All inputs are provided with clamping diodes, exemplified by D1 and D2 in *Figure 1-1.* These diodes conduct when an input signal goes negative (*Figure 1-5),* which limits undershoot and helps to control ringing on long signal lines following a HIGH-to-LOW transition. These diodes are intended only for the suppression of transient currents and should not be used as steady-state clamps in interface applications. A clamp current exceeding 2 mA and with a duration greater than 500 ns can activate a parasitic lateral NPN transistor, which in turn can steal current from internal nodes of the LS circuit and thus cause logic errors.

The effective capacitance of an LSTTL input is 5 pF for DIP and 4 pF for Flatpak. For an input that serves more than one internal function, each additional function adds 1.5 pF.

OUTPUT CONFIGURATION. The output circuitry of LSTTL have several features not found in conventional TTL. A few of these features are discussed below.

Referring to *Figure 1-1,* the base of the pull-down output transistor Q5 is returned to ground through Q3 and a pair of resistors instead of through a simple resistor. This arrangement is called a squaring network since it squares up the transfer characteristics *(Figure 1-2)* by preventing conduction in the phase splitter Q1 until the input voltage rises high enough to allow Q1 to supply base current to Q5. The squaring network also improves the propagation delay by providing a low resistance path to discharge capacitance at the base of Q5 during turn-off.

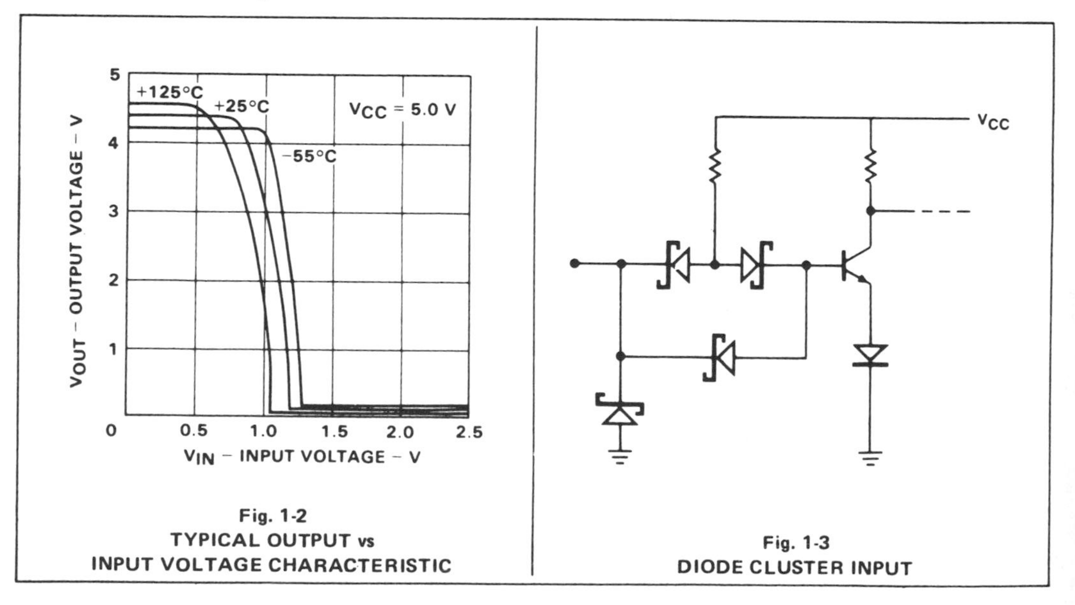

Fig. 1-2
TYPICAL OUTPUT vs
INPUT VOLTAGE CHARACTERISTIC

Fig. 1-3
DIODE CLUSTER INPUT

The output pull-up circuit is a 2-transistor Darlington circuit with the base of the output transistor returned through a 5 K resistor to the output terminals, unlike 74H and 74S where it is returned to ground which is a more power consuming configuration. This configuration allows the output to pull up to one V_{BE} below V_{CC} for low values of output current.

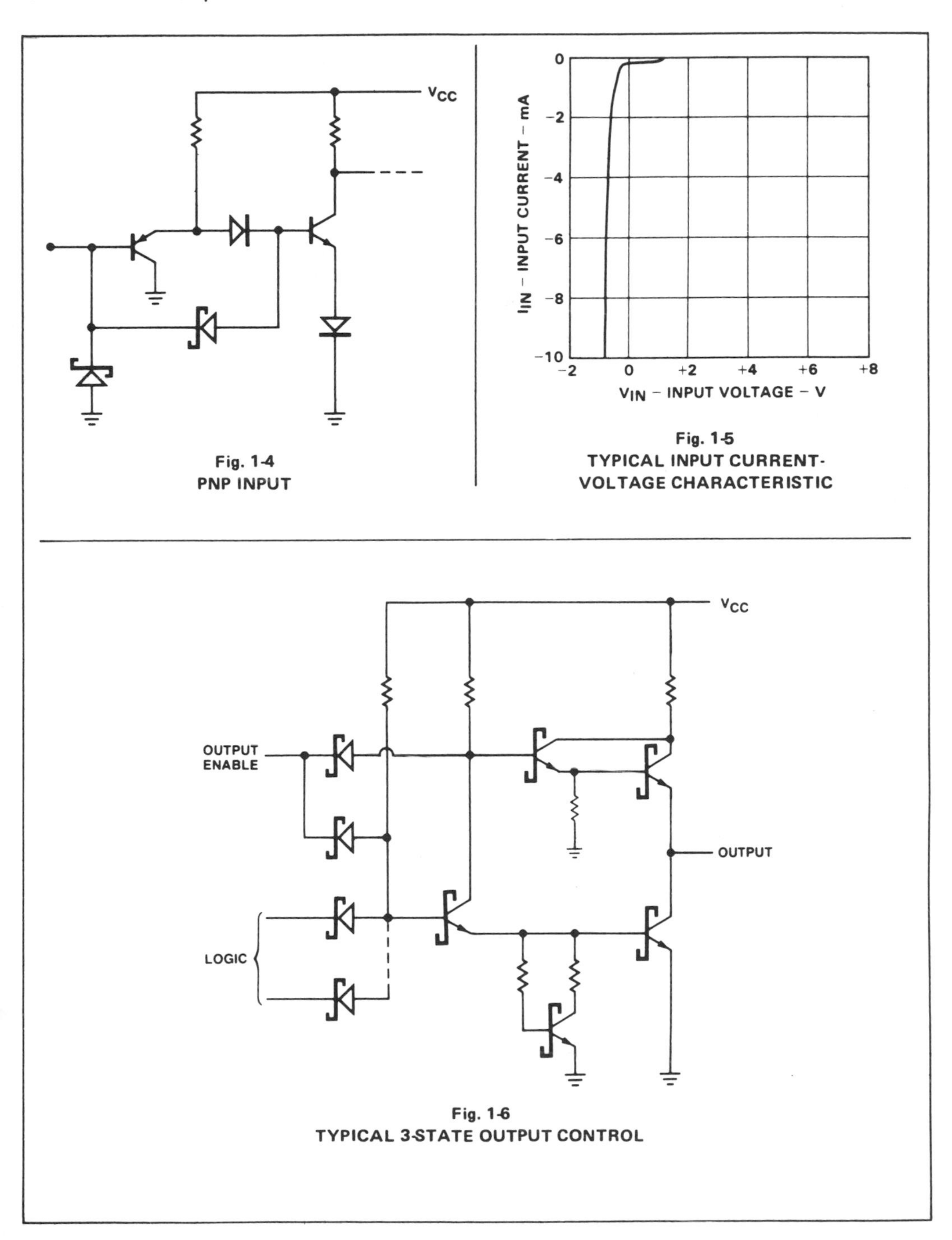

Fig. 1-4
PNP INPUT

Fig. 1-5
TYPICAL INPUT CURRENT-VOLTAGE CHARACTERISTIC

Fig. 1-6
TYPICAL 3-STATE OUTPUT CONTROL

Figure 1-6 shows the extra circuitry used to obtain the "high Z" condition in 3-state outputs. When the Output Enable signal is LOW, both the phase splitter and the Darlington pull-up are turned off. In this condition the output circuitry is non-conducting, which allows the outputs of 2 or more such circuits to be connected together in a bus application wherein only one output is enabled at any particular time.

OUTPUT CHARACTERISTICS. *Figure 1-7* shows the LOW-state output characteristics. For LOW I_{OL} values, the pull-down transistor is clamped out of deep saturation to shorten the turn-off delay. The curves also show the clamping effect when I_{OL} tends to go negative, as it often does due to reflections on a long interconnection after a negative-going transition. This clamping effect helps to minimize ringing.

The waveform of a rising output signal resembles an exponential, except that the signal is slightly rounded at the beginning of the rise. Once past this initial rounded portion, the starting-edge rate is approximately 0.5 V/ns with a 15 pF load and 0.25 V/ns with a 50 pF load. For analytical purposes, the rising waveform can be approximated by the following expression.

$$V(t) = V_{OL} + 3.7\left[1 - \exp(-t/T)\right]$$

where $T = 8$ ns for $C_L = 15$ pF and 16 ns for $C_L = 50$ pF

The waveform of a falling output signal resembles that part of a cosine wave between angles of 0° and 180°. Fall times from 90% to 10% are approximately 4.5 ns with a 15 pF load and 8.5 ns with a 50 pF load. Equivalent edge rates are approximately 0.8 and 0.4 V/ns respectively. For analytical purposes, the falling waveform can be approximated by the following.

$$V(t) = V_{OL} + 1.9\,\mu(t)\left[1 + \cos\omega t\right] - 1.9\,\mu(t-a)\left[1 + \cos\omega(t-a)\right]$$

where $\mu(t) = 0$ for $t < 0$, $= 1$ for $t > 0$ and $\mu(t-a) = 0$ for $t < a$, $= 1$ for $t > a$

For t in nanoseconds and $C_L = 15$ pF, $a = 7.5$ ns, $\omega = 0.42$

For $C_L = 50$ pF, $a = 14$ ns, $\omega = 0..23$

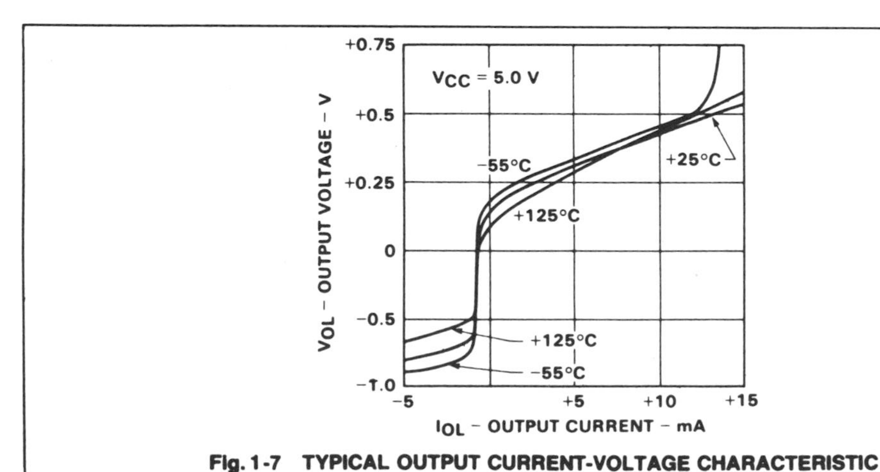

Fig. 1-7 TYPICAL OUTPUT CURRENT-VOLTAGE CHARACTERISTIC

AC SWITCHING CHARACTERISTICS. The average propagation delay of a Low Power Schottky gate is 5 ns at a load of 15 pF as shown in *Figure 1-8.* The delay times increase at an average of 0.08 ns/pF for larger values of capacitance load. These delay times are relatively insensitive to variations in power supply and temperature. The average propagation delay time changes less than 1.0 ns over temperature and less than 0.5 ns with V_{CC} for the military temperature and voltage ranges. (See *Figures 1-10* and *1-11*).

The power versus frequency characteristics of Motorola's LS family, as shown in *Figure 1-9,* indicate that at operating frequencies above 1 MHz the Low Power Schottky devices are more efficient than CMOS for most applications.

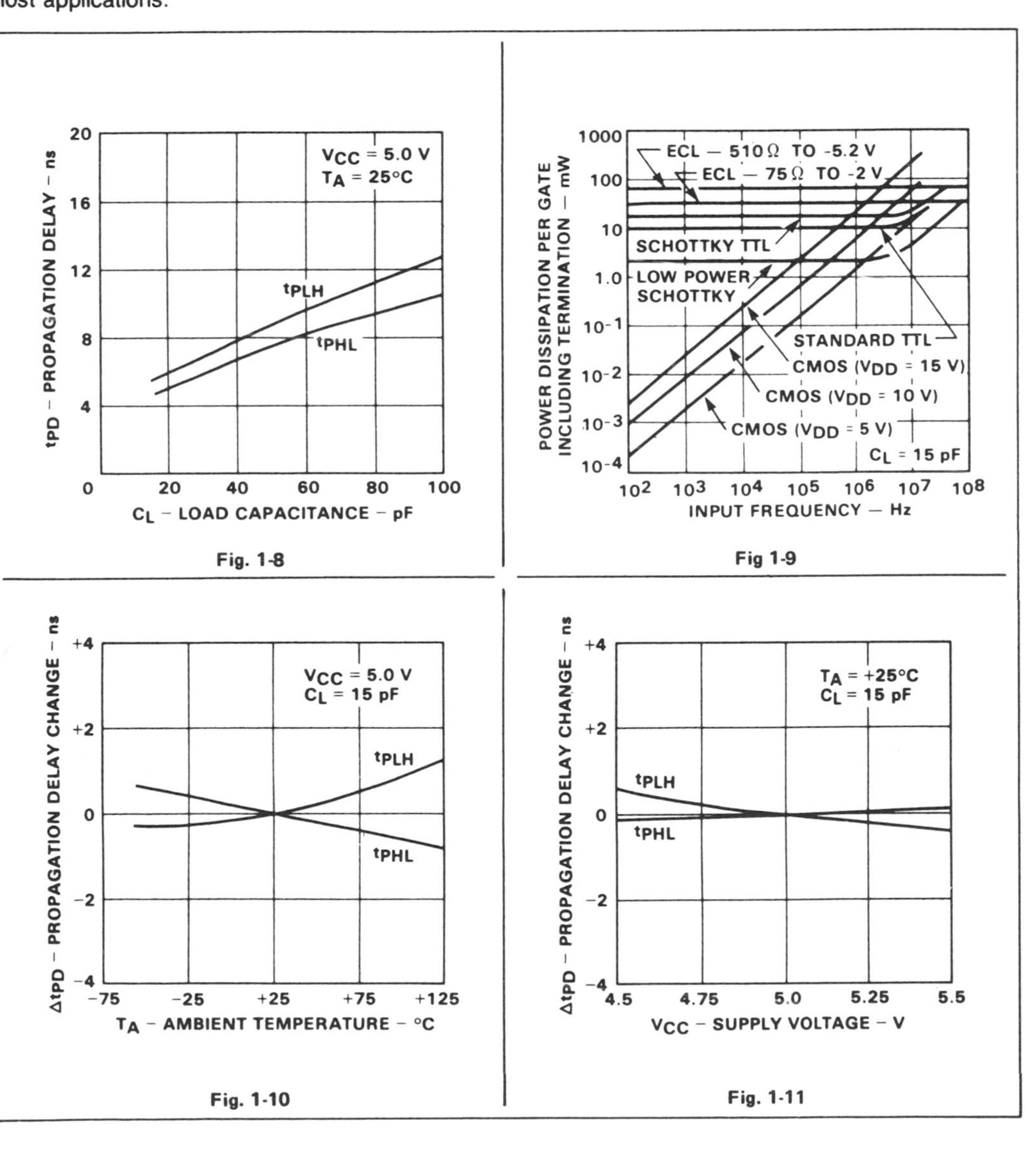

Fig. 1-8

Fig 1-9

Fig. 1-10

Fig. 1-11

AC TEST CIRCUITS AND WAVEFORMS

The following test circuits and conditions represent Motorola's typical AC test procedures. The output loading for standard Low Power Schottky devices is a 15 pF capacitor. Experimental evidence shows that test results using the additional diode-resistor load are within 0.2 ns of the capacitor only load. The capacitor only load also has the advantage of repeatable, easily correlated test results. The input pulse rise and fall times are specified at 6 ns to closely approximate the Low Power Schottky output transitions through the active threshold region. The specified propagation delay limits can be guaranteed with a 15 ns input rise time on all parameters except those requiring narrow pulse widths. Any frequency measurement over 15 MHz or pulse width less than 30 ns must be performed with a 6 ns input rise time.

Test Circuit for Standard Output Devices

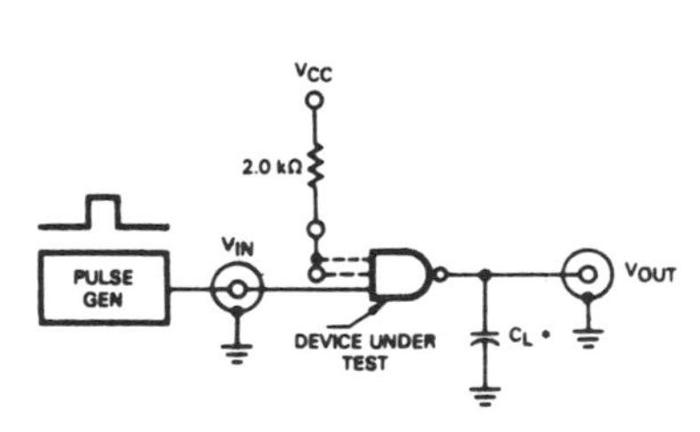

*Includes all probe and jig capacitance

Optional Load (Guaranteed—Not Tested)

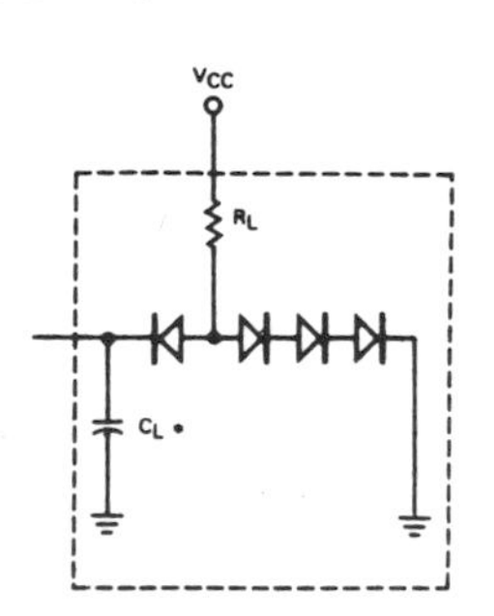

Test Circuit for Open Collector Output Devices

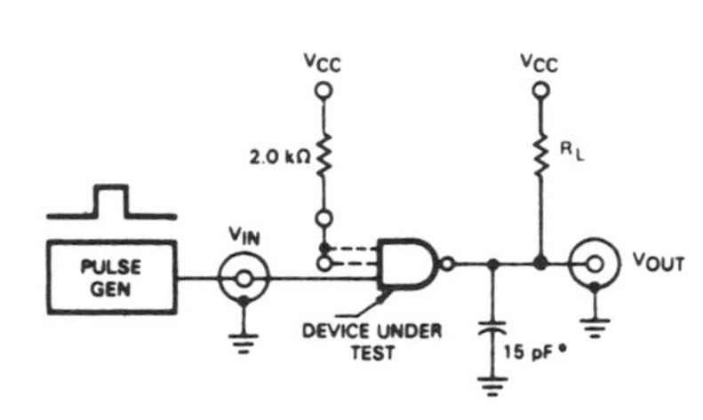

*Includes all probe and jig capacitance

Pulse Generator Settings
(unless otherwise specified)

Frequency = 1 MHz
Duty Cycle = 50%
t_{TLH} (t_r) = 6 ns
t_{THL} (t_f) = 6 ns
Amplitude = 0 to 3 V

Waveform for Inverting Outputs

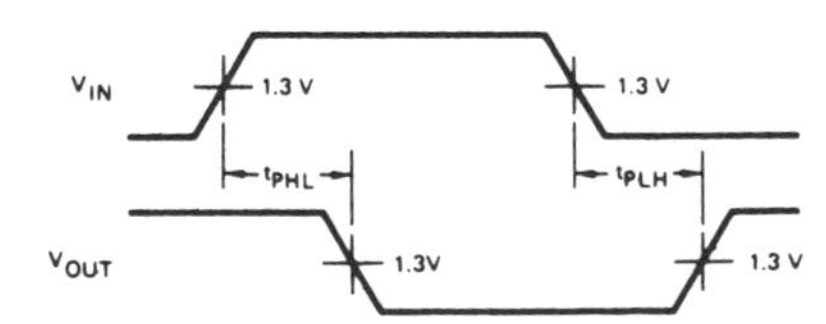

Waveform for Non-inverting Outputs

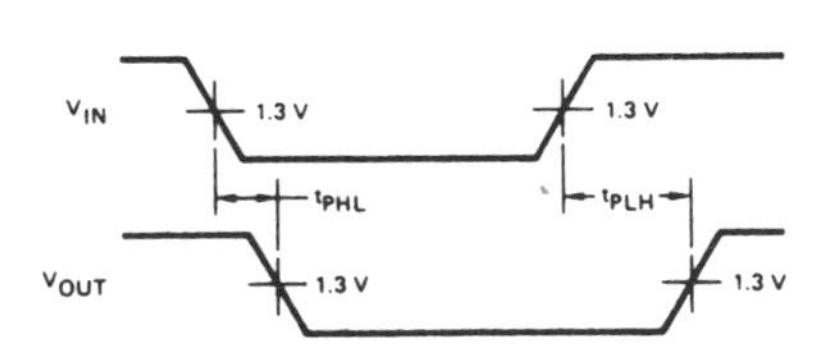

ABSOLUTE MAXIMUM RATINGS
(above which the useful life may be impaired)

Storage Temperature	−65°C to +150°C
Temperature (Ambient) Under Bias	−55°C to +125°C
Vcc Pin Potential to Ground Pin	−0.5 V to +7.0 V
*Input Voltage (dc) Diode Inputs	−0.5 V to 15 V
*Input Current (dc)	−30 mA to +5.0 mA
*Input Voltage (dc) Emitter Inputs	−0.5 V to +5.5 V
Voltage Applied to Outputs (Output HIGH)	−0.5 V to + 5.5 V
High Level Voltage Applied to Disabled 3-State Output	7.0 V
Open Collector Outputs	−0.5 V to +10 V

*Either input voltage limit or input current limit is sufficient to protect the inputs—Circuits with 5.5 V maximum limits are listed below.

Device types having inputs limited to 5.5 V are as follows:

SN74LS196, SN74LS197—Emitter inputs on clock terminals.
SN74LS390/393 —Emitter inputs on clock terminals.
SN74LS242/243, SN74LS245—Inputs connected to outputs.
SN74LS640/641/642/645 —Inputs connected to outputs.

F. General Information on MECL ICs*

SECTION I — HIGH-SPEED LOGICS

High speed logic is used whenever improved system performance would increase a product's market value. For a given system design, high-speed logic is the most direct way to improve system performance and emitter-coupled logic (ECL) is today's fastest form of digital logic. Emitter-coupled logic offers both the logic speed and logic features to meet the market demands for higher performance systems.

MECL PRODUCTS

Motorola introduced the original monolithic emitter-coupled logic family with MECL I (1962) and followed this with MECL II (1966). These two families are now obsolete and have given way to the MECL III (MC1600 series), MECL 10,000, MECL 10800, and PLL (MC12000 series) families.

Chronologically the third family introduced, MECL III (1968) is a higher power, higher speed logic. Typical 1 ns edge speeds and propagation delays along with greater than 500 MHz flip-flop toggle rates, make MECL III useful for high-speed test and communications equipment. Also, this family is used in the high-speed sections and critical timing delays of larger systems. For more general purpose applications, however, trends in large high-speed systems showed the need for an easy-to-use logic family with propagation delays on the order of 2 ns. To match this requirement, the MECL 10,000 Series was introduced in 1971.

An important feature of MECL 10,000 is its compatibility with MECL III to facilitate using both families in the same system. A second important feature is its significant power economy — MECL 10,000 gates use less than one-half the power of MECL III. Finally, low gate power and advanced circuit design techniques have permitted a new level of complexity for MECL 10,000 circuits. For example, the complexity of the MC10803 Memory Interface Function compares favorably to that of any bipolar integrated circuit on the market.

The basic MECL 10,000 Series has been expanded by a subset of devices with even greater speed. This additional series provides a selection of MECL 10,000 logic functions with flip-flop repetition rates up to 200 MHz min. The MECL 10,200 Series is meant for use in critical timing chains, and for clock distribution circuits. MECL 10,200 parts are otherwise identical to their 10,000 Series counterparts (subtract 100 from the MECL 10,200 part number to obtain the equivalent standard MECL 10,000 part number).

Continuing technical advances led more recently to the development of the M10800 LSI processor family. The M10800 family combines the performance of ECL with the system advantages of LSI density. Architectural features of the M10800 family significantly reduce the component count of a high-performance processor system. The M10800 LSI family is fully compatible with the MECL 10,000 and MECL III logic families for a complete selection of system design components.

MECL FAMILY COMPARISONS

Feature	MECL 10,000			MECL III
	10,100 Series 10,500 Series	10,200 Series 10,600 Series	10,800 LSI*	
1. Gate Propagation Delay	2 ns	1.5 ns	1 – 2.5 ns	1 ns
2. Output Edge Speed	3.5 ns	2.5 ns	3.5 ns	1 ns
3. Flip-Flop Toggle Speed	160 MHz	250 MHz	N.A.	300 – 500 MHz
4. Gate Power	25 mW	25 mW	2.3 mW	60 mW
5. Speed Power Product	50 pJ	37 pJ	4.6 pJ	60 pJ

*Average for Equivalent LSI Gate.

FIGURE 1a – GENERAL CHARACTERISTICS

*Courtesy Motorola

Ambient Temperature Range	MECL 10,000	M10800	MECL III	PLL
0° to 75°C	MCM10100 Series	–	MC1697P	MC12000 Series
-30°C to +85°C	MC10100 Series MC10200 Series	MC10800 Series	MC1600 Series	MC12000 Series
-55°C to 125°C	MC10500 Series MC10600 Series MCM10500 Series	–	MC1648M	MC12500 Series

FIGURE 1b – OPERATING TEMPERATURE RANGE

Package Style	MECL 10,000	M10800	MECL III	PLL
16-Pin Plastic DIP	MC10100P Series MC10200P Series	–	MC1658P	MC12000P Series
16-Pin Ceramic DIP	MC10100L Series MC10200L Series MC10500L Series MC10600L Series MCM10100L Series MCM10500L Series	MC10804L MC10807L	MC1600L Series	MC12000L Series MC12500L Series
16-Pin Flat Package	MC10500F Series MC10600F Series MCM10500F Series	–	MC1600F Series	MC12513F
20-Pin Ceramic DIP	–	MC10805L	–	–
24-Pin Plastic Package	MC10181P	–	–	–
24-Pin Ceramic DIP	MC10181L, MC10581L	MC10802L	–	–
24-Pin Flat Package	MC10581F	–	–	–
48-Pin Ceramic Quil	–	MC10800L Series	–	–
14-Pin Plastic DIP	–	–	MC1648P	MC12000P MC12002P MC12020P MC12040P
14-Pin Ceramic DIP	–	–	MC1648L	MC12000L MC12002L MC12020L MC12040L
14-Pin Flat Package	–	–	MC1648F	MC12540F
8-Pin Plastic DIP	–	–	MC1697P	–

For package information see page 1-28.

FIGURE 1c – PACKAGE STYLES

MECL IN PERSPECTIVE

In evaluating any logic line, speed and power requirements are the obvious primary considerations. Figure 1 provides the basic parameters of the MECL 10,000, M10800, and MECL III families. But these provide only the start of any comparative analysis, as there are a number of other important features that make MECL highly desirable for system implementation. Among these:

Complementary Outputs cause a function and its complement to appear simultaneously at the device outputs, without the use of external inverters. It reduces package count by eliminating the need for associated invert functions and, at the same time, cuts system power requirements and reduces timing differential problems arising from the time delays introduced by inverters.

High Input Impedance and Low Output Impedance permit large fan out and versatile drive characteristics.

Insignificant Power Supply Noise Generation, due to differential amplifier design which eliminates current spikes even during signal transition period.

Nearly Constant Power Supply Current Drain simplifies power-supply design and reduces costs.

Low Cross-Talk due to low-current switching in signal path and small (typically 850 mV) voltage swing, and to relatively long rise and fall times.

Wide Variety of Functions, including complex functions facilitated by low power dissipation (particularly in MECL 10,000 series). A basic MECL 10,000 gate consumes less than 8 mW in on-chip power in some complex functions.

Wide Performance Flexibility due to differential amplifier design which permits MECL circuits to be used as linear as well as digital circuits.

Transmission Line Drive Capability is afforded by the open emitter outputs of MECL devices. No "Line Drivers" are listed in MECL families, because *every* device is a line driver.

Wire-ORing reduces the number of logic devices required in a design by producing additional OR gate functions with only an interconnection.

Twisted Pair Drive Capability permits MECL circuits to drive twisted-pair transmission lines as long as 1000 feet.

Wire-Wrap Capability is possible with MECL 10,000 and the M10800 LSI family because of the low rise and fall time characteristic of the circuits.

Open Emitter-Follower Outputs are used for MECL outputs to simplify signal line drive. The outputs match any line impedance and the absence of internal pulldown resistors saves power.

Input Pulldown Resistors of approximately 50kΩ permit unused inputs to remain unconnected for easier circuit board layout.

MECL APPLICATIONS

Motorola's MECL product lines are designed for a wide range of systems needs. Within the computer market, MECL 10,000 is used in systems ranging from special purpose peripheral controllers to large mainframe computers. Big growth areas in this market include disk and communication channel controllers for larger systems and high performance minicomputers.

The industrial market primarily uses MECL for high performance test systems such as IC or PC board testers. However, the high bandwidths of MECL 10,000, MECL III, and MC12,000 are required for many frequency synthesizer systems using high speed phase lock loop networks. MECL will continue to grow in the industrial market through complex medical electronic products and high performance process control systems.

MECL 10,000 and MECL III have been accepted within the Federal market for numerous signal processors and navigation systems. Full military temperature range MECL 10,000 is offered in the MC10500 and MC10600 Series, and in the PLL family as the MC12500 Series.

BASIC CONSIDERATIONS FOR HIGH-SPEED LOGIC DESIGN

High-speed operation involves only four considerations that differ significantly from operation at low and medium speeds:

1. Time delays through interconnect wiring, which may have been ignored in medium-speed systems, become highly important at state-of-the-art speeds.

2. The possibility of distorted waveforms due to reflections on signal lines increases with edge speed.

3. The possibility of "crosstalk" between adjacent signal leads is proportionately increased in high-speed systems.

4. Electrical noise generation and pick-up are more detrimental at higher speeds.

In general, these four characteristics are speed- and frequency-dependent, and are virtually independent of the type of logic employed. The merit of a particular logic family is measured by how well it compensates for these deleterious effects in system applications.

The interconnect-wiring time delays can be reduced only by reducing the length of the interconnecting lines. At logic speeds of two nanoseconds, an equivalent "gate delay" is introduced by every foot of interconnecting wiring. Obviously, for functions interconnected within a single monolithic chip, the time delays of signals travelling from one function to another are insignificant. But for a great many externally interconnected parts, this can soon add up to an appreciable delay time. Hence, the greater the number of functions per chip, the higher the system speed. *MECL circuits, particularly those of the MECL 10,000 Series are designed with a propensity toward complex functions to enhance overall system speed.*

Waveform distortion due to line reflections also becomes troublesome principally at state-of-the-art speeds. At slow and medium speeds, reflections on interconnecting lines are not usually a serious problem. At higher speeds, however, line lengths can approach the wavelength of the signal and improperly terminated lines can result in reflections that will cause false triggering (see Figure 2). The solution, as in RF technology, is to employ "transmission-line" practices and properly terminate each signal line with its characteristic impedance at the end of its run. *The low-impedance, emitter-follower outputs of MECL circuits facilitate transmission-line practices without upsetting the voltage levels of the system.*

The increased affinity for crosstalk in high-speed circuits is the result of very steep leading and trailing edges (fast rise and fall times) of the high-speed signal. These steep wavefronts are rich in harmonics that couple readily to adjacent circuits. *In the design of MECL 10,000, the rise and fall times have been deliberately slowed. This reduces the affinity for crosstalk without compromising other important performance parameters.*

From the above, it is evident that the MECL logic line is not simply capable of operating at high speed, but has been specifically designed to reduce the problems that are normally associated with high-speed operation.

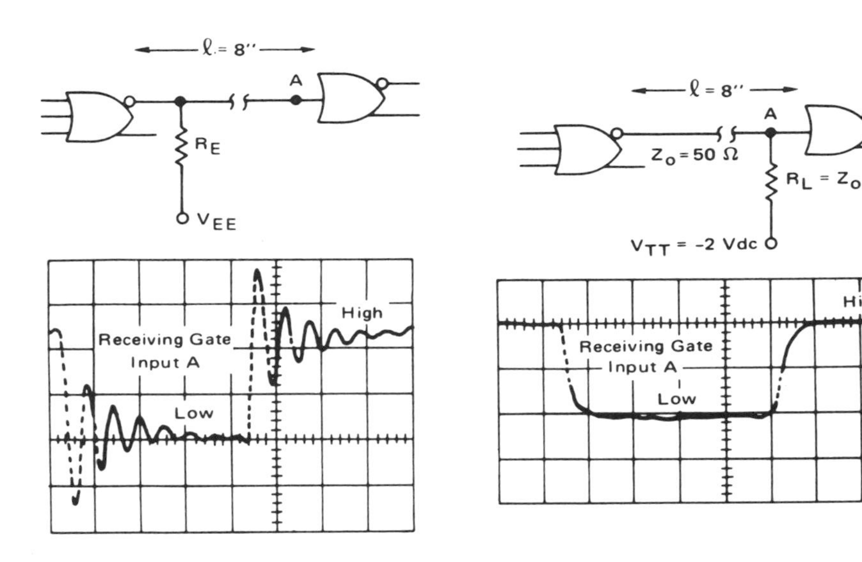

FIGURE 2a – UNTERMINATED TRANSMISSION LINE (No Ground Plane Used)

FIGURE 2b – PROPERLY TERMINATED TRANSMISSION LINE (Ground Plane Added)

FIGURE 3 – MECL GATE STRUCTURE AND SWITCHING BEHAVIOR

CIRCUIT DESCRIPTION

The typical MECL circuit, Figure 3, consists of a differential-amplifier input circuit, a temperature and voltage compensated bias network, and emitter-follower outputs to restore dc levels and provide buffering for transmission line driving. High fan-out operation is possible because of the high input impedance of the differential amplifier input and the low output impedance of the emitter follower outputs. Power-supply noise is virtually eliminated by the nearly constant current drain of the differential amplifier, even during the transition period. Basic gate design provides for simultaneous output of both the OR function and its complement, the NOR function.

Power-Supply Connections – Any of the power supply levels, V_{TT}, V_{CC}, or V_{EE} may be used as ground; however, the use of the V_{CC} node as ground results in best noise immunity. In such a case: V_{CC} = 0, V_{TT} = -2.0 V, V_{EE} = -5.2 V.

System Logic Specifications – The output logic swing of 0.85 V, as shown by the typical transfer characteristics curve, varies from a LOW state of V_{OL} = -1.75 V to a HIGH state of V_{OH} = -0.9 V with respect to ground.

Positive logic is used when reference is made to logical "0's" or "1's." Then

"0" = -1.75 V = LOW
typical
"1" = -0.9 V = HIGH

Circuit Operation -- Beginning with all logic inputs LOW (nominal -1.75 V), assume that Q1 through Q4 are cut off because their P-N base-emitter junctions are not conducting, and the forward-biased Q5 is conducting. Under these conditions, with the base of Q5 held at -1.29 V by the V_{BB} network, its emitter will be one diode drop (0.8 V) more negative than its base, or -2.09 V. (The 0.8 V differential is a characteristic of this P-N junction.) The base-to-emitter differential across Q1 – Q4 is then the difference between the common emitter voltage (-2.09 V) and the LOW logic level (-1.75 V) or 0.34 V. This is less than the threshold voltage of Q1 through Q4 so that these transistors will remain cut off.

When any one (or all) of the logic inputs are shifted upward from the -1.75 V LOW state to the -0.9 V HIGH state, the base voltage of that transistor increases beyond the threshold point and the transistor turns on. When this happens, the voltage at the common-emitter point rises from -2.09 V to -1.7 (one diode drop below the -0.9 V base voltage of the input transistor), and since the base voltage of the fixed-bias transistor (Q5) is held at -1.29 V, the base-emitter voltage Q5 cannot sustain conduction. Hence, this transistor is cut off.

This action is reversible, so that when the input signal(s) return to the LOW state, Q1 – Q4 are again turned off and Q5 again becomes forward biased. The collector voltages resulting from the switching action of Q1 – Q4 and Q5 are transferred through the output emitter-follower to the output terminal. Note that the differential action of the switching transistors (one section being off when the other is on) furnishes simultaneous complementary signals at the output. This action also maintains constant power supply current drain.

DEFINITIONS OF LETTER SYMBOLS AND ABBREVIATIONS

Current:

I_{CC} — Total power supply current drawn from the positive supply by a MECL unit under test.

I_{CBO} — Leakage current from input transistor on MECL devices without pulldown resistors when test voltage is applied.

I_{CCH} — Current drain from V_{CC} power supply with all inputs at logic HIGH level.

I_{CCL} — Current drain from V_{CC} power supply with all inputs at logic LOW level.

I_E — Total power supply current drawn from a MECL test unit by the negative power supply.

I_F — Forward diode current drawn from an input of a saturated logic-to-MECL translator when that input is at ground potential.

I_{in} — Current into the input of the test unit when a maximum logic HIGH ($V_{IH\ max}$) is applied at that input.

*I_{INH} — HIGH level input current into a node with a specified HIGH level ($V_{IH\ max}$) logic voltage applied to that node. (Same as I_{in} for positive logic.)

*I_{INL} — LOW level input current, into a node with a specified LOW level ($V_{IL\ min}$) logic voltage applied to that node.

I_L — Load current that is drawn from a MECL circuit output when measuring the output HIGH level voltage.

*I_{OH} — HIGH level output current: the current flowing into the output, at a specified HIGH level output voltage.

*I_{OL} — LOW level output current: the current flowing into the output, at a specified LOW level output voltage.

I_{OS} — Output short circuit current.

I_{out} — Output current (from a device or circuit, under such conditions mentioned in context).

Current (cont.) :

I_R — Reverse current drawn from a transistor input of a test unit when V_{EE} is applied at that input.

I_{SC} — Short-circuit current drawn from a translator saturating output when that output is at ground potential.

Voltage:

V_{BB} — Reference bias supply voltage.

V_{BE} — Base-to-emitter voltage drop of a transistor at specified collector and base currents.

V_{CB} — Collector-to-base voltage drop of a transistor at specified collector and base currents.

V_{CC} — General term for the most positive power supply voltage to a MECL device (usually ground, except for translator and interface circuits).

V_{CC1} — Most positive power supply voltage (output devices). (Usually ground for MECL devices.)

V_{CC2} — Most positive power supply voltage (current switches and bias driver). (Usually ground for MECL devices.)

V_{EE} — Most negative power supply voltage for a circuit (usually -5.2 V for MECL devices).

V_F — Input voltage for measuring I_F on TTL interface circuits.

V_{IH} — Input logic HIGH voltage level (nominal value).

*$V_{IH\ max}$ — Maximum HIGH level input voltage: The most positive (least negative) value of high-level input voltage, for which operation of the logic element within specification limits is guaranteed.

V_{IHA} — Input logic HIGH threshold voltage level.

$V_{IHA\ min}$ — Minimum input logic HIGH level (threshold) voltage for which performance is specified.

*$V_{IH\ min}$ — Minimum HIGH level input voltage: The least positive (most negative) value of HIGH level input voltage for which operation of the logic element within specification limits is guaranteed.

V_{IL} — Input logic LOW voltage level (nominal value).

*$V_{IL\ max}$ — Maximum LOW level input voltage: The most positive (least negative) value of LOW level input voltage for which operation of the logic element within specification limits is guaranteed.

V_{ILA} — Input logic LOW threshold voltage leve

$V_{ILA\ max}$ — Maximum input logic LOW level (thre hold) voltage for which performance specified.

*$V_{IL\ min}$ — Minimum LOW level input voltage: Th least positive (most negative) value c LOW level input voltage for which opera tion of the logic element within specifica tion limits is guaranteed.

V_{in} — Input voltage (to a circuit or device).

V_{max} — Maximum (most positive) supply voltag permitted under a specified set conditions.

*V_{OH} — Output logic HIGH voltage level: Th voltage level at an output terminal for specified output current, with th specified conditions applied to establish HIGH level at the output.

V_{OHA} — Output logic HIGH threshold volta level.

$V_{OHA\ min}$ — Minimum output HIGH threshold volta level for which performance is specifie

$V_{OH\ max}$ — Maximum output HIGH or high-lev voltage for given inputs.

$V_{OH\ min}$ — Minimum output HIGH or high-level vo age for given inputs.

*V_{OL} — Output logic LOW voltage level: Th voltage level at the output terminal for specified output current, with th specified conditions applied to establish LOW level at the output.

V_{OLA} — Output logic LOW threshold voltage lev

$V_{OLA\ max}$ — Maximum output LOW threshold volta level for which performance is specifie

$V_{OL\ max}$ — Maximum output LOW level voltage f given inputs.

$V_{OL\ min}$ — Minimum output LOW level voltage f given inputs.

V_{TT} — Line load-resistor terminating voltage f outputs from a MECL device.

V_{OLS1} — Output logic LOW level on MECL 10,0 line receiver devices with all inputs V_{EE} voltage level.

V_{OLS2} — Output logic LOW level on MECL 10,0 line receiver devices with all inputs op

*JEDEC, EIA, NEMA standard definition

Time Parameters:

t+	Waveform rise time (LOW to HIGH), 10% to 90%, or 20% to 80%, as specified.
t-	Waveform fall time (HIGH to LOW), 90% to 10%, or 80% to 20%, as specified.
t_r	Same as t+
t_f	Same as t-
t+-	Propagation Delay, see Figure 9.
t-+	Propagation Delay, see Figure 9.
t_{pd} $t_{x\pm y\pm}$	Propagation delay, input to output from the 50% point of the input waveform at pin x (falling edge noted by – or rising edge noted by +) to the 50% point of the output waveform at pin y (falling edge noted by – or rising edge noted by +). (Cf Figure 9.)
t_{x+}	Output waveform rise time as measured from 10% to 90% or 20% to 80% points on waveform (whichever is specified) at pin x with input conditions as specified.
t_{x-}	Output waveform fall time as measured from 90% to 10% or 80% to 20% points on waveform (whichever is specified) at pin x, with input conditions as specified.
Tog	Toggle frequency of a flip-flop or counter device.
shift	Shift rate for a shift register.

Read Mode (Memories)

t_{ACS}	Chip Select Access Time
t_{RCS}	Chip Select Recovery Time
t_{AA}	Address Access Time

Write Mode (Memories)

t_W	Write Pulse Width
t_{WSD}	Data Setup Time Prior to Write
t_{WHD}	Data Hold Time After Write
t_{WSA}	Address setup time prior to write
t_{WHA}	Address hold time after write
t_{WSCS}	Chip select setup time prior to write
t_{WHCS}	Chip select hold time after write
t_{WS}	Write disable time
t_{WR}	Write recovery time

Temperature:

T_{stg}	Maximum temperature at which device may be stored without damage or performance degradation.
T_J	Junction (or die) temperature of an integrated circuit device.
T_A	Ambient (environment) temperature existing in the immediate vicinity of an integrated circuit device package.
θ_{JA}	Thermal resistance of an IC package, junction to ambient.
θ_{JC}	Thermal resistance of an IC package, junction to case.
lfpm	Linear feet per minute.
θ_{CA}	Thermal resistance of an IC package, case to ambient.

Miscellaneous:

e_g	Signal generator inputs to a test circuit.
TP_{in}	Test point at input of unit under test.
TP_{out}	Test point at output of unit under test.
D.U.T.	Device under test.
C_{in}	Input capacitance.
C_{out}	Output capacitance.
Z_{out}	Output impedance.
*P_D	The total dc power applied to a device, not including any power delivered from the device to a load.
R_L	Load Resistance.
R_T	Terminating (load) resistor.
R_p	An input pull-down resistor (i.e., connected to the most negative voltage).
P.U.T.	Pin under test.

*JEDEC, EIA, NEMA standard definition

SECTION II — TECHNICAL DATA

GENERAL CHARACTERISTICS and SPECIFICATIONS

(See pages 1-6 through 1-8 for definitions of symbols and abbreviations.)

In subsequent sections of this Data Book, the important MECL parameters are identified and characterized, and complete data provided for each of the functions. To make this data as useful as possible, and to avoid a great deal of repetition, the data that is common to all functional blocks in a line is not repeated on each individual sheet. Rather, these common characteristics, as well as the application information that applies to each family, are discussed in this section.

In general, the common characteristics of major importance are:

Maximum Ratings, including both dc and ac characteristics and temperature limits;

Transfer Characteristics, which define logic levels and switching thresholds;

DC Parameters, such as output levels, threshold levels, and forcing functions.

AC Parameters, such as propagation delays, rise and fall times and other time dependent characteristics.

In addition, this section will discuss general layout and design guides that will help the designer in building and testing systems with MECL circuits.

LETTER SYMBOLS AND ABBREVIATIONS

Throughout this section, and in the subsequent data sheets, letter symbols and abbreviations will be used in discussing electrical characteristics and specifications. The symbols used in this book, and their definitions, are listed on the preceding pages.

MAXIMUM RATINGS

The limit parameters beyond which the life of the devices may be impaired are given in Figure 4a. In addition, Table 4b provides certain limits which, if exceeded, will not damage the devices, but could degrade the performance below that of the guaranteed specifications.

MECL TRANSFER CURVES

For MECL logic gates, the dual (complementary) outputs must be represented by two transfer curves: one to describe the OR switching action and one to describe the NOR switching action. A typical transfer curve and associated data for all MECL families is shown in Figure 5.

It is not necessary to measure transfer curves at all points of the curves. To guarantee correct operation it is sufficient merely to measure two sets of min/max logic level parameters.

FIGURE 4a — LIMITS BEYOND WHICH DEVICE LIFE MAY BE IMPAIRED

Characteristic	Symbol	Unit	MECL 10,000	M10800 LSI	MECL III
Characteristic	V_{EE}	Vdc	−8.0 to 0	−8.0 to 0	−8.0 to 0
Supply Voltage (V_{CC} = 0)	V_{TT}	Vdc	—	−4.0 to 0	—
Input Voltage (V_{CC} = 0)	V_{in}	Vdc	0 to V_{EE}	0 to V_{EE}	0 to V_{EE}
Input Voltage Bus (V_{CC} = 0)	V_{in}	Vdc	—	0 to −2.0 (1)	—
Output Source Current Continuous	I_{out}	mAdc	50	50	40
Output Source Current Surge	I_{out}	mAdc	100	100	—
Storage Temperature	T_{stg}	°C	−55 to +150	−55 to +150	−55 to +150
Junction Temperature Ceramic Package (2)	T_J	°C	165	165	165 (3)
Junction Temperature Plastic Package	T_J	°C	150	—	150

NOTES: (1) Input voltage limit is V_{CC} to −2 volts when bus is used as an input and the output drivers are disabled.

(2) Maximum T_J may be exceeded (≤ 250°C) for short periods of time (≤ 240 hours) without significant reduction in device life.

(3) Except MC1666 — MC1670 which have maximum junction temperatures = 145°C.

FIGURE 4b – LIMITS BEYOND WHICH PERFORMANCE MAY BE DEGRADED

Characteristics	Symbol	Unit	MECL 10,000	M10800 LSI	MECL III
Operating Temperature Range Commercial ①	T_A	°C	MC: −30 to +85 MCM: 0 to 75	−30 to +85	−30 to +85
Operating Temperature Range MIL ①	T_A	°C	−55 to +125	–	−55 to +125 (MC1648M)
Supply Voltage (V_{CC} = 0)②	V_{EE}	Vdc	MC: −4.68 to −5.72 MCM: −4.94 to −5.46	−4.68 to −5.72	−4.68 to −5.72
Supply Voltage (V_{CC} = 0)	V_{TT}	Vdc	–	−1.9 to −2.2	–
Output Drive Commercial	–	Ω	50 Ω to −2.0 Vdc	50 Ω to −2.0 Vdc	50 Ω to −2.0 Vdc④
Output Drive MIL	–	Ω	100 Ω to −2.0 Vdc	100 Ω to −2.0 Vdc	–
Maximum Clock Input Rise and Fall Time (20% to 80%)	t_r, t_f	ns	–	10	③

NOTES: ① With airflow ≥ 500 lfpm.
② Functionality only. Data sheet limits are specified for −5.2 V ± 0.010 V.
③ 10 ns maximum limit for MC1690, MC1697, and MC1699.
④ Except MC1648 which has an internal output pulldown resistor.

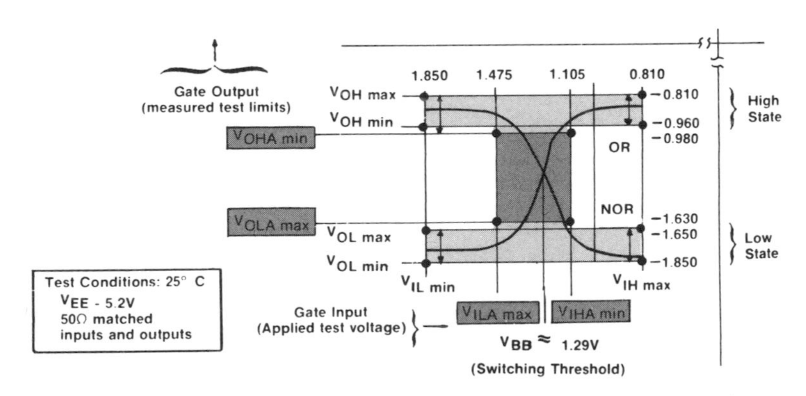

FIGURE 5 – MECL TRANSFER CURVES (MECL 10,000 EXAMPLE) and SPECIFICATION TEST POINTS

The first set is obtained by applying test voltages, $V_{IL\ min}$ and $V_{IH\ max}$ (sequentially) to the gate inputs, and measuring the OR and NOR output levels to make sure they are between V_{OL} max and V_{OL} min, and V_{OHmax} and V_{OHmin} specifications.

The second set of logic level parameters relates to the switching thresholds. This set of data is distinguished by an "A" in symbol subscripts. A test voltage, $V_{ILA\ max}$, is applied to the gate and the NOR and OR outputs are measured to see that they are above the $V_{OHA\ min}$ and below the $V_{OLA\ max}$ levels, respectively. Similar checks are made using the test input voltage $V_{IHA\ min}$.

The result of these specifications insures that:

a) The switching threshold (≈ V_{BB}) falls within the darkest rectangle; i.e. switching does not begin outside this rectangle;

b) Quiescent logic levels fall in the lightest shaded ranges;

c) Guaranteed noise immunity is met.

Figure 6 shows the guaranteed MECL 10,000 and MECL III logic levels and switching thresholds over specified temperature ranges. As shown in the Figure 6a Typical Transfer Curves, MECL outputs rise with increasing ambient temperature. All circuits in each family have the same worst-case output level specifications regardless of power dissipation or junction temperature differences to reduce loss of noise margin due to thermal differences.

All of these specifications assume −5.2 V power supply operation. Operation at other power-supply voltages is possible, but will result in further transfer curve changes. Transfer characteristic data obtained for a variety of supply voltages are shown in Figure 7. The table accompanying these graphs indicates the change rates of output voltages as a function of power supply voltages.

TRANSFER DATA FOR TEMPERATURE VARIATIONS

FIGURE 6a – TYPICAL TRANSFER CHARACTERISTICS AS A FUNCTION OF TEMPERATURE
(See tables below for data)

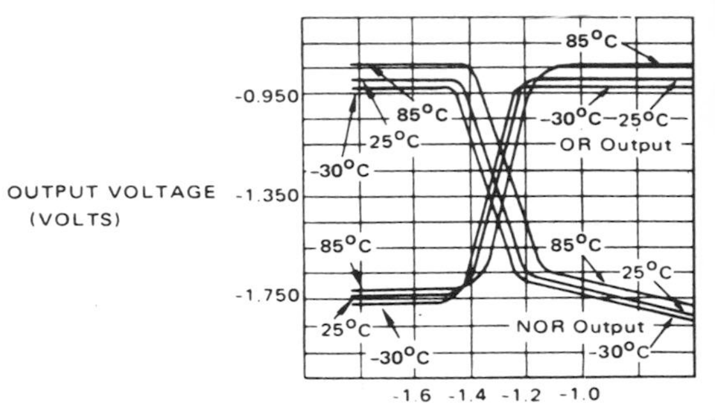

Forcing Function	Parameter	-55°C①	-30°C②	0°C③	25°C②	25°C①	75°C③	85°C②	125°C①
		MC10500 MC10600 MCM10500	MC10100 MC10200 MC10800	MCM10100	MC10100 MC10200 MC10800	MC10500 MC10600 MCM10500	MCM10100	MC10100 MC10200 MC10800	MC10500 MC10600 MCM1050
V_{IHmax} =	V_{OHmax}	-0.880	-0.890	-0.840	-0.810	-0.780	-0.720	-0.700	-0.630
	V_{OHmin}	-1.080	-1.060	-1.000	-0.960	-0.930	-0.900	-0.890	-0.825
	V_{OHAmin}	-1.100	-1.080	-1.020	-0.980	-0.950	-0.920	-0.910	-0.845
V_{IHAmin}		-1.255	-1.205	-1.145	-1.105	-1.105	-1.045	-1.035	-1.000
V_{ILAmax}		-1.510	-1.500	-1.490	-1.475	-1.475	-1.450	-1.440	-1.400
	V_{OLAmax}	-1.635	-1.655	-1.645	-1.630	-1.600	-1.605	-1.595	-1.525
	V_{OLmax}	-1.655	-1.675	-1.665	-1.650	-1.620	-1.625	-1.615	-1.545
V_{ILmin} =	V_{OLmin}④	-1.920	-1.890	-1.870	-1.850	-1.850	-1.830	-1.825	-1.820
V_{ILmin}	I_{INLmin}	0.5	0.5	0.5	0.5	0.5	0.3	0.3	0.3

NOTES: ① MC10500, MC10600, and MCM10500 series specified driving 100 Ω to -2.0 V.
② MC10100, MC10200, and MC10800 series specified driving 50 Ω to -2.0 V.
③ Memories (MCM10100) specified 0-75°C for commercial temperature range, 50 Ω to -2.0 V. Military temperature range memories (MCM10500) specified per Note 1.
④ Special circuits such as MC10123, MC10118, MC10119, and MC10800 family bus outputs have lower than normal V_{OLmin}. See individual data sheets for specific values.

Each MECL 10,000 series device has been designed to meet the dc specifications shown in the test table, after thermal equilibirum has been established. The circuit is in a test socket or mounted on a printed circuit board and transverse airflow greater than 500 linear fpm is maintained. V_{EE} = -5.2 V ± 0.010 V.

FIGURE 6b – MECL 10,000 DC TEST PARAMETERS

Forcing Function	Parameter	-30°C	25°C	85°C
V_{IHmax} =	V_{OHmax}	-0.875	-0.810	-0.700
	V_{OHmin}	-1.045	-0.960	-0.890
	V_{OHAmin}	-1.065	-0.980	-0.910
V_{IHAmin}		-1.180	-1.095	-1.025
V_{ILAmax}		-1.515	-1.485	-1.440
	V_{OLAmax}	-1.630	-1.600	-1.555
	V_{OLmax}	-1.650	-1.620	-1.575
V_{ILmin} =	V_{OLmin}	-1.890	-1.850	-1.830
V_{ILmin}	I_{INLmin}	0.5	0.5	0.3

NOTE: All outputs loaded 50 Ω to -2.0 Vdc except MC1648 which has an internal output pulldown resistor.

ELECTRICAL CHARACTERISTICS

Each MECL III series device has been designed to meet the dc specification shown in the test table, after thermal equilibrium has been established. The circuit is in a test socket or mounted on a printed circuit board and transverse airflow greater than 500 linear fpm maintained. V_{EE} = -5.2 V ± 0.10 V.

FIGURE 6c – MECL III DC TEST PARAMETERS

TRANSFER DATA FOR POWER SUPPLY VARIATIONS

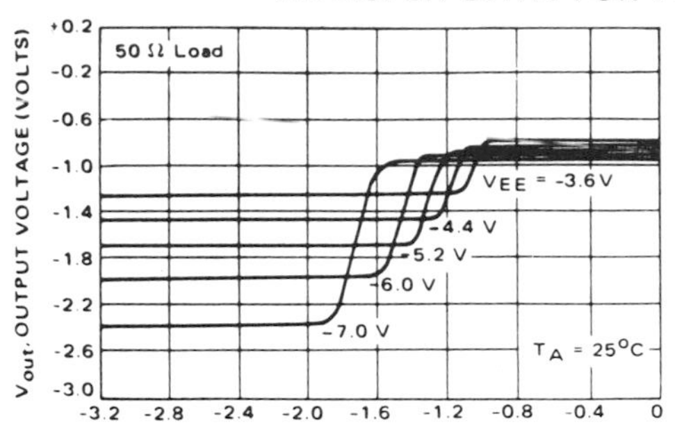

FIGURE 7a – MECL III/10,000 "OR"

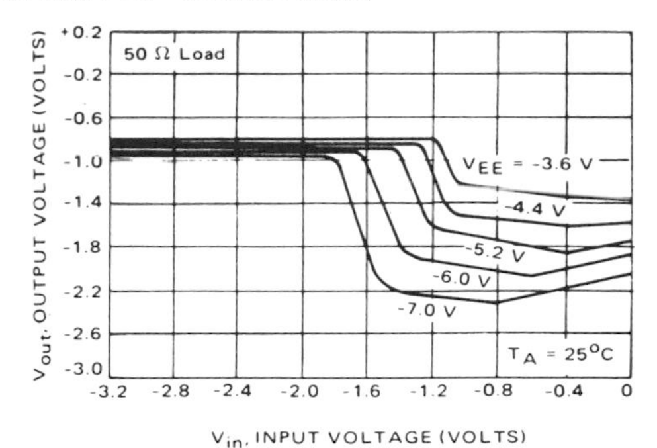

FIGURE 7b – MECL III/10,000 "NOR"

Voltage	MECL 10,000*	MECL III	M10800 LSI
$\Delta V_{OH}/\Delta V_{EE}$	0.016	0.033	0.016
$\Delta V_{OL}/\Delta V_{EE}$	0.250	0.270	0.030
$\Delta V_{BB}/\Delta V_{EE}$	0.148	0.140	0.015

*and subsets: 10,200; 10,500; 10,600.

FIGURE 7C – TYPICAL LEVEL CHANGE RATES

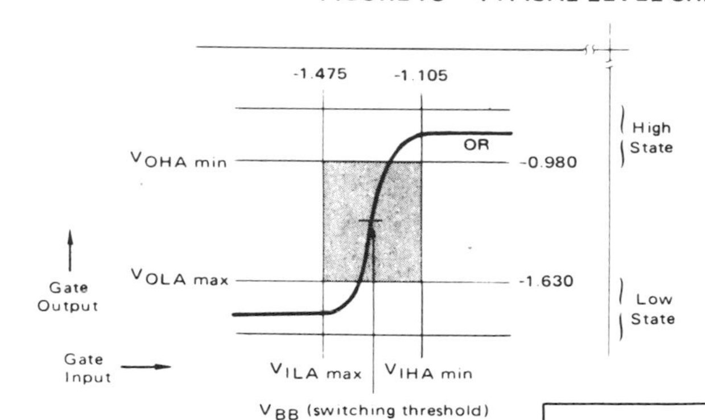

Specification Points for Determining Noise Margin

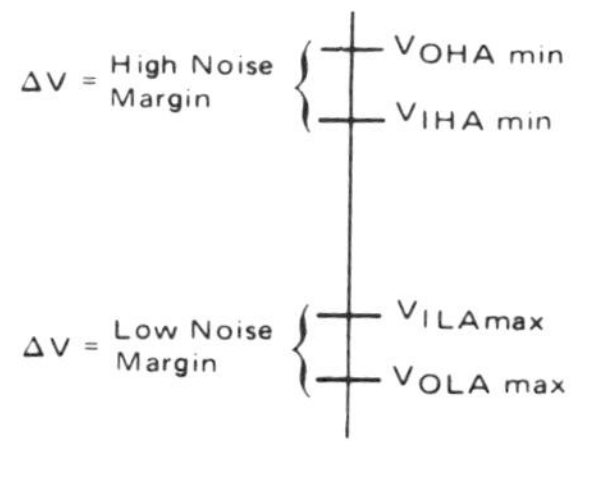

Noise Margin Computations

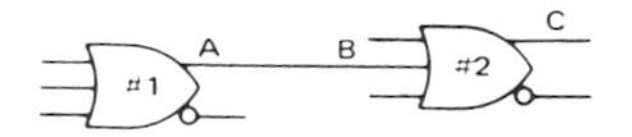

Family	Guaranteed Worst-Case dc Noise Margin	Typical dc Noise Margin
All MECL 10,000	0.125	0.210
MECL III	0.115	0.200

FIGURE 8 – MECL Noise Margin Data

NOISE MARGIN

"Noise margin" is a measure of a logic circuit's resistance to undesired switching. MECL noise margin is defined in terms of the specification points surrounding the switching threshold. The critical parameters of interest here are those designated with the "A" subscript ($V_{OHA\ min}$, $V_{OLA\ max}$, $V_{IHA\ min}$, $V_{ILA\ max}$) in the transfer characteristic curves.

Guaranteed noise margin (NM) is defined as follows:

$$NM_{HIGH\ LEVEL} = V_{OHA\ min} - V_{IHA\ min}$$
$$NM_{LOW\ LEVEL} = V_{ILA\ max} - V_{OLA\ max}$$

To see how noise margin is computed, assume a MECL gate drives a similar MECL gate, Figure 8.

At a gate input (point B) equal to $V_{ILA\ max}$, MECL gate #2 can begin to enter the shaded transition region.

This is a "worst case" condition, since the $V_{OLA\ max}$ specification point guarantees that no device can enter the transition region before an input equal to $V_{ILA\ max}$ is reached. Clearly then, $V_{ILA\ max}$ is one critical point for noise margin computation, since it is the edge of the transition region.

To find the other critical voltage, consider the output from MECL gate #1 (point A). What is the most positive value possible for this voltage (considering worst case specifications)? From Figure 8 it can be observed that the $V_{OLA\ max}$ specification insures that the LOW state OR output from gate #1 can be no greater than $V_{OLA\ max}$.

Note that $V_{OLA\ max}$ is more negative than $V_{ILA\ max}$. Thus, with $V_{OLA\ max}$ at the input to gate #2, the transition region is not yet reached. (The input voltage to gate #2 is still to the left of $V_{ILA\ max}$ on the transfer curve.)

In order to ever run the chance of switching gate #2, we would need an additional voltage, to move the input from $V_{OLA\ max}$ to $V_{ILA\ max}$. This constitutes the "safety factor" known as noise margin. It can be calculated as the magnitude of the difference between the two specification voltages, or for the MECL 10,000 levels shown:

$$NM_{LOW} = V_{ILA\ max} - V_{OLA\ max} = -1.475\ V - (-1.630\ V) = 155\ mV.$$

Similarly, for the HIGH state:

$$NM_{HIGH} = V_{OHA\ min} - V_{IHA\ min} = -0.980\ V - (-1.105\ V) = 125\ mV$$

Analogous results are obtained when considering the "NOR" transfer data.

Note that these noise margins are absolute worst case conditions. The lesser of the two noise margins is that for the HIGH state, 125 mV. This then, constitutes the guaranteed margin against signal undershoot, and power or thermal disturbances.

As shown in the table, typical noise margins are usually better than guaranteed — by about 75 mV.

Noise margin is a dc specification that can be calculated, since it is defined by specification points tabulated on MECL data sheets. However, by itself, this specification does not give a complete picture regarding the noise immunity of a system built with a particular set of circuits. Overall system noise immunity involves not only noise-margin specifications, but also other circuit-related factors that determine how difficult it is to apply a noise signal of sufficient magnitude and duration to cause the circuit to propagate a false logic state. In general, then, noise immunity involves line impedances, circuit output impedances, and propagation delay in addition to noise-margin specifications. This subject is discussed in greater detail in Application Note AN-592.

AC OR SWITCHING PARAMETERS

Time-dependent specifications are those that define the effects of the circuit on a specified input signal, as it travels through the circuit. They include the time delay involved in changing the output level from one logic state to another. In addition, they include the time required for the output of a circuit to respond to the input signal, designated as propagation delay, or access time, in the case of memories. Since this terminology has varied over the years, and because the "conditions" associated with a particular parameter may differ among logic families, the common MECL waveform and propagation delay terminologies are depicted in Figure 9. Specific rise, fall, and propagation delay times are given on the data sheet for each specific functional block, but like the transfer characteristics, ac parameters are temperature and voltage dependent. Typical variations for MECL 10,000 are given in the curves of Figure 10.

SETUP AND HOLD TIMES

Setup and hold times are two ac parameters which can easily be confused unless clearly defined. For MECL logic devices, t_{setup} is the minimum time (50% — 50%) before the positive transition of the clock pulse (C) that information must be pres-

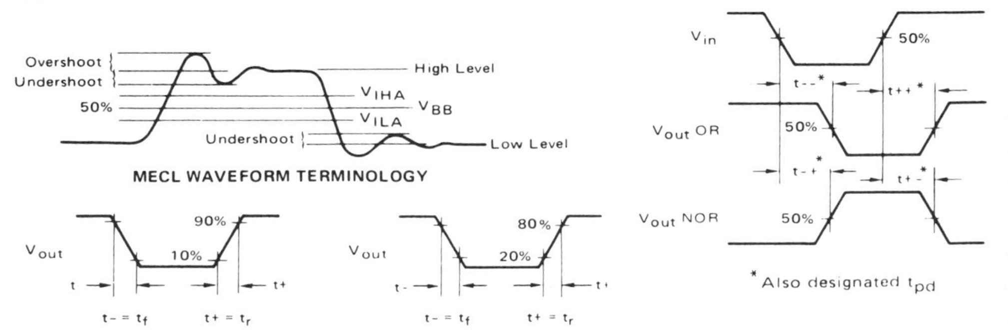

MECL III Rise and Fall Times MECL 10,000 Rise and Fall Times MECL Propagation Delay

FIGURE 9a — TYPICAL LOGIC WAVEFORMS

FIGURE 9b – MEMORY CHIP SELECT ACCESS TIME WAVEFORM

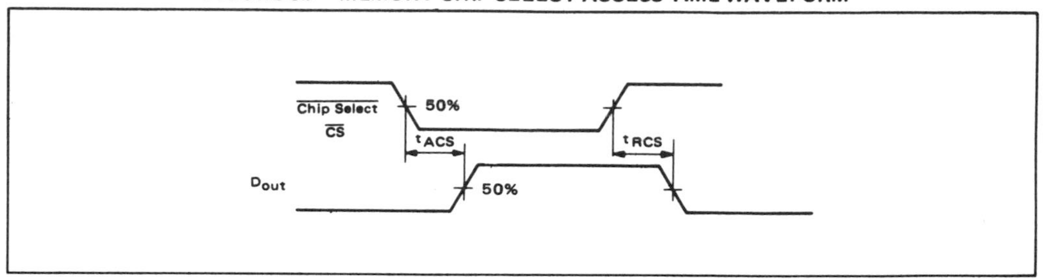

FIGURE 9c – MEMORY ADDRESS ACCESS TIME WAVEFORM

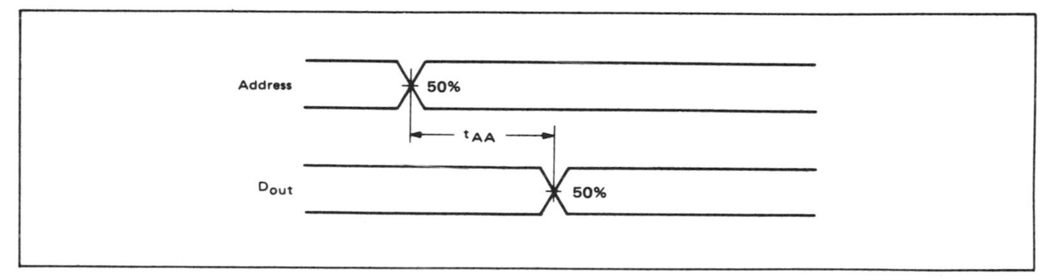

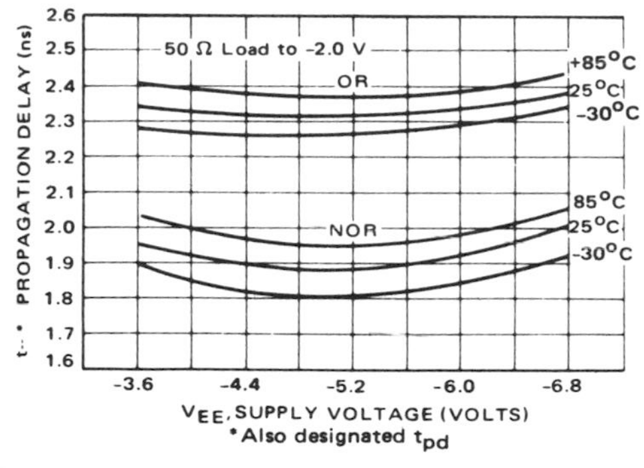

URE 10a – TYPICAL PROPAGATION DELAY t– – versus V_{EE} AND TEMPERATURE (MECL 10,000)

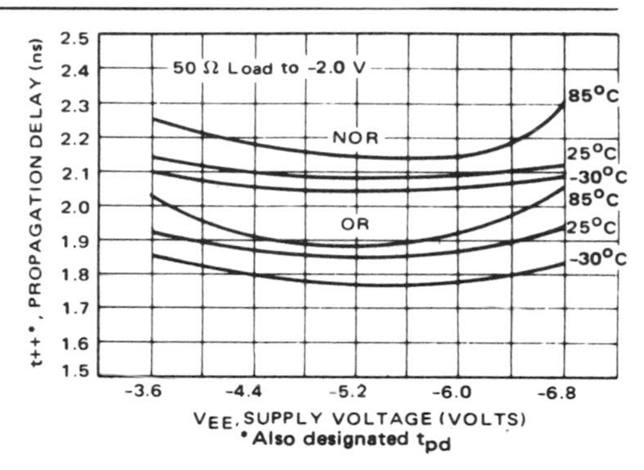

FIGURE 10b – TYPICAL PROPAGATION DELAY t++ versus V_{EE} AND TEMPERATURE (MECL 10,000)

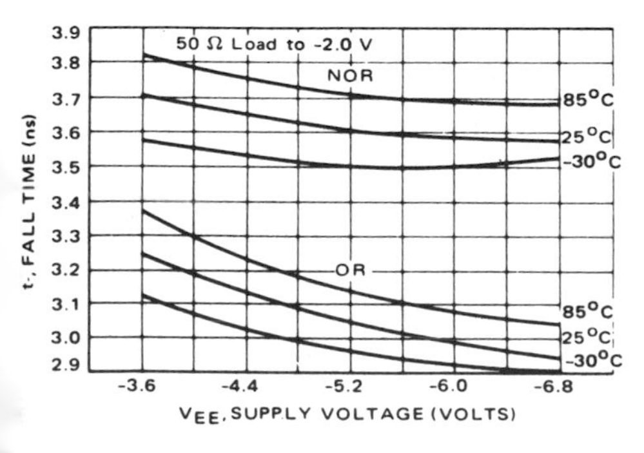

IGURE 10c – TYPICAL FALL TIME (90% to 10%) versus EMPERATURE AND SUPPLY VOLTAGE (MECL 10,100)

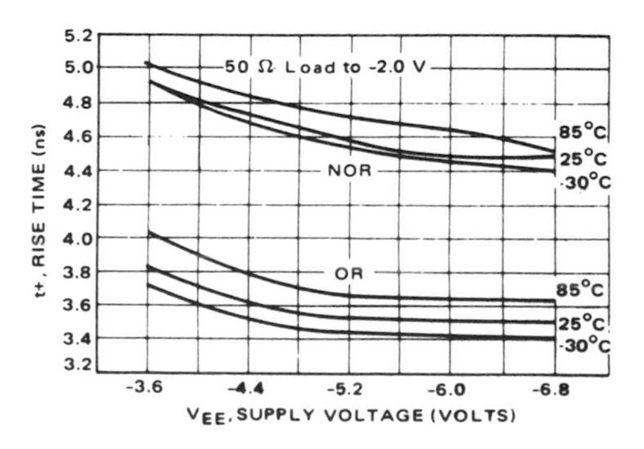

FIGURE 10d – TYPICAL RISE TIME (10% to 90%) versus TEMPERATURE AND SUPPLY VOLTAGE (MECL 10,100)

sent at the Data input (D) to insure proper operation of the device. The t_{hold} is defined similarly as the minimum time after the positive transition of the clock pulse (C) that the information must remain unchanged at the Data input (D) to insure proper operation. Setup and hold waveforms for logic devices are shown in Figure 11a.

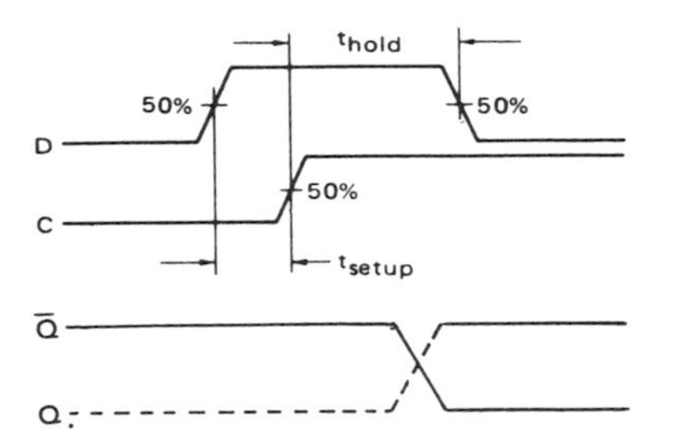

FIGURE 11a – SETUP AND HOLD WAVEFORMS FOR MECL LOGIC DEVICES

For MECL memory devices, t_{setup} is the minimum time before the negative transition of the write enable pulse ($\overline{WE}$) that information must be present at the chip select ($\overline{CS}$), Data (D), and address (A) inputs for proper writing of the selected cell. Similarly t_{hold} is the minimum time after the positive transition of the write enable pulse ($\overline{WE}$) that the information must remain unchanged at the inputs to insure proper writing. Memory setup and hold waveforms are shown in Figure 11b.

In specifying devices, Motorola establishes and guarantees values (shown as minimums on the data sheets) for t_{setup} and t_{hold}. For most MECL circuits, proper device operation typically occurs with the inputs present for somewhat less time than that specified for t_{setup} and t_{hold}.

TESTING MECL 10,000 and MECL III

To obtain results correlating with Motorola circuit specifications certain test techniques must be used. A schematic of a typical gate test circuit is shown in Figure 12a, and a typical memory test circuit in Figure 12b.

A solid ground plane is used in the test setup, and capacitors bypass V_{CC1}, V_{CC2}, and V_{EE} pins to ground. All power leads and signal leads are kept as short as possible.

The sampling scope interface runs directly to the 50-ohm inputs of Channel A and B via 50-ohm coaxial cable. Equal-length coaxial cables must be used between the test set and the A and B scope inputs. A 50-ohm coax cable such as RG58/U or RG188A/U, is recommended.

Interconnect fittings should be 50 ohm GR, BNC, Sealectro Conhex, or equivalent. Wire length should be < ¼ inch from TP_{in} to input pin and TP_{out} to output pin.

FIGURE 11b – SETUP AND HOLD WAVEFORMS FOR MECL MEMORIES (WRITE MODE)

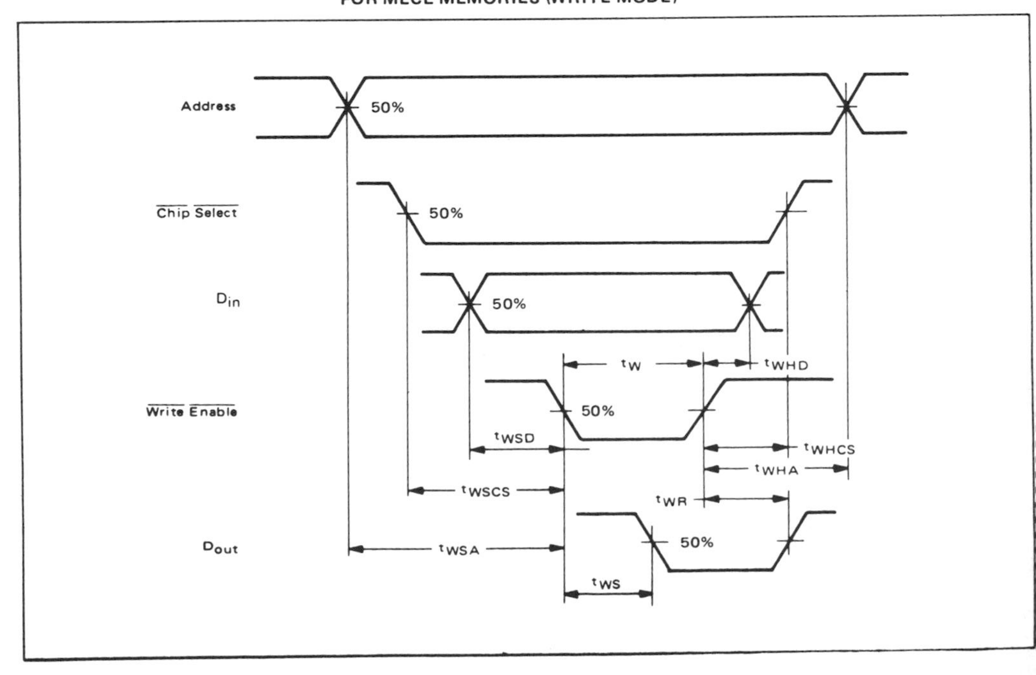

The pulse generator must be capable of 2.0 ns rise and fall times for MECL 10,000 and 1.5 ns for MECL III. In addition, the generator voltage must have an offset to give MECL signal swings of ≈ ±400 mV about a threshold of ≈+0.7 V when V_{CC} = +2.0 V and V_{EE} = −3.2 V for ac testing of logic devices.

The power supplies are shifted +2.0 V, so that the device under test has only one resistor value to load into – the precision 50-ohm input impedance of the sampling oscilloscope. Use of this technique yields a close correlation between Motorola and customer testing. Unused outputs are loaded with a 50-ohm resistor (100-ohm for MIL temp devices) to ground. The positive supply (V_{CC}) should be decoupled from the test board by RF type 25μF capacitors to ground. The V_{CC} pins are bypassed to ground with 0.1 μF, as is the V_{EE} pin.

Additional information on testing MECL 10,000 and understanding data sheets is found in Application Notes AN-579 and AN-701.

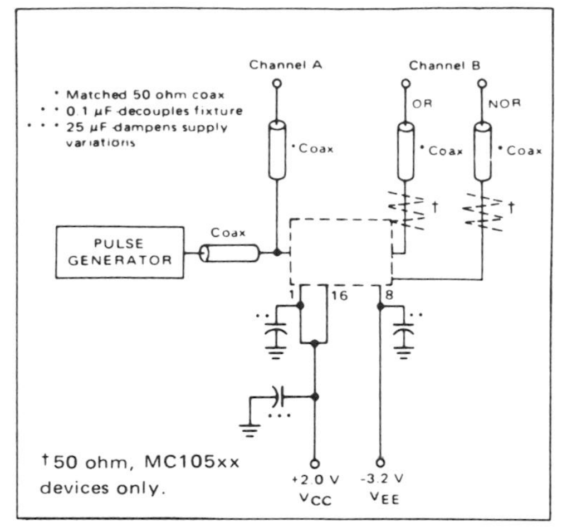

NOTE: All power supply levels are shown shifted 2 volts positive.

FIGURE 12a – MECL LOGIC SWITCHING TIME TEST SETUP

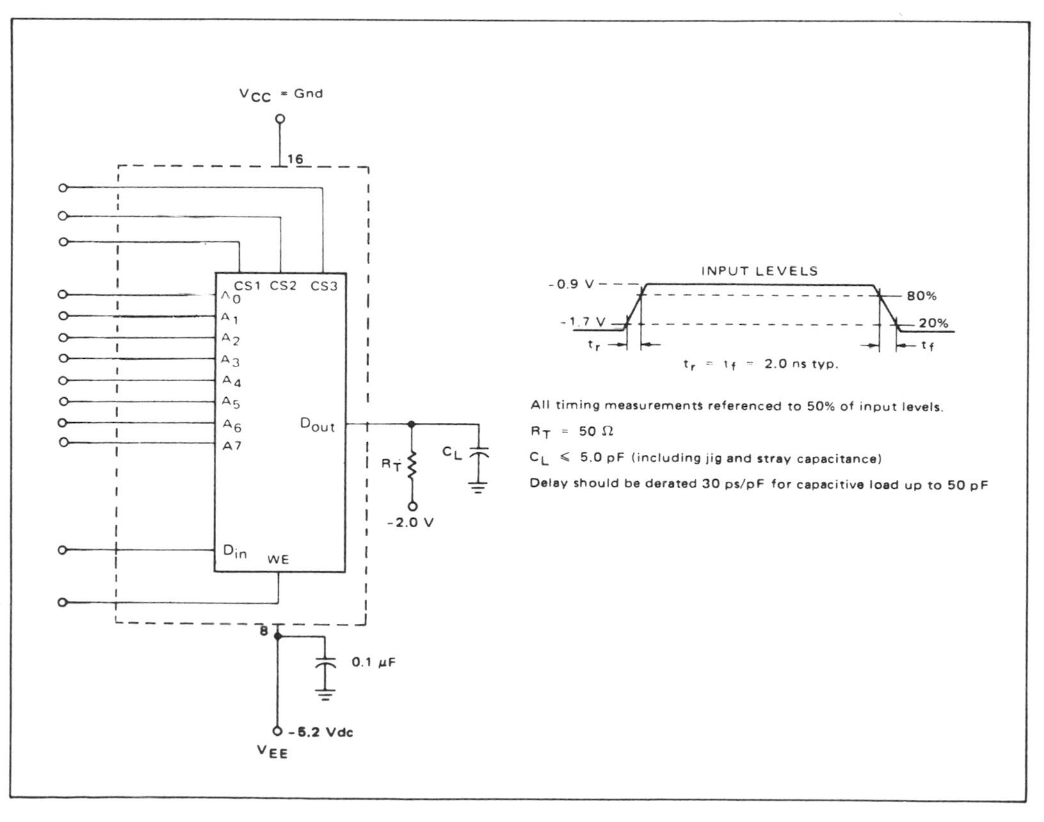

FIGURE 12b – MECL MEMORY SWITCHING TIME TEST CIRCUIT

G. Glossary, TTL Terms and Definitions†

Introduction

These symbols, terms, and definitions are in accordance with those currently agreed upon by the JEDEC Council of the Electronic Industries Association (EIA) for use in the USA and by the International Electrotechnical Commission (IEC) for international use.

PART I—OPERATING CONDITIONS AND CHARACTERISTICS (INCLUDING LETTER SYMBOLS)

Clock Frequency

Maximum Clock Frequency, f_{max}

The highest rate at which the clock input of a bistable circuit can be driven through its required sequence while maintaining stable transitions of logic level at the output with input conditions established that should cause changes of output logic level in accordance with specification.

Current

High-level input current, I_{IH}
The current into* an input when a high-level voltage is applied to that input.

High-level output current, I_{OH}
The current into* an output with input conditions applied that according to the product specification will establish a high level at the output.

Low-level input current, I_{IL}
The current into* an input when a low-level voltage is applied to that input.

Low-level output current, I_{OH}
The current into* an output with input conditions applied that according to the product specification will establish a low level at the output.

Off-state output current, $I_O(off)$
The current flowing into* an output with input conditions applied that according to the product specification will cause the output switching element to be in the off state.

Note: This parameter is usually specified for open-collector outputs intended to drive devices other than logic circuits.

---

†Texas Instruments Incorporated
*Current out of a terminal is given as a negative value.

Off-state (high-impedance-state) output current (of a three-state output), I_{OZ}

The current into* an output having three-state capability with input conditions applied that according to the product specification will establish the high-impedance state at the output.

Short-circuit output current, I_{OS}

The current into* an output when that output is short-circuited to ground (or other specified potential) with input specifications applied to establish the output logic level farthest from ground potential (or other specified potential).

Supply current, I_{CC}

The current into* the V_{CC} terminal of an integrated circuit.

Hold Time

Hold time, t_h

The interval during which a signal is retained at a specified input terminal after an active transition occurs at another specified input terminal.

NOTES: 1. The hold time is the actual time between two events and may be insufficient to accomplish the intended result. A minimum value is specified that is the shortest interval for which correct operation of the logic element is guaranteed.

2. The hold time may have a negative value in which case the minimum limit defines the longest interval (between the release of data and the active transition) for which correct operation of the logic element is guaranteed.

Output Enable and Disable Time

Output enable time (of a three-state output) to high level, t_{PZH} (or low level, t_{PZL})†

The propagation delay time between the specified reference points on the input and output voltage waveforms with the three-state output changing from a high-impedance (off) state to the defined high (or low) level.

Output enable time (of a three-state output) to high or low level, t_{PZX}†

The propagation delay time between the specified reference points on the input and output voltage waveforms with the three-state output changing from a high-impedance (off) state to either of the defined active levels (high or low).

Output disable time (of a three-state output) from high level, t_{PHZ} (or low level, t_{PLZ})†

The propagation delay time between the specified reference points on the input and output voltage waveforms with the three-state output changing from the defined high (or low) level to a high-impedance (off) state.

---

*Current out of a terminal is given as a negative value.

†On older data sheets, similar symbols without the P subscript were used; i.e. t_{ZH}, t_{ZL}, t_{HZ}, and t_{LZ}.

Output disable time (of a three-state output) from high or low level, t_{PXZ}†
The propagation delay time between the specified reference points on the input and output voltage waveforms with the three-state output changing from either of the defined active lengths (high or low) to a high-impedance (off) state.

Propagation Time

Propagation delay time, t_{PD}
The time between the specified reference points on the input and output voltage waveforms with the output changing from one defined level (high or low) to the other defined level.

Propagation delay time, low-to-high-level output, t_{PLH}
The time between the specified reference points on the input and output voltage waveforms with the output changing from the defined low level to the defined high level.

Propagation delay time, high-to-low-level output, t_{PHL}
The time between the specified reference points on the input and output voltage waveforms with the output changing from the defined high level to the defined low level.

Pulse Width

Pulse width, t_w
The time interval between specified reference points on the leading and trailing edges of the pulse waveform.

Recovery Time

Sense recovery time, t_{SR}
The time interval needed to switch a memory from a write mode to a read mode and to obtain valid data signals at the output.

Release Time

Release time, $t_{release}$
The time interval between the release from a specified input terminal of data intended to be recognized and the occurrence of an active transition at another specified input terminal.

Note: When specified, the interval designated "release time" falls within the setup interval and constitutes, in effect, a negative hold time.

Setup Time

Setup time, t_{su}
The time interval between the application of a signal that is maintained at a specified input terminal and a consecutive active transition at another specified input terminal.

---

†On older data sheets, similar symbols without the P subscript were used; i.e. t_{ZH}, t_{ZL}, t_{HZ}, and t_{LZ}.

NOTES: 1. The setup time is the actual time between two events and may be insufficient to accomplish the setup. A minimum value is specified that is the shortest interval for which correct operation of the logic element is guaranteed.

2. The setup time may have a negative value in which case the minimum limit defines the longest interval (between the active transition and the application of the other signal) for which correct operation of the logic element is guaranteed.

Transition Time

Transition time, low-to-high-level, t_{TLH}
The time between a specified low-level voltage and a specified high-level voltage on a waveform that is changing from the defined low level to the defined high level.

Transition time, high-to-low-level, t_{THL}
The time between a specified high-level voltage and a specified low-level voltage on a waveform that is changing from the defined high level to the defined low level.

Voltage

High-level input voltage, V_{IH}
An input voltage within the more positive (less negative) of the two ranges of values used to represent the binary variables.
NOTE: A minimum is specified that is the least positive value of high-level input voltage for which operation of the logic element within specification limits is guaranteed.

High-level output voltage, V_{OH}
The voltage at an output terminal with input conditions applied that according to the product specification will establish a high level at the output.

Input clamp voltage, V_{IK}
An input voltage in a region of relatively low differential resistance that serves to limit the input voltage swing.

Low-level input voltage, V_{IL}
An input voltage level within the less positive (more negative) of the two ranges of values used to represent the binary variables.
NOTE: A maximum is specified that is the most positive value of low-level input voltage for which operation of the logic element within specification limits is guaranteed.

Low-level output voltage, V_{OL}
The voltage at an output terminal with input conditions applied that according to the product specification will establish a low level at the output.

Negative-going threshold voltage, V_{T-}
The voltage level at a transition-operated input that causes operation of the logic element according to specifications as the input voltage falls from a level above the positive-going threshold voltage, V_{T+}.

Off-state output voltage, $V_{O(OFF)}$
The voltage at an output terminal with input conditions applied that according to the product specification will cause the output switching element to be in the off state.

Note: This characteristic is usually specified only for outputs not having internal pull-up elements.

On-state output voltage, $V_{O(on)}$
The voltage at an output terminal with input conditions applied that according to the product specification will cause the output switching element to be in the on state.

NOTE: This characteristic is usually specified only for outputs not having internal pull-up elements.

Positive-going threshold voltage, V_{T+}
The voltage level at the transition-operated input that causes operation of the logic element according to specifications as the input voltage rises from a level below the negative-going threshold voltage, V_{T-}.

PART II—CLASSIFICATION OF CIRCUIT COMPLEXITY

Gate Equivalent Circuit

A basic unit-of-measure of relative digital-circuit complexity. The number of gate equivalent circuits is that number of individual logic gates that would have to be interconnected to perform the same function.

Large-Scale Integration, LSI

A concept whereby a complete major subsystem or system function is fabricated as a single microcircuit. In this context a major subsystem or system, whether digital or linear, is considered to be one that contains 100 or more equivalent gates or circuitry of similar complexity.

Medium-Scale Integration, MSI

A concept whereby a complete subsystem or system function is fabricated as a single microcircuit. The subsystem or system is smaller than for LSI, but whether digital or linear, is considered to be one that contains 12 or more equivalent gates or circuitry of similar complexity.

Small-Scale Integration, SSI

Integrated circuits of less complexity than medium-scale integration (MSI).

Very-Large-Scale Integration, VLSI

A concept whereby a complete system function is fabricated as a single microcircuit. In this context, a system, whether digital or linear, is considered to be one that contains 1000 or more gates or circuitry of similar complexity.

TTL — EXPLANATION OF FUNCTION TABLES

The following symbols are now being used in function tables on TI data sheets:

H	= high level (steady state)
L	= low level (steady state)
↑	= transition from low to high level
↓	= transition from high to low level
X	= irrelevant (any input, including transitions)
Z	= off (high-impedance) state of a 3-state output
a..h	= the level of steady-state inputs at inputs A through H respectively
Q_O	= level of Q before the indicated steady-state input conditions were established
$\overline{Q}_O$	= complement of Q_O or level of $\overline{Q}$ before the indicated steady-state input conditions were established
Q_n	= level of Q before the most recent active transition indicated by ↓ or ↑
_⎍_	= one high-level pulse
‾⊔‾	= one low-level pulse
TOGGLE	= each output changes to the complement of its previous level on each active transition indicated by ↓ or ↑

If, in the input columns, a row contains only the symbols H, L, and/or X, this means the indicated output is valid whenever the input configuration is achieved and regardless of the sequence in which it is achieved. The output persists so long as the input configuration is maintained.

If, in the input columns, a row contains H, L, and/or X together with ↑ and/or ↓, this means the output is valid whenever the input configuration is achieved but the transition(s) must occur following the achievement of the steady-state levels. If the output is shown as a level (H, L, Q_O, or $\overline{Q}_O$), it persists so long as the steady-state input levels and the levels that terminate indicated transitions are maintained. Unless otherwise indicated, input transitions in the opposite direction to those shown have no effect at the output. (If the output is shown as a pulse, _⎍_ or ‾⊔‾, the pulse follows the indicated input transition and persists for an interval dependent on the circuit.)

Among the most complex function tables in this book are those of the shift registers. These embody most of the symbols used in any of the function tables, plus more. Following is the function table of a 4-bit bidirectional universal shift register, e.g., type SN74194.

FUNCTION TABLE

INPUTS										OUTPUTS			
CLEAR	MODE		CLOCK	SERIAL		PARALLEL				Q_A	Q_B	Q_C	Q_D
	S1	S0		LEFT	RIGHT	A	B	C	D				
L	X	X	X	X	X	X	X	X	X	L	L	L	L
H	X	X	L	X	X	X	X	X	X	Q_{A0}	Q_{B0}	Q_{C0}	Q_{0}
H	H	H	↑	X	X	a	b	c	d	a	b	c	d
H	L	H	↑	X	H	X	X	X	X	H	Q_{An}	Q_{Bn}	Q_{Cn}
H	L	H	↑	X	L	X	X	X	X	L	Q_{An}	Q_{Bn}	Q_{Cn}
H	H	L	↑	H	X	X	X	X	X	Q_{Bn}	Q_{Cn}	Q_{Dn}	H
H	H	L	↑	L	X	X	X	X	X	Q_{Bn}	Q_{Cn}	Q_{Dn}	L
H	L	L	X	X	X	X	X	X	X	Q_{AO}	Q_{BO}	Q_{CO}	Q_{DO}

The first line of the table represents a synchronous clearing of the register and says that if clear is low, all four outputs will be reset low regardless of the other inputs. In the following lines, clear is inactive (high) and so has no effect.

The second line shows that so long as the clock input remains low (while clear is high), no other input has any effect and the outputs maintain the levels they assumed before the steady-state combination of clear high and clock low was established. Since on other lines of the table only the rising transition of the clock is shown to be active, the second line implicitly shows that no further change in the outputs will occur while the clock remains high or on the high-to-low transition of the clock.

The third line of the table represents synchronous parallel loading of the register and says that if S1 and S0 are both high then, without regard to the serial input, the data entered at A will be at output Q_A, data entered at B will be at Q_B, and so forth, following a low-to-high clock transition.

The fourth and fifth lines represent the loading of high- and low-level data, respectively, from the shift-right serial input and the shifting of previously entered data one bit; data previously at Q_A is now at Q_B, the previous levels of Q_B and Q_C are now at Q_D respectively, and the data previously at Q_D is no longer in the register. This entry of serial data and shift takes place on the low-to-high transition of the clock when S1 is low and S0 is high and the levels at inputs A through D have no effect.

The sixth and seventh lines represent the loading of high- and low-level data, respectively, from the shift-left serial input and the shifting of previously entered data one bit; data previously at Q_B is now at Q_A, the previous levels of Q_C and Q_D are now at Q_B and Q_C, respectively, and the data previously at Q_A is no longer in the register. This entry of serial data and shift takes place on the low-to-high transition of the clock when S1 is high and S0 is low and the levels at inputs A through D have no effect.

The last line shows that as long as both mode inputs are low, no other input has any effect and, as in the second line, the outputs maintain the levels they assumed before the steady-state combination of clear high and both mode inputs low was established.

INDEX

A

B

C

D

E

I

J

K

L

M

N

O

T

U

V